**Sovereignty and Command in Canada–US
Continental Air Defence, 1940–57**

Studies in Canadian Military History

Series editor: Andrew Burtch, Canadian War Museum

The Canadian War Museum, Canada's national museum of military history, has a threefold mandate: to remember, to preserve, and to educate. Studies in Canadian Military History, published by UBC Press in association with the Museum, extends this mandate by presenting the best of contemporary scholarship to provide new insights into all aspects of Canadian military history, from earliest times to recent events. The work of a new generation of scholars is especially encouraged, and the books employ a variety of approaches – cultural, social, intellectual, economic, political, and comparative – to investigate gaps in the existing historiography. The books in the series feed immediately into future exhibitions, programs, and outreach efforts by the Canadian War Museum. A list of the titles in the series appears at the end of the book.

CANADIAN WAR MUSEUM
MUSÉE CANADIEN DE LA GUERRE

Sovereignty and Command in Canada–US Continental Air Defence, 1940–57

Richard Goette

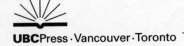

UBCPress · Vancouver · Toronto

© UBC Press 2018

All rights reserved. No part of this publication may be reproduced, stored in a retrieval system, or transmitted, in any form or by any means, without prior written permission of the publisher, or, in Canada, in the case of photocopying or other reprographic copying, a licence from Access Copyright, www.accesscopyright.ca.

27 26 25 24 23 22 21 20 19 18   5 4 3 2 1

Printed in Canada on FSC-certified ancient-forest-free paper
(100% post-consumer recycled) that is processed chlorine- and acid-free.

Library and Archives Canada Cataloguing in Publication

Goette, Richard Evan, author
    Sovereignty and command in Canada – US continental air defence, 1940-57 / Richard Goette.

(Studies in Canadian military history)
Includes bibliographical references and index.
Issued in print and electronic formats.

ISBN 978-0-7748-3687-6 (hardcover). – ISBN 978-0-7748-3689-0 (PDF)
ISBN 978-0-7748-3690-6 (EPUB). – ISBN 978-0-7748-3691-3 (Kindle)

1. Air defenses – Canada – History – 20th century. 2. Air defenses – United States – History – 20th century. 3. Command and control systems – Canada – History – 20th century. 4. Command and control systems – United States – History – 20th century. 5. Canada – Military relations – United States – History – 20th century. 6. United States – Military relations – Canada – History – 20th century. I. Title. II. Series: Studies in Canadian military history

UG735.C3G64 2018   358.4'160971   C2018-900885-7
                                                        C2018-900886-5

Canada

UBC Press gratefully acknowledges the financial support for our publishing program of the Government of Canada (through the Canada Book Fund), the Canada Council for the Arts, and the British Columbia Arts Council.

This book has been published with the help of a grant from the Canadian Federation for the Humanities and Social Sciences, through the Awards to Scholarly Publications Program, using funds provided by the Social Sciences and Humanities Research Council of Canada.

Publication of this book has been financially supported by the Canadian War Museum.

Printed and bound in Canada by Friesens
Set in Minion Pro by Marquis Interscript
Copy editor: Matthew Kudelka
Proofreader: Judith Earnshaw
Indexer: Cathy Murphy
Cover designer: David Drummond
Cartographer: Mike Bechthold

UBC Press
The University of British Columbia
2029 West Mall
Vancouver, BC V6T 1Z2
www.ubcpress.ca

To my family

Contents

List of Illustrations / viii

Foreword / ix
By Lieutenant-General Pierre St-Amand

Acknowledgments / xi

List of Abbreviations / xv

Introduction / 3

1 Command and Control, Sovereignty, Civil–Military Relations, and the Profession of Arms / 21

2 Command and Control Culture and Systems / 36

3 Wartime Planning for Command and Control / 71

4 Wartime Operational Level Command and Control / 86

5 Replacing ABC-22 / 106

6 Organizing and Coordinating Canada–US Air Defences / 132

7 The US Northeast Command / 151

8 Integrating North American Air Defences under Operational Control / 170

Conclusion / 193

Appendices / 201

Notes / 211

Select Bibliography / 275

Illustrations

MAPS
1 The Canadian Northwest Atlantic Command / 103
2 Command Boundaries, RCAF Air Defence Command and US Northeast Command, April 1953 / 167
3 The Pinetree, Mid-Canada, and Distant Early Warning (DEW) radar lines, 1958 / 173
4 RCAF and USAF Interceptor Deployment and Coverage, 1958 / 190

Foreword

CANADA'S SECURITY IN the context of our relationship with the United States and the way we think about North American defence is a subject of ongoing discussion and debate. Through this book, Richard Goette offers us a first-row seat to an era that was instrumental in the development of military command and control principles and resulting arrangements between Canada and the United States that endure to this day. These principles not only apply to NORAD but also guide us in renewed discussions about the best way to defend the North American continent while preserving Canadian sovereignty.

NORAD was formed initially in response to the single threat of long-range nuclear-armed bombers; sixty years later we now face a far more complete and difficult defence problem. If we only look at modern capabilities that can reach out and threaten North America (loosely defined as the territories, airspace, and maritime approaches to Canada and the United States), it is clear that the three oceans surrounding our continent no longer represent the obstacles to an adversary that they once did. Ballistic missiles, precise and very long-range air- and sea-launched cruise missiles, and cyber and space capabilities are all redefining and multiplying potential threats to both our nations. These modern security challenges and the potential solutions they bring raise the very same concerns about sovereignty and command and control arrangements that were dealt with from the onset of the Second World War to the creation of NORAD in 1957.

Canadians assigned to NORAD live Canadian sovereignty every day; in defending Canada and the United States we exercise and defend Canadian sovereignty as well. As in the past, our relationship with our US counterparts within the NORAD enterprise remains one of mutual respect and service to both our nations. The fact that we still maintain this type of relationship some sixty years after NORAD's creation is revealing: our history, so eloquently described through this book, is testimony to our achievements and is also an essential compass for the future as we continue to seek solutions to the potential threats that affect both nations.

Pierre St-Amand
Lieutenant-General
NORAD Deputy Commander, RCAF

Acknowledgments

I AM DEEPLY grateful to several individuals and institutions for making this book possible. This study began as a doctoral dissertation at Queen's University, and I appreciate all the assistance I have received in crafting it into a viable book manuscript. Allan English opened many doors for me and also opened my eyes to many different perspectives and approaches during my time under his tutelage at Queen's University, and for that I will always be grateful.

Funding for the doctoral dissertation from which this book derived was provided by generous Department of National Defence (DND) Security and Defence Forum doctoral scholarships, Queen's Graduate Awards, Western Ontario Fellowships, and R. Samuel McLaughlan Fellowships. Travel for research purposes was made possible by a Graduate Dean's Doctoral Field Travel Grant and a Donald S. Rickerd Fellowship in Canadian–American Studies. My thanks go out to the institutions that administer these awards for their financial support. In transforming the dissertation into a book, I am grateful for funding from a DND Security and Defence Forum post-doctoral fellowship and from the Canadian Forces College (CFC). I also thank the Federation for the Humanities and Social Sciences' Awards to Scholarly Publications Program for a financial contribution that helped facilitate this publication.

One of the best advantages of being a historian is the joy of doing research at archival institutions. At the United States National Archives and Records Administration in College Park, Maryland, I thank Beth Lipford, Will Mahoney, and Ken Schlessinger for their assistance in facilitating my research. My appreciation also goes to Paul Banfield and Heather Home at the Queen's University Archives, Carol Reid at the Canadian War Museum Archives, the staff at the Massey Library at the Royal Military College of Canada, Cathy Murphy and her staff at the Canadian Forces College Information Resource Centre, and the Reading Room staff at Library and Archives Canada. I reserve special recognition and appreciation to Steve Harris and his excellent staff at DND's Directorate of History and Heritage in Ottawa. This book would not have been possible without access to DHH's excellent collection of Canadian military history documents and the support and encouragement of the superb people who work in the C.P. Stacey Building. In particular, I thank Mat Joost, the late Charles Rhéaume,

Warren Sinclair, Isabel Campbell, Bob Caldwell, Jean Martin, Kurt Grant, Mike Whitby, John McFarlane, and Bill Johnston.

For the past few years I have greatly benefited from my experience working at the Canadian Forces College: it has been the best tutorial on defence studies and the Canadian Armed Forces that one could ask for. I am particularly grateful and deeply indebted to all my academic colleagues in the Department of Defence Studies at CFC for their support, encouragement, and collegiality, and for broadening my methodological approaches and perspectives. I have also greatly benefited from my many discussions with the military Directing Staff and Curriculum Development Officers at CFC, and my thanks go to them as well. I would like to express my sincere gratitude to CFC Chief Librarian Cathy Murphy not only for indexing this book but also for her unending support, reassurance, friendship, and genuine enthusiasm for all things RCAF and air power. I also thank one of my air power mentors and colleague in the National Security Program at the Canadian Forces College, former Commander of Air Command, Lieutenant-General (ret.) Fred Sutherland, for imparting to me the value of building professional relationships and its vital link to the "human element." At CFC I learn as much from my students as (hopefully) they learn from me. I am grateful for their enthusiasm, professionalism, and service so that people like I and others have the freedom and opportunity to research and write on subjects we are passionate about. Special thanks go to my DS 501 Modern Joint Air Campaigns students, who unfailingly reinforce the fact that air power is, indeed, awesome.

I also thank the organizations that have given me the opportunity to present research from this book at their various conferences and symposia: the Canadian Forces Aerospace Warfare Centre, RCAF History and Heritage, the Centre for Military and Strategic Studies at the University of Calgary, the Canadian Historical Association, the Organization for the History of Canada, the Society for Military History, the Canadian Aviation Historical Society, the Queen's Centre for International Relations, and the Association for Canadian Studies in the United States. Special thanks go to the Laurier Centre for Military Strategic and Disarmament Studies (LCMSDS) for supporting my post-doctoral studies and especially for giving young historians like myself the opportunity to present research at their annual military history symposium. In particular, I am grateful to Mike Bechthold, Roger Sarty, Terry Copp, and Mark Humphries at LCMSDS.

Several people have commented on my passion for the RCAF, and so I would like to thank Canada's professional air power institution for its support. My appreciation goes to the Canadian Forces Aerospace Warfare Centre and especially Colonel Kelvin Truss, Colonel Shane Elder, Lieutenant-Colonel Pux Barnes,

Brad Gladman, Lieutenant-Colonel Doug Moulton, Lieutenant-Colonel Chris Roy, and Lieutenant-Colonel Bruno Paulhus. Special thanks go to Richard Mayne and Major Bill March at RCAF History and Heritage for their friendship, collegiality, and sounding boards as fellow RCAF historians. I also thank RCAF NORAD Deputy Commander Lieutenant-General Pierre St-Amand for writing the foreword to this book. My gratitude goes to my *Airforce* magazine and RCAF Association colleagues Dean Black and Terry Leversedge for their support and encouragement.

I thank the anonymous reviewers of my manuscript for their useful suggestions, which helped make this book a more thorough study. My appreciation also goes to Emily Andrew, Randy Schmidt, Nadine Pedersen, and Megan Brand at UBC Press for shepherding my manuscript through the publication and production process. I am also grateful to Tim Cook and Andrew Burtch at the Canadian War Museum for supporting my book as part of the Studies in Canadian Military History series. I thank Mary Metcalfe from Laskin Publishing for proofreading the manuscript and my friend and colleague Mike Bechthold for drawing the maps. Thanks also go to Brian Laslie at the NORAD History Office in Colorado Springs for supplying photographs.

Undertaking a study like this is not possible without the support of many friends and colleagues. I therefore express my thanks to: Bill March, Matt Trudgen, Howard Coombs, Richard Mayne, Craig Mantle, Andrew Godefroy, Rachel Lea Heide, Danielle Metcalfe-Chenail, Mike Bechthold, Jim Wood, Jeff Noakes, Andrew Burtch, Gary Williams, Tim Cook, Jim Bell, Ken Struthers, Brian Borsc, Mike Borsc, Adam Chapnick, Al Okros, Paul Mitchell, Eric Ouellet, Chris Madsen, Pierre Pahlavi, Chris Spearin, Barbara Falk, Craig Stone, Miloud Chennouffi, Walter Dorn, Dan Eustace, Paul Johnston, Jen Menard, Martin Rosenthal, Andy Torrance, Matt Lillibridge, Carlton "Ike" Turner, Bryan Bedard, Paul Earnshaw, Hugues Canuel, Jay Ballard, MGen (ret) Fraser Holman, Pierre Viens, MGen (ret) Daniel Gosselin, Tanya Grodzinski, Ross Pigeau, Carol McCann, Dan Heidt, Mat Joost, Randy Wakelam, Bert Frandsen, Oliver Haller, Ray Stouffer, Rich Gimblett, and Geoff Hayes. My humblest apologies if I have missed anyone: there are so many who have helped me on this journey and I am most grateful to them all. Any errors that appear in this book are my responsibility alone.

My biggest debt of gratitude goes to my postdoctoral supervisor, Whitney Lackenbauer. He not only read the entire manuscript and provided sage advice throughout the journey of publishing this book, but was (and continues to be) extremely generous with his time, compassion, insights, exuberance, and genuine kindness. He is an outstanding scholar and an even better person, and I am honoured and privileged to call him my friend.

Acknowledgments

Most of all I thank my family who provided encouragement, love, understanding, support, and patience throughout the research, writing, and production of this book. It was during many visits to my grandparents Ina and Erich Goette's house in Hamilton that I fell in love with aviation and air force history while seeing the airplanes from the nearby Hamilton Airport and Canadian Warplane Heritage Museum fly overhead. They, my grandmother Marg Smith, and my Aunt Janis always encouraged my passion for history and I miss them deeply. Special thanks also go to my parents Marlene and Manard Goette, my brother Steve, his wife Clara and their daughter Annabella, my Aunt Anne and Uncle Dan, my cousins Andrew and Marianne and her partner Lisa, and my mother-in-law Ermelinda Carreiro. Most especially, I thank my greatest supporters: my wife Sandra and my daughters Sarah and Kaitlyn. Words simply cannot express my appreciation for the amount of love, joy, and happiness they bring to my life.

I am grateful for permission to incorporate material from the following publications:

Richard Goette, "The Acid Test of Sovereignty: Canada, the United States, and the Command and Control of Combined Forces for Continental Defence, 1940–1945," in *New Perspectives on Canada in the Second World War*, ed. Abe Roof and Christine Leppard (Calgary: University of Calgary Papers in Military and Strategic Studies, Occasional Paper no. 6, 2012), 23–43.
–. "The Command and Control of Canadian and American Maritime Air Power in the Northwest Atlantic, 1941–1943," *Canadian Military History*, 26, No. 2 (Summer/Autumn 2017): 1–27.
–. "Guarding Canada's Approaches: Air Defence Command," *Airforce* 38, nos. 2–3 (2014): 37–42.

Abbreviations

AAA	Anti-Aircraft Artillery
ABDA	American–British–Dutch–Australian Command
A/C	Air Commodore
ACGS	Assistant Chief of the General Staff
ACHQ	Area Combined Headquarters
ADC	Air Defence Command (RCAF)/Air Defense Command (USAF)
ADCC	Air Defence Control Centre
ADSG	Air Defence Study Group
AFHQ	Air Force Headquarters, Ottawa
AHSG	Canada–US Ad Hoc Study Group on air defence
AIAW	Air Interceptor and Air Warning
Air	Air Ministry File, Public Record Office, London
AMAP	Air Member for Air Personnel
AMAS	Air Member for Air Staff
AOC	Air Officer Commanding
AOCinC	Air Officer Commanding-in-Chief
ATIP	Access to Information and Privacy
AVM	Air Vice-Marshal
BSP	Basic Security Plan
CAB	War Cabinet File, Public Record Office, London
CAF	Canadian Armed Forces
CANUSA	Canada–United States Chief of Staff
CAS	Chief of the Air Staff
CCS	Combined Chiefs of Staff
CDC	Cabinet Defence Committee
CFC IRC	Canadian Forces College Information Resource Centre
CG	Commanding General
CGS	Chief of the General Staff

CinC	Commander-in-Chief
CINCADCANUS	Commander-in-Chief Air Defence Canada–United States
CJS	Canadian Joint Staff, Washington
CMSS	Centre for Military and Strategic Studies
CNS	Chief of the Naval Staff
COAC	Commanding Officer Atlantic Coast
ConAC	Continental Air Command
CONAD	Continental Air Defense Command
CSC	Chiefs of Staff Committee
CUSRPG	Canada–US Regional Planning Group
DCAS	Deputy Chief of the Air Staff
DCER	Documents on Canadian External Relations
DCinC	Deputy Commander-in-Chief
DEA	Department of External Affairs
DEW	Distant Early Warning
DHH	Directorate of History and Heritage, Ottawa
DMO&P	Director of Military Operations and Planning
DND	Department of National Defence
EAC	Eastern Air Command
EADP	Emergency Air Defence Plan
EDP	Emergency Defence Plan
FONF	Flag Officer Newfoundland Force
G/C	Group Captain
GOCinC	General Officer Commanding-in-Chief
HQ	Headquarters
JCS	Joint Chiefs of Staff
JPC	Joint Planning Committee
JSPC	Joint Strategic Plans Committee
JWPC	Joint War Plans Committee
LAC	Library and Archives Canada
Maj. Gen.	Major-General (USAF)
MCC	Canada–US Military Co-operation Committee
MND	Minister of National Defence
MSG	Canada–US Military Study Group

NARA	United States National Archives and Records Administration
NATO	North Atlantic Treaty Organization
NDA	National Defence Act
NEF	Newfoundland Escort Force
NORAD	North American Air Defense Command
OPCOM	Operational Command
OPCON	Operational Control
PJBD	Permanent Joint Board on Defence
PRO	Public Record Office, London
RAF	Royal Air Force
RCAF	Royal Canadian Air Force
RCN	Royal Canadian Navy
RG	Record Group
RMC	Royal Military College of Canada
RN	Royal Navy
ROE	Rules of Engagement
RPG	Regional Planning Group
SAC	Strategic Air Command
SACEUR	Supreme Allied Commander Europe
SSEA	Secretary of State for External Affairs
TAC	Tactical Air Command
TACON	Tactical Control
TNA	The National Archives (Britain)
UCP	Unified Command Plan
US	United States
USAAF	United States Army Air Forces
USAF	United States Air Force
USN	United States Navy
USSEA	Under-Secretary of State for External Affairs
VCAS	Vice-Chief of the Air Staff
VCGS	Vice-Chief of the General Staff
WPD	US Army War Plans Division

Sovereignty and Command in Canada–US Continental Air Defence, 1940–57

Introduction

CANADA HAS NEVER fought a war alone. Entering conflicts allied with another country or as a member of a coalition is part of the Canadian way of war.[1] This principle holds equally true for Canada's participation in expeditionary operations and continental air defence. Before the Second World War, Canadian armed forces primarily fought with or within British forces. After 1939, the Canadian military continued this trend, but gradually during the war its forces increasingly operated with American forces. This was particularly apparent in the area of continental air defence, where Canada and the United States worked together to protect North America first from Axis aggression and then from the Soviet bomber threat. In developing their continental air defence partnership, the two countries tackled the central issue of managing command and control over the operations of their combined forces.

This book offers a detailed assessment of the Canada–United States continental air defence command and control relationship from 1940 to 1957, starting with the Ogdensburg Agreement and ending with the establishment of the North American Air Defense Command (NORAD). In particular, it critically examines the degree of command and control authority that each nation was willing to grant commanders to exercise over their combined military forces to accomplish a mission. This authority was embodied in specific command and control principles, the definitions of which varied depending on the degree of authority and responsibility exercised. This book shows that the effort to find compromises between the unique joint command and control cultures of Canada, the United States, and Britain had a significant effect on the growth of the Canadian–American defence relationship. Equally important, history reveals that Canada upheld its sovereignty through this continental air defence command and control relationship.

National Command

> *The concern for command arrangements is not entirely based on emotion or pedantry. They are real concerns of national sovereignty.*
> – Douglas Bland[2]

In 1941, Lieutenant-General Andrew McNaughton, Canada's top soldier and future Minister of National Defence, remarked that "the acid test of sovereignty is the control of the armed forces."[3] This study on Canadian–American command and control arrangements encourages revision of McNaughton's statement to reflect the reality of the bilateral defence relationship. Definitions of command and control principles continually evolved from 1940 to 1957, and military personnel used terms with less precision than do members of today's Canadian Armed Forces (CAF).[4]

As McNaughton's quote demonstrates, military personnel often used the words "control" and "command" interchangeably. This is problematic, particularly as Canada's degree of command and control authority over its military forces has re-emerged as an important issue in the twenty-first century. Recent continental defence policy and organizational developments include the establishment of the United States Northern Command in 2002; the ongoing debate regarding Canadian participation in American ballistic missile defence; the establishment of Canadian unified commands in 2005 and Canadian Joint Operations Command (CJOC) in 2012; NORAD's expanded maritime surveillance mission and renewal "in perpetuity" in 2006; the possible enlargement of NORAD to include all continental defence responsibilities in addition to its aerospace defence role; and the 2017 Canadian Defence Policy's emphasis on a Canada "secure in North America" by being "active in a renewed defence partnership in NORAD and with the United States."[5] In several of these cases, some commentators fear that Canada's command and control relationship with the United States is undermining Canadian sovereignty.

For instance, in anticipation of the possible creation of a new Canadian–American North American Continental Defence Command, international law professor Michael Byers stresses that the NORAD operational control arrangement (discussed in Chapter 8) erodes Canadian sovereignty. In his report "Canadian Armed Forces under US Command," Byers asserts that the NORAD definition of "operational control" is too broad and includes "many of the powers civilians would envisage as falling within command," leading him to assert that "concerns about sovereignty cannot be overcome by technical distinction between 'command and operational control.'"[6]

This book offers an alternative hypothesis based on a careful examination of the historical evolution of command and control principles, including operational control. Whereas Byers asserts that "control over one's armed forces is regarded as a central quality of a sovereign state," this study demonstrates that the retention of *command* over a country's armed forces is a central quality of a sovereign state.[7] The difference between command and operational control is not mere semantics. It represents a fundamental feature of Canadian military tradition and command and control culture: to retain command when operating with American armed forces in continental defence.[8] The purpose of this study, therefore, is to demonstrate that the key issue for Canada in its continental defence relationship with the United States was not control but *command*: command over Canadian air defence forces was the actual "acid test of sovereignty." Canada was able to maintain command of its forces in all continental air defence command and control arrangements with the United States, thus securing Canadian sovereignty.

The Canada–United States continental air defence relationship deliberately excluded administration and discipline over Canadian forces in bilateral command and control arrangements. Instead, authority over these matters has remained a national prerogative.[9] It formed part of the command or "national command" that Canada's civilian government maintained over Canadian forces, exercised through its military chiefs of staff – a practice that continues. From 1940 to 1957, these national military authorities were the Canadian Chiefs of Staff Committee (CSC) and, more specifically, the RCAF's Chief of the Air Staff (CAS) for air force operations. All aspects of national command were a service (i.e., air force, navy, and army) prerogative, including authority over the assignment and original composition of forces, logistics, and administration and discipline. These powers could not be assigned or delegated to a commander from a foreign nation; the retention of national command in command and control arrangements with other countries was an effective guarantee of Canadian sovereignty.[10] Moreover, as Chapter 2 demonstrates, Canada's retention of national command as a service prerogative was consistent with American command and control practice and culture.

Chapter 1 examines the link between command and sovereignty, demonstrating how the source of the Canadian military's national command was (and remains) the Crown and flows down the military chain of command. The National Defence Act (NDA) granted the three service chiefs of staff responsibility for overall control and direction of their respective services under the direction of the Minister of National Defence (MND).[11] The heads of Canada's air force, navy, and army thus had the authority to assign forces to a command organization and to delegate operational command of them to operational-level commanders for specific tasks. This arrangement ensured a clear chain of command up from the operational-level commanders to the service chiefs of staff and onward to the defence minister.[12] In 1941, Brigadier-General Maurice Pope explained that the function of the service chiefs in Ottawa was

> to assign missions and to provide the means necessary thereto. It is for the local [i.e., operational-level] commanders ... to execute the missions they receive. The Department of National Defence can only exercise its true function by means of directives. Any action on its part to take charge of operations as such, would simply hamper the responsible commander in the field.[13]

A service chief could technically exercise operational command of his service's forces, though this was not normal practice because he "would simply hamper the responsible commander in the field."[14] The exercise of operational-level command and control authority by a service chief also threatened to draw attention away from his national command responsibilities at the strategic level in Ottawa.[15]

Instead, a service chief usually delegated operational command to his service's operational-level commanders. For example, the RCAF's Chief of the Air Staff delegated operational command to the Air Officer Commanding (AOC) Eastern Air Command (EAC) in Halifax during the Second World War and to the AOC RCAF Air Defence Command (ADC) in St-Hubert during the 1950s. A service chief could also grant operational command (or operational control) to a commander from a different service for joint operations or, if necessary, to a foreign commander in command of bilateral or multilateral forces.[16] As the research in this book reveals, the subject of delegating operational command or operational control of RCAF forces to an American commander was a key issue in the Canada–US continental air defence command and control relationship.

Service culture and operational factors therefore played a large part in the decision to permit one's forces to come under the operational command or operational control of a foreign commander. From 1940 to 1957, national command and also operational command of Canadian air forces defending North America remained with the Canadian Chiefs of Staff, specifically the Chief of the Air Staff, and through him RCAF operational commanders. In this way, the Canadian military was able to protect Canada's sovereignty in its continental air defence relationship with the United States.

This book is written as Canadian military history, focusing specifically on the Royal Canadian Air Force's role in the maintenance of Canadian sovereignty from 1940 to 1957. The central aim is to show how Canadian military professionals ensured effective command and control arrangements for the prosecution of the Canada–US continental air defence effort while maintaining Canadian sovereignty by retaining command over air defence forces. This study takes a more military functional approach to the maintenance of Canadian sovereignty – a perspective that needs to be explored in further detail in order to properly understand Canada–US continental air defence cooperation between 1940 and 1957.

A Functional Approach

Borrowing a designation from American political scientist Joseph Jockel, this book takes a 1957 "functional view" instead of a 1958 "political view" of the Canada–US continental air defence relationship and the founding of NORAD. The former concerns the September 1957 establishment of NORAD as a binational Canada–US command with operational control authority over the two nations' air defence forces. This perspective, based on the functional military professional imperative (discussed in greater detail in Chapter 1), focuses on the purely military or "functional" aspects of the Canada–US continental air defence relationship, particularly the efforts of officers from both countries to ensure effective coordination at the operational level to defend North America. By contrast, the political perspective (based on the May 1958 exchange of diplomatic notes as NORAD's founding date) focuses on intergovernmental relations. It places greater weight on officials from Canada's Department of External Affairs (DEA), who emphasized the "political" features of the continental air defence relationship such as diplomatic procedure and enhanced strategic consultation.[17]

"Functionalism" or the "functional principle" was a popular and much discussed Canadian foreign policy subject from the mid-1940s to the early 1950s. Representative of Canada's internationalism and greater engagement in world affairs during this "golden age" of Canadian foreign policy, it entailed "the idea that state participation in international affairs be determined on an issue by issue basis, contingent upon the state's interests, its prior contribution and its capacity for further involvement."[18] While the military functionalism examined in this book is similar to the foreign policy functional principle in that both emphasize Canadians contributing to international endeavours and increasing Canadian influence, it also has important differences.

The functional 1957 perspective highlights that Canadian forces played an active and integral part of the overall air defence effort with the United States. Whereas the 1958 political perspective focused on Canada having a "seat at the table" to ensure a say in continental defence policy and strategy, the 1957 functional perspective allowed for a more intimate operational relationship between Canadian and American military personnel. This exemplifies what Canadian political scientist Joel Sokolsky has called a "seat at the console" – a concept of equal if not more importance than "a seat at the table" because it allows Canadian officers at the operational level, working hand in hand with their American colleagues, to exercise and safeguard Canadian sovereignty while at the same time fulfilling an important operational role.[19]

Through a functional approach, the Canadian armed forces also negated a "defence against help" situation with the United States. As W.A.B. Douglas observed, "so often in coalition warfare, large and powerful allies sometimes seemed to pose the greatest threat."[20] The concept of "defence against help" thus dictates that a country has to establish and maintain military credibility in order to avoid unwanted "help" from its larger neighbours. As theorist Nils Ørvik explained: "One credible objective for small states would be, while not attempting military resistance against a large neighbour, to persuade him that they are strong enough to defend themselves against any of the large neighbour's potential enemies. This could help avoid the actual military presence of the great neighbour on one's territory for reasons of military 'help' and assistance."[21] Although Canada was (and still is) the larger partner geographically in the North American defence relationship, the United States was (and remains) much greater in terms of population, economic power, and military might. These factors, coupled with the strategic reality that Canada's geography placed it between the United States and its enemies, meant that the "threat" that Douglas referred to put Canada in a classic potential defence-against-help situation from 1940 to 1957.

The concept of defence against help stressed that Canada maintain a certain credible level (in American eyes) of defensive capabilities. Canada could not take a "free ride" when it came to defence: if the United States lost confidence in Canada's ability to defend itself, it might usurp Canadian sovereignty by taking independent unilateral action in Canadian territory, waters, or airspace to protect US security and defence interests.[22] General Pope articulated the dilemma succinctly in a 1944 letter to a colleague on the topic of postwar planning with the United States: "To the Americans the defence of the United States is continental defence, which includes us [Canada], and nothing that I can think will ever drive that idea out of their heads ... What we have to fear is more a lack of confidence in the United States as to our security, rather than enemy action."[23] When Canada began collaborating with the Americans in continental defence during the Second World War it faced a defence-against-help situation; and that situation was only enhanced during the Cold War period with the rising threat posed by the Soviet Union. "Canadian actions would be determined not by 'what we think the Soviets might do,'" Louis Grimshaw explained, "but rather by 'what we think the Americans think the Soviets might do.'" Preventing American unilateral action in Canadian territory to defend its northern flank became a key concern of the Canadian government.[24] From 1940 onwards, Canada had to ensure its security from the enemy through close collaboration with the United States while at the same time safeguarding its sovereignty in the context of bilateral defence cooperation.[25]

While some literature on Canada's early defence relationship with the United States has a negative tone, this study pursues a more positive approach. Pessimistic defence commentators have argued that the military situation during the Second World War and the Cold War placed Canada in the unenviable position of having to bend to American pressures and become a US "satellite" or "protectorate."[26] By contrast, this study shows that in its continental air defence command and control relations with the United States, Canada was able to protect its sovereignty by avoiding a defence-against-help situation. Canadian steadfastness was important, but so too was the accommodating and respectful approach that the United States took with its northern neighbour. Granted, some American officials were annoyed by Canadian resoluteness – what one called Canada's "pride and little brother attitude."[27] Nonetheless, the Americans were also quite cognizant of Canadian sensibilities regarding sovereignty and command. For this reason, they did not try to coerce Canada into accepting the United States' position.[28]

The Americans did not violate or undermine Canadian sovereignty by bullying Canada and imposing undesirable command and control arrangements. Instead, respectful of Canada's opinions, the Americans did what they have always done and continue to do: they sat down and negotiated mutually agreeable solutions that respected Canadian sovereignty and that placed limits on American command and control over Canadian forces. This was entirely consistent with the historical record of Canadian–American relations as they relate to sovereignty (especially in the Arctic) in that Canada was able to balance sovereignty and security interests, the United States did not undermine Canada's sovereignty, and collaboration and compromise remained an effective and viable approach.[29]

After 1940, the Canadian military (and specifically the RCAF during the early Cold War period) also aided the cause of Canadian sovereignty through its attitude towards the continental defence relationship. Instead of taking a belligerent or defensive stance, the Canadian military advocated playing an active role by developing effective and efficient bilateral command and control arrangements with the Americans. In other words, Canadian officers chose what historian Whitney Lackenbauer has called a "piece of the action" approach: participating in continental defence efforts with the United States enabled Canada to protect its sovereignty from American intervention.[30] Early Cold War RCAF officers definitely understood this requirement. Writing in 1954, Air Commodore W.I. Clements observed:

> I feel that one of *the most important* considerations is that now, in peacetime, we have a good opportunity of getting a set-up that would suit us or come somewhere near it ("us" being Canadians – government, services and civilians). If nothing is done until war comes we might find things moving with great rapidity and the Americans might, on the excuse of national survival, suddenly take over everything overnight and if New York, etc., were being hydrogen bombed Canada's complaints about national sovereignty might not be heard above the other noises. I feel Canada should take the initiative now in view of what we stand to gain.[31]

What Canada stood to gain was a "seat at the console" for the RCAF. Canadian airmen would have a "piece of the action" in terms of an important operational role in continental air defence while simultaneously safeguarding Canadian sovereignty.

Active participation with the United States in air defence also promised another, often overlooked advantage for Canada. Not only would Canadian sovereignty be protected, but working with the Americans would also help protect Canadian territory and people from enemy attack.[32] Canada's relationship with the United States was not simply a choice between sovereignty *or* security: by engaging with the Americans in the defence of North America, Canada could have both.[33]

In summary, this book fleshes out Lackenbauer's "piece of the action" hypothesis. Despite the overwhelming power of the United States, Canada took an active role in arranging an effective continental air defence command and control relationship with the Americans, which helped ensure that Canadian forces did not come under US command. Canada was able to avoid a defence-against-help situation with the United States, maintain Canadian sovereignty, and provide effective air defence of the continent.

Distinguishing between Continental Defence and Coalitions

There are unique circumstances in a continental defence situation compared to coalition undertakings that dictate a different approach to command and control. In the experience of Canada and the United States, a coalition effort, whether it is a formal military alliance such as NATO or a collection of "two or more nations' militaries working together to support a specific objective," is typically expeditionary and almost always consists of various nations. In this case, the term "combined" (denoting the multilateral interaction of forces from two or more countries) applies.³⁴ Command and control of operations in coalitions can be very complex and inefficient due to varying and often conflicting national interests. Martha Maurer describes this best:

> Within a coalition, common cause and mutual interest are balanced against minority views and national interests. One body has one head and one perspective, but a coalition has many heads and many national views reflecting economic, cultural, and institutional differences. The motivation and self-interest that underline the development of a coalition must be powerful enough to counter the forces of separation. Yet, divisiveness remains part of the nature of the coalition and that tension must be acknowledged.³⁵

Some differences can be avoided in formal military alliances such as NATO, where there is greater standardization of equipment, doctrine, and approaches to operations.³⁶ This is more difficult in coalitions consisting of nations with greater cultural differences. Again, Maurer's observations are illustrative: "any coalition can be overlaid with regional variations of politics, ethnic and cultural values, and religious influences. These differences may extend into the command and control arena. Different philosophies of life or world view (Western, Asian, Arab) may influence national theories of command and control and, therefore[,] of military doctrine."³⁷ Although Canada–US continental defence has some similarities to coalition command and control, there are also important differences.

The close regional Canada–US defence partnership makes the North American continental security relationship unique. Although the term "combined" could technically apply to the Canada–US continental defence effort, the terms "bilateral" or "binational" are more accurate and are accepted in the lexicon of the two countries. Whereas bilateral denotes cooperation between two separate national military chains of command, binational entails (in addition to two separate chains of command) the integration of two nations' military forces into a single, unified entity. In the case of Canada and the United States, that organization has been institutionalized in the form of NORAD, the world's most successful binational military command.³⁸

The most important distinctive feature of continental defence is that it is the defence of one's sovereign territory *in partnership* with another nation in the defence of their territory. There are correspondingly important command and control considerations. In national (domestic) defence operations one has full national command over one's forces and there are no restrictions on the authority of one's operational commanders.³⁹ In coalitions there are multiple restrictions and caveats. Continental defence is different because it is neither purely national defence nor is it coalition warfare. Because part of continental defence involves defending one's own territory, authorities and inherent sovereignty concerns need to be taken into account in command and control arrangements. It was therefore predictable that Canadian military leaders adopted a harder line to ensure Canadian command over its forces in its continental air defence relationship with the United States.

One reality of both coalition (combined) and continental defence (bilateral) command and control is that the more substantial the contribution by a nation – in terms of both number of forces provided and financial responsibility assumed – the greater influence it will have over the command and control of operations.[40] The nation that contributes the most will in all likelihood provide the overall commander exercising command and control authority over the assigned (combined or bilateral) forces. The size of contribution became a key consideration for Canadian and American planners and commanders during the Second World War (notably in Newfoundland, as Chapter 4 demonstrates) and was one of the main reasons why the USAF took a leading role when Canada and the United States began coordinating and then integrating their air defence forces during the early Cold War period. It is also the primary reason why the NORAD Commander has always been an American four-star officer (with a Canadian air force three-star deputy).

Since the air defence mission dominated the problem of continental defence in North America by the early Cold War, this book focuses on the Canada–US continental *air* defence command and control relationship. During the Second World War, however, air defence formed an integral part of the overall continental defence effort – army, navy as well as air force – in both countries. When examining that time period, therefore, this study analyzes the bilateral air defence command and control relationship within more general continental defence efforts. The air forces that Canada and the United States devoted to continental defence during the war also played an important part in the defence of Allied shipping. Since both countries' air forces conducted most operations in a maritime trade defence role, the issue of command and control over maritime air forces became intimately involved with the overall Canada–US debate over continental defence forces. So too did military culture.

Military Culture, Institutional Service Identity, and Motivations

Military culture embodied in institutional service identities, interests, and other motivations related to the profession of arms played an important role in Canadian–American negotiations for continental air defence command and control arrangements. The military profession possesses what political scientist Eliot Cohen calls a "corporateness or a sense of community and commitment to members of one's group."[41] The Canadian planners who conducted the command and control negotiations identified themselves as being part of a unique profession and as an integral part of an important national institution. In addition to identifying with the profession of arms writ large, Canadian military officers identified with their particular operational environment or service. This was embodied in service culture, and identifiers included specific customs and traditions associated with membership in the army, navy, or air force, but also particular institutional knowledge and skills of the service – military expertise.[42]

Professional military identity was also transnational. As American political scientist Samuel B. Huntington noted, "the possession of a common professional skill is a bond among military officers cutting across other differences," including nationalities and borders.[43] Canadian officers saw themselves as Canadian military professionals but also as part of the community of Western military professionals. This is what Joel Sokolsky calls the "international fraternity of the uniform." It emphasizes the military functional imperative and consists of a "set of institutional and personal ties, which can exist almost independently of governments. In this fraternity, allegiance is to the common goal and the military means of implementing it."[44] Traditionally, Canadian officers identified most with the British military and thus adopted British military culture and traditions. During the early Cold War period, however, they increasingly came to identify with their American allies. This was especially so for Canadian airmen.

An important part of the identity of Canadian air force officers was the RCAF's institutional knowledge of and skill in air power. Because these officers identified themselves as airmen, service concerns about the RCAF as a professional air power institution – including the future of that institution[45] – were key features of continental air defence command and control arrangements with the United States. During the early Cold War period in particular, this institutional air power service culture also meant that RCAF officers began to increasingly identify with their USAF brethren. Airmen on both sides of the border saw air defence as a common continental problem and agreed on the interrelated goals of defending people, cities, and industries in North America from Soviet strategic bombers armed with atomic bombs and protecting the deterrent value of the USAF's Strategic Air Command (SAC) as a means to prevent the outbreak of a nuclear war.[46] As James Fergusson notes, "for the two air forces, strategic air defence cooperation made simple functional sense, reinforced by the natural ties that exist via organizational culture between functionally identical military services."[47]

The RCAF dedicated a significant amount of its resources and attention to the air defence role in conjunction with the USAF. This included designing and fielding all-weather jet interceptors such as the CF-100 Canuck, building American F-86 Sabre fighter aircraft in Canada, and working in conjunction with the United States to build a series of radar lines in Canada's North.[48] The RCAF also sought to achieve greater standardization, interoperability, and integration with the USAF. In 1947, the Canada–US Permanent Joint Board on Defence (PJBD) released a Joint Statement calling for "common designs and standards in arms, equipment, organization, methods of training and new developments" with the United States, and in 1948 the RCAF was a signatory to the Air Standardization Coordination Committee, the forerunner of today's Air Space Interoperability Council.[49] The RCAF's increasing identification with the United States Air Force (USAF) was also an important cultural paradigm shift for Canadian airmen.

During the early Cold War period, Canadian airmen began to identify less with Britain's Royal Air Force (RAF) and more with the USAF. The RCAF adopted much of the USAF's doctrine and methods, sent some of its officers to American professional military education institutions such as the National Defence College and the Air University, and began posting officers to the USAF Air Defense Command to ensure greater consistency in practices and a common operational picture. Historian Ray Stouffer notes that Canadian and American airmen "came from similar social backgrounds" and that working closely with the USAF "was less [of] a cultural affront" to the RCAF because American airmen treated the Canadians as allies, unlike the British, who had treated them as "colonials" – which the Canadians resented – during the Second World War, when Canadian airmen deployed overseas served under the RAF (see Chapter 2).[50] Here the value of the "human element" was readily apparent.

This study demonstrates the importance of the "human element" of individual professional interaction in the Canada–US continental air defence relationship. Canadian military personnel had to foster common views and cordial and effective working relationships with individuals from their own service, members from the other Canadian services, and American personnel as they planned and worked together at the operational level. The relationship between Canadian and American personnel could be difficult and confrontational at times, but the development of positive, personal working relationships between like-minded people promoted close bonds and encouraged common views on air defence, especially during the early Cold War. From an air force perspective, professional interactions, mutual respect, and cordial working relationships enhanced the bonds between Canadian and American airmen and the interoperability between their services.[51] Indeed, this shared identity as airmen working together in the continental air defence effort strengthened the functional imperative and was a key factor in the increased coordination, integration, and centralization of Canadian–American air defence command and control during the 1950s, which culminated in the formation of NORAD in 1957.[52]

The cultural bonds between Canadian and American airmen also served the RCAF's institutional interests. Canadian airmen were able to convince their political masters of the primacy of air power in the nuclear age and thereby further the RCAF's institutional goals.[53] As James Eayrs has noted, during the early Cold War the RCAF's "role was more easily defined, its status more prestigious, its connections more powerful. Its funds, in consequence, were more plentiful and its future more assured."[54] It was no surprise that during these "golden years" the RCAF garnered the lion's share of the Canadian defence budget, a significant portion of which was dedicated to air defence. With this increased funding came increased responsibility and accountability as well as difficult choices regarding priorities, but the RCAF leadership was able to successfully advocate air power (especially the air defence role) and leverage its growing bond with the USAF to further its institutional goals.[55] Moreover, by focusing on the air defence role and greater integration and interoperability with the USAF in carrying out this role, the RCAF was also able to reinforce its independence from the Canadian Army and Royal Canadian Navy.[56] Yet by no means did the growing military bond between Canada and the United States mean that Canadian officers submitted entirely to American wishes.

Canadian nationalism was also an important motivation in the Canada–US continental air defence command and control relationship. It was not just politicians and officials from the Department of External Affairs who, anxious to consolidate gains made during the interwar period in securing greater Canadian autonomy, were desirous of freedom of action and weary of American domination of the Canada–US defence relationship. Canadian officers identified as Canadians and expressed nationalist passions in their efforts to retain command of Canada's military forces.[57] For personal and nationalistic reasons, no Canadian officer indicated that he was keen to be forced under foreign command.[58] Chapters 3 and 4 demonstrate this nationalism most clearly in the context of the negotiations for the ABC-22 Canada–US defence plan and the operational relationship in Newfoundland during the Second World War. Different national cultural approaches to military matters between Canada and the United States affected their command and control relationship. This included conflicting national strategic estimates of the enemy threat to North America as well as differing strategic, organizational, and doctrinal approaches to air power, all of which were instrumental to Canadian success at resisting American unity of command during the Second World War. Canadian officers went to great lengths to retain Canadian command when negotiating command and control arrangements with the United States, affirming their nationalist motives and concerns about protecting Canadian sovereignty.

Nationalism and differing national approaches to military matters also extended into the early Cold War period. During this time, American strategic culture was very offensively oriented towards the nuclear-armed Strategic Air Command and strategic defence was thus not as much of a priority for the United States as it was for Canada. This factor would play into discussions on the future of air defence and Canada's role and relative importance. The main purpose of the RCAF's air defence system was to protect the SAC deterrent and provide warning time for USAF strategic bombers to launch a counterstrike against the Soviet Union – a role that was in direct support of (and therefore made an important functional contribution to) the overall American offensive strategic posture. Nonetheless, this RCAF role in the continental air defence system was still doctrinally a defensive one that appealed to the Canadian public and politicians and served Canadian security interests.[59] Moreover, Canadian airmen were also not as engrossed in SAC nuclear bomber theory and doctrine as their USAF counterparts. To be sure, the RCAF grew closer to the USAF in terms of its culture and identity, but it is notable that the RCAF adopted a "fighters first" emphasis on air defence and retained its distinctive British-style rank structure and uniform during the early Cold War period.[60] Furthermore, James Eayrs credits Canadian airmen for their cautious approach of "capitalizing on (but capitulating to) doctrines of air power sedulously propagated in the United States" and "reject[ing] the simplistic SAC approach to international problems."[61] James Jackson, a professor at the RCAF Staff College during the 1950s, echoes this sentiment, noting that the RCAF "avoid[ed] the USAF's excesses."[62] The RCAF's functional approach to air defence found favour in Ottawa, which gave it a high priority in terms of political and material commitment. Ottawa devoted greater relative political and military attention to continental defence arrangements than Washington, a factor that helps explain how and why Canada was able to use its functional approach to protect Canadian sovereignty.

Chapter Summary

This book has eight chapters. Chapter 1 delves into the issue of command and control, articulating the various theories and approaches on the subject, and shows the link between command and sovereignty. It also situates this study in civil–military relations (CMR) theory and Canadian Profession of Arms doctrine. Chapter 2 discusses command and control culture by examining specific command and control principles, concepts, organization, and ideas, demonstrating that the Canada–US continental air defence relationship was based on a combination of British, American, and Canadian command and control culture and practice, categorized into five individual "systems." Chapter 3 argues that the unique Canadian and American command and control cultures conflicted once the two nations began planning together for the defence of North America in the early years of the Second World War. Nonetheless, the two countries were eventually able to put their differences aside and find middle ground by establishing a compromise cooperation/unity of command system in ABC-22. Chapter 4 examines the relationship between Canadian and American operational-level commanders during the war. Focusing on Newfoundland, where the forces of both nations were stationed, and on the Pacific coast after Pearl Harbor, it shows that national disagreements persisted among commanders regarding what the command and control relationship between their forces should be. When it became apparent that the Canadian system was not working for combined Canada–US maritime air operations in the Northwest Atlantic, Canadian and American military officials implemented the more effective British operational control system, establishing an important precedent for the postwar period.

Chapter 5 outlines the development of a new Soviet aerial threat to North America in the early postwar period and Canadian efforts to avoid a potential defence-against-help situation with the United States. Canada–US planning changed from a generalized continental defence focus to concentration on air defence, though the Canada–US Basic Security Plan (BSP) largely retained the same command and control arrangements. Chapter 6 examines bilateral emergency defence planning and argues that Canada and the United States began to organize their national air defences under centralized command and control in response to the growing strategic threat from Soviet bombers, now armed with nuclear weapons, in the late 1940s and early 1950s. The two countries agreed to new arrangements for cross-border fighter interception and mutual reinforcement, which had important command and control implications.

Chapter 7 offers a case study of the RCAF's command and control relationship with the US Northeast Command. The United States and Canada formalized the principle of operational control in national airspace through an arrangement that allowed the RCAF to exercise operational control over USAF aircraft operating in Newfoundland airspace. This arrangement established a crucial precedent for NORAD, the culmination of the Canada–US continental air defence command and control relationship, which is critically re-examined in Chapter 8. When USAF and RCAF officers sought to integrate North American air defences and centralize them under one overall commander, Canadian officials expressed deep concern that they might be surrendering sovereignty by giving command of Canadian air forces to the Americans. This final chapter argues that sovereignty concerns were allayed by deciding, at the suggestion of the RCAF leadership, to formalize operational control as the command and control principle exercised by the NORAD Commander-in-Chief (and his RCAF deputy in his absence).

Terminology and Language

Precise terminology is important. Definitions of command and control principles were still evolving in the 1940–57 period, and military personnel often used terms such as "control" and "command" interchangeably to mean the same thing. This study reveals how modern definitions of command and control terms arose from years of evolution and change. For the sake of clarity and consistency, this book adopts modern definitions of command and control principles from current Canadian Armed Forces doctrine. In quotations that contain command and control terminology that does not adhere to current definitions, the modern command and control term is placed in square brackets.

It is also important to define the levels of war or conflict. The strategic level is the one "at which a nation, or group of nations, determines national or alliance security aims and objectives and develops and uses national resources to attain them." The operational level is the one "at which campaigns are planned, conducted and sustained to achieve military strategic objectives within an area of responsibility." The tactical level is where "military actions are planned and executed to achieve the military objectives assigned to tactical formations and units."[63] The operational level thus connects the strategic and tactical levels and focuses on the use of military forces to accomplish missions. Command and control of Canadian air forces at the strategic level in Ottawa was quite distinct from the authority that Canadian or foreign commanders exercised at the operational level, which is where the most important command and control interaction between Canada and the United States took place from 1940 to 1957. Consequently, this book focuses on the evolution of operational-level command and control arrangements between the two countries.

In the 1940–57 period (and especially during the Second World War), the operational level of conflict remained a relatively new concept in the English-speaking world. Joe Sharpe and Allan English explain that Commonwealth aircrews during the war saw the term "operational" as "indicat[ing] that someone was ready to go on 'ops' as opposed to still being in training."[64] The exact meaning of the term "operational" and the relationship of the operational level to the other two levels of warfare continued to evolve during the period studied. Military personnel used the term "operational" when referring to the operational level of warfare but also freely used the terms "strategic" and "tactical" interchangeably to mean the same thing.

Service personnel also used the terms "joint" and "combined" interchangeably, and oftentimes even in reverse of today's usage. To be consistent with current military terminology, the term "joint" in this book will mean two or more services, while "combined" will refer to multilateral (two or more nations) interaction, usually in a coalition or an alliance.[65] This study also uses the most recent Canadian Armed Forces doctrine to serve both as a basis of understanding for the reader and as a comparison to historical military terminology.

1
Command and Control, Sovereignty, Civil–Military Relations, and the Profession of Arms

AN UNDERSTANDING OF COMMAND and control and the relationship between these two military concepts is important to theoretically and doctrinally situate Canada's 1940–57 continental air defence relationship with the United States. Furthermore, in order to demonstrate that Canada's retention of command safeguarded Canadian sovereignty, it is necessary to examine the link between sovereignty and command. By linking command to sovereignty, this study makes a new contribution to the discourse on the role that a nation's military plays in the preservation and exercise of a state's sovereignty. As subsequent chapters show, the maintenance of national command and limiting the degree of authority given to a particular commander in bilateral command and control arrangements was the key concern for planners. Finally, examination of civil–military relations theory and Canadian profession of arms doctrine demonstrates that command and control arrangements, though definitely concerned with functional issues of military efficiency and effectiveness, also had broader political implications related to sovereignty that planners had to consider.

What Is Command and Control?

Command and control refers to the degree of authority that a nation grants to a commander to be exercised over assigned military forces to accomplish a mission.[1] This authority is embodied in specific command and control principles, which vary depending on the degree of authority and responsibility exercised. Command is a unique aspect of the unique military profession. The most important aspect of the military profession that distinguishes it from other professions is unlimited liability: the requirement to put oneself – or order (or "command") others – into harm's way.[2] There is a requirement for a system to control this uniqueness – what Samuel B. Huntington calls the "management of violence" – and this is a role that military command and control plays. Because the military is a collective profession, it is the collective as a whole that acts, and this requires a higher degree of organization and specialization that is very unique to the military profession. Command is therefore part of the military profession's acquisition and use of specialized knowledge, one that becomes part of its unique expertise.[3] Control derives from command (see below), and both are formally and legally defined in doctrine.

The most commonly recognized conceptualization of command and control – especially among those in military uniform – is what can be described as "legal" or "doctrinal" command and control.[4] Usually articulated in military doctrine manuals, planning documents, and official government publications, doctrinal command and control specifically outlines the degree of authority that a commander can exercise over assigned forces. With regard to the command and control of continental air defence examined in this book, doctrinal command and control consisted of the principles that were the subject of Canadian–American negotiations. As it was Canada's retention of command that ensured sovereignty, it is important to distinguish the differences between command and control.

Command refers to the legal and personal authority derived from a nation's governing institutions for the proper management, administration, deployment, and performance of the military.[5] The official definition in current Canadian military doctrine – which is based on classic NATO doctrine – reflects this premise, placing emphasis on authority vested in one individual: "the authority vested in an individual of the armed forces for the direction, co-ordination and control of military forces."[6] The emphasis on the individual is an important feature of the legal/doctrinal definition of command in terms of personal authority and responsibility; as Kenneth Allard notes, "one of the most striking characteristics" is how it evokes "the personal nature of command itself, especially the fact that it is vested in an individual who, being responsible for the 'direction, coordination, and control of military forces,' is then legally and professionally accountable for everything those forces do or fail to do."[7]

Control derives from command but is subordinate to it, focusing on a commander's use of forces for operational and tactical purposes. It is officially defined in doctrine as "the authority exercised by a commander over part of the activities of subordinate organizations, or other organizations not normally under his command, which encompasses the responsibility for implementing orders or directions. All or part of this authority may be transferred or delegated."[8] The distinction between command and control is essential. As defence scientist Ross Pigeau has noted, "whereas command constitutes formal authority, control derives by delegation from command; whereas command provides oversight, unifying all action, control supports command in detail; whereas command is focused on establishing common intent, control is focused upon the details of execution."[9] In other words, command entails formal authority while control provides direction or a *means* of exercising effective command.

Together, command and control in doctrine is defined as "the exercise of authority and direction by a commander over assigned, allocated and attached forces in the accomplishment of a mission."[10] The distinction between command entailing formal authority and control consisting of execution is the reason the retention of command and not control is central to the protection of sovereignty. The link between command and sovereignty will be discussed in greater detail later.

This book documents how Canadian and American officers debated and negotiated doctrinal definitions of command and control as the basis of the Canada–US continental air defence relationship from 1940 to 1957. Indeed, modern doctrinal command and control principles originated during the Second World War and evolved and matured significantly during the Cold War. Nonetheless, since multiple concepts and theories of command and control other than the doctrinal/legal ones have been developed in recent years, a brief discussion of some of the most pertinent ones is in order.

One of the main theories examines command and control as a decision-making exercise, exemplified in the work of Thomas Coakley. Coakley attempted to demystify the arcane command and control world by explaining command and control in plain language, using common sense and ordinary historical illustrations for the general reader. He highlights that command and control exists in different contexts: terminology focused, in the technological realm, in the human context, and organizationally.[11] Coakley's emphasis is on command and control processes and systems, and for an analogy he turns to the central nervous system. He describes command and control "as a nervous system which coordinates the muscles of [the] national security system." The sensory nerves (control functions) detect what is going on within and outside the body and sends findings to the brain. The brain (the commander) interprets the findings, compares the existing condition to the desired condition, decides on a course of action, and sends the appropriate orders to the muscles by way of the motor nerves.[12]

Coakley emphasizes the importance of having a command and control system of radars, computers, headquarters staff, and so on, to effectively carry out the battle. He argues that having a superior command and control system can negate numerical inferiority and that *not* having such a command and control system – being in the dark – can be crippling. The example he gives is the 1991 Gulf War, when US-led coalition forces destroyed or incapacitated Iraqi command and control systems, leaving Saddam Hussein's forces disoriented. In other words, it is important to target the enemy's command and control, but it is also important to anticipate how the enemy will target one's own command

and control and be sufficiently flexible and agile to adapt in order to minimize the impact on one's own decision-making.[13]

Joel S. Lawson's model of command and control also emphasizes decision-making, but his accommodates five functions. The *sense* function gathers data on the environment – the world "out there," including friendly and enemy forces, allied forces, terrain, weather, and so on. The *process* function draws together and correlates the data to give the commander information about the environment. The *compare* function juxtaposes the existing state of the environment – the relative strengths, weaknesses, positions, and so on – with the desired state, that is, the commander's view of what the state of the environment should be. The *decide* function chooses among available courses of action for reconciling the existing state of the environment with the desired state. Finally, the *act* function translates the decision into action.[14]

Another important decision-making command and control theory, especially in air force circles, is that of USAF fighter pilot and air power theorist John Boyd. His OODA (Observe, Orient, Decide, Act) Loop theory stresses that the key to military victory is "getting inside" the enemy's decision cycle: to work through the sequence of functions that constitute command and control faster than the enemy does. Having faster command and control processes allows time to figure out what enemy commanders are trying to do, as well as time to cut off their opportunities for doing those things. It permits one to determine what indicators enemy commanders will key on, so that one can manipulate those indicators to mislead the enemy. In other words, faster command and control is the key to playing with the enemy's mind.[15]

The drawback of many decision-making command and control models is that they represent a highly mechanistic approach. They give the illusion of dealing with complexity but devalue the command or authority aspect of command and control by relegating the human commander to "user" status – that is, a "user in the loop," not the initiator of action. As a consequence, command and control as a decision-making perspective degenerates into a race for more accurate and timely information, with a heavy emphasis on technology – which can be problematic (e.g., during the Battle of Britain the command and control system was more important than the technology used). Thus, mechanistic models that are overly technology-based explain control functions but not command – the technical performance of the command and control system but not the intellectual performance of the commander. Indeed, too much focus on technology is at the expense of the human cultural element of command and control – in particular, the cognitive process of the commander as the initiator of action. It neglects the fact that command is vested in the individual commander and is carried out based on the commander's vision or intent.[16]

Command and control can also be analyzed from an organizational perspective, as demonstrated in Eliot Cohen and John Gooch's seminal book *Military Misfortunes*.[17] Their implicit theory of command and control examines failures in decision-making at all levels of command based on a variety of variables and outside influences instead of narrowly focusing on single elements such as a particular individual or cause for military failure (i.e., a scapegoat). Cohen and Gooch base their model on the core assumption that armed forces or militaries are large and complex organizations and are therefore prone to organizational failure. In other words, their approach more closely examines the organizational systems within which the individuals and institutions have to operate, with emphasis on the environment around them that led to the specific decisions being taken that led to failure. Cohen and Gooch take all these factors into account in their organizational model of military failure, which shows that organizations that become complacent often cannot read the environment because they have blinders on in the form of shared organizational mental models that are stuck – what they call organization ossification or hardening. This can lead to three different types or degrees of failure: simple failures (oftentimes easy to recover from and not usually leading to a catastrophe); aggregate failures (possible though very difficult to recover from); and catastrophic failures (fatal, with no chance of recovery).[18] Humans play a large part in such organizations and are the main focus in the next command and control theory examined.

A command and control theory that has strongly influenced the Canadian military is based on the human-centred approach of defence scientists Ross Pigeau and Carol McCann. Derived from and developed in a Canadian context through an analysis of Canadian military operations and doctrine, the Pigeau–McCann theory is "consistent with Canada's culture, national values, and military ethos."[19] Unlike the theories discussed above, which focus on technology, systems, and organizations, Pigeau and McCann focus on the human element of command. They recognize human uniqueness, advantages, and imperfections, and their theories truly represent the following observation from the 1997 Somalia Inquiry report: "Military command is of course a human activity, fashioned by creative imagination and therefore beset by the frailties of human nature."[20]

Pigeau and McCann began their study of command and control by examining the differences between control and command. They assert that only humans command, emphasizing that the "buck stops" at the human commander. Command is thus first and foremost a *human* endeavour based on human *will*. Accordingly, they define command as "the creative expression of human will necessary to accomplish the mission."[21]

Command is not enough, however, because commanders need a way to exercise their will – to actually invoke action in the military environment. This is provided by control. Control cannot exist without command; in fact, control *derives* from command, again reinforcing the "buck stops here" concept of command. Specifically, Pigeau and McCann define control as "those structures and processes devised by command to enable it and to manage risk."[22] Control manages uncertainty in the military domain and is therefore a tool of command. Control structures and processes range from the abstract – for example, a policy statement or general guidelines contained in doctrine – to things that are quite concrete and even physical, such as software and hardware technologies and military equipment. Because command initiates control and command is able to modify the procedures and structures of control, Pigeau and McCann assert that the human-centred concept of command will always retain priority over control. According to them, therefore, the essential difference between command and control is that command is a human attribute and control is a process, system, or organization. Command is thus paramount.[23]

Pigeau and McCann posit that there are three dimensions to command: competency, authority, and responsibility – or CAR. Each of these dimensions is further divided into constituent parts. They define *competency* as "the skills and abilities so that missions can be accomplished successfully" and maintain that these abilities fall into four general classes: physical, intellectual, emotional, and interpersonal.[24] *Authority* refers to command's domain of influence and consists of "the degree to which a commander is empowered to act, the scope of this power and the resources available for enacting his or her will." Pigeau and McCann stress that authority comes from two sources that an individual earns by virtue of personal credibility: legal authority (the power to act as assigned by a formal agency outside the military, typically the government), and personal authority (given informally to an individual by peers and subordinates).[25] Legal authority, as the next section of this chapter demonstrates, is the essential link between command and sovereignty, while personal authority (also related to emotional and interpersonal competency) is a vital aspect of the "human element" (described in the Introduction) that is so essential for ensuring an effective command and control relationship. *Responsibility* consists of "the degree to which an individual accepts the legal and moral liability commensurate with command" and has two components: extrinsic responsibility and intrinsic responsibility.[26] Responsibility is essentially the cornerstone of CAR. Extrinsic responsibility's emphasis on accountability as a result of legal authority bestowed upon an individual was an important consideration during the 1940–57 continental air defence command and control negotiations because of the degree of authority that one nation's commander would exercise over another nation's forces. The Pigeau–McCann theory also addresses the responsibility attribute of the Canadian military ethos (discussed below). Overall, the dimensions of command in the Pigeau–McCann theory underscore the fact that command is a unique part of a unique profession.[27]

The focus of this book will primarily be on the doctrinal command and control principles. Canadian and American military planners were concerned primarily with them when negotiating command and control arrangements for continental air defence. The doctrinal command and control principles evolved significantly between 1940 and 1957, and the definitions that appear today in NATO and other professional military doctrine often reflect the results of that evolution. It is worth focusing on doctrinal command and control principles because they are the ones that today's military professionals recognize and utilize the most. Likewise, as indicated by Allard's observations (see above) about the doctrinal definitions evoking the "personal nature of command," it is essential to be cognizant of the "human element" aspects of command and control. Accordingly, this study will also utilize Pigeau and McCann's human-centred theories to emphasize the importance of individual interactions. The above theories, especially Pigeau and McCann's, place foremost emphasis on command, and indeed, command has an important link to sovereignty that bears examination.

Sovereignty and Command

Sovereignty is a complex and contested concept without a universally accepted definition. Particularly since the end of the Cold War, the proliferation of non-state actors, non-governmental organizations, transnational terrorism, and failed states has meant there are varying concepts of sovereignty that relate to individuals and human rights, and to human security concerns therein. This is in addition to the more traditional, state-centred interpretation of sovereignty as a "world of autonomous, internationally recognized, and well-governed states" – what political scientist Stephen Krasner has termed "conventional sovereignty."[28] Given the context of the 1940–57 time frame in which Canada–US continental air defence command and control arrangements were negotiated, and the parties involved (i.e., Canadian and American military and government officials representing their respective states), the conventional sovereignty concept as it applies to states is the most relevant one for this study.[29]

What are the links between command and sovereignty that entail the maintenance of command as protecting sovereignty? Discernibly at the heart of most conventional definitions of sovereignty is an emphasis on supreme or ultimate authority to be obeyed.[30] This characteristic of sovereignty bears a striking resemblance to the concept of command: here, too, authority and obedience are pre-eminent. Whereas sovereignty emphasizes supreme authority over a land and its people, command emphasizes supreme authority over military forces.[31] The key link between sovereignty and command is therefore authority, which is vital to both.

Possession of the capacity to exercise effective control over one's territory has long been an important characteristic of state sovereignty. It is also related to command. In a legal and political sense, sovereignty is a matter of jurisdiction: the right of the state to dictate the law, so "those who exercise supreme jurisdiction exercise sovereignty and can be said to be sovereign."[32] Similarly, the concept of command – especially full or national command – entails the sovereign or the commander-in-chief having complete authority over the personnel under his or her command. Command, not control, is central to the protection of sovereignty. Although control is an important aspect of command, it is subordinate to command; control derives from command and is thus a tool of command. It is therefore the distinction between command (which entails formal authority) and control (which is more supportive, and focused on execution) that points to why command rather than control is the link to sovereignty. This raises the question: Where does command authority come from in the United States and Canada?

The constitution of the United States, as a republic, designates the president the commander-in-chief of American military forces.[33] Acting in this capacity, the president transmits orders to the Secretary of Defense (during the Second World War these were transmitted to the secretaries of the War and Navy departments) for execution by the chiefs of staff (by 1942 the Joint Chiefs of Staff) of the US Army, Navy, and (after 1947) Air Force. From there command authority was delegated to the various operational commanders in the US military service chain of command.[34]

In Canada, military command authority is more complicated because the parliamentary system combines the executive and legislative branches and because the Canadian head of state is a monarch. Nonetheless, there is an obvious constitutional link between sovereignty and command in Canada. The term "sovereignty" itself derives from the word "sovereign." As political scientist Philippe Lagassé notes, "strictly speaking, the state in Canada is the Crown, the locus of Canadian sovereignty," and "the authority for maintaining and commanding military forces flows from the sovereign power."[35]

Legal authority of command over the military comes from Section 15 of the Constitution Act (1867), which vests command-in-chief of Canada's military in the British Crown. Quite literally, the "sovereign" has command over Canada's armed forces: the Crown is "the executive power" and "the fount of

Canada's supreme military command authority."[36] In the absence in Canada of the sitting monarch of England, it is the designated Canadian executive that exercises supreme authority. This consists of the prime minister and his or her cabinet, empowered by and acting on behalf of the Governor General (i.e., the representative of the Crown in Canada) and the Queen, in their capacity as the Crown's cabinet ministers (i.e., not as elected Members of Parliament[37]), who have command authority over the military. The executive subsequently delegates command authority down to professional military officers (i.e., the military chain of command via regulations) as servants of (and commissioned by) the Crown.[38] Put simply, command authority in Canada flows from the Crown (the executive) down the military chain of command. When one revisits the formal doctrinal definition of command and in particular the words "the authority vested in an individual of the armed forces," the source of this vesting in Canada is the Crown as the fount of command authority.

The military chain of command is essential to the flow of command authority. The chain of command should never be bypassed, for it links all those in uniform from the lowest to the highest rank and is also the essential and formal link between the civil authorities (i.e., the executive) and commanders at the operational and tactical levels. Gosselin emphasizes that "a superior commander can only command the commanders on the next subordinate level of authority. Accordingly, the chain of command is an important instrument of command, exercised through the flow of orders and information, and acting as a hierarchy of individual commanders, with delegated authority, who take decisions within their connected military formations and units."[39] Not only does the chain of command have command authority vested in it from the Crown via the executive (Pigeau and McCann's legal authority), but it also has a responsibility back to the executive and the Crown for accountability and an obligation to subordinates and military efficiency (Pigeau and McCann's extrinsic and intrinsic responsibility and the professional military ethos attribute of responsibility[40]). Simply put, in exercising command authority, the executive and military chain of command are responsible to the sovereign as servants of the Crown.

Having established the link between command and sovereignty, we can also state that when they negotiate command and control arrangements with their American counterparts that place specific restrictions on the degree of authority employed, Canadian military planners are exercising sovereignty.

The focus of sovereignty is not always territorial: it can also be *functional* in terms of a state's *exercise* of its sovereignty. State sovereignty is usually divided into internal sovereignty and external sovereignty. Internal sovereignty emphasizes the domestic or internal affairs of a state, and in particular the supreme authority of a state in terms of its "liberty of action ... inside its borders." External sovereignty emphasizes a state's "liberty of action outside its borders" in its interactions with other sovereign states.[41] The key aspect of internal and external sovereignty as it relates to command is liberty of action. Internally or externally, states in some ways are always constrained, but liberty of action is possible in the sense of the freedom or right to decide.

Sovereignty can be limited to some degree in terms of a state's functional exercise of its supreme authority. A state can yield some of this authority and at the same time functionally exercise its sovereignty. Joining alliances, entering into schemes for regional economic integration, and even treaty-making are the most cited examples of this exercise of sovereignty.[42] It also applies to agreeing on command and control arrangements with other nations. Again we turn to the supreme authority link between sovereignty and command. At issue is a state delegating a certain degree of command and control authority over a part of its armed forces to a foreign commander. The key sovereignty concern is finding the right balance in terms of how much command and control authority a foreign commander is to exercise over the state's forces assigned to him or her. It is technically yielding limited sovereignty when a state grants a degree of command and control authority over its forces to a foreign commander.[43] Although a loss of state sovereignty in such instances can be interpreted negatively (as some have done[44]), there is a more positive interpretation. Entering freely (i.e., not being coerced) into command and control arrangements with other countries that voluntarily place specific limits on the degree of authority that a foreign commander can exercise over a state's military forces is a functional exercise of sovereignty by the state that, it is important to note, at the same time also ensures the benefit of military efficiency (the military functional imperative – see below). Put another way, a state loses part of its sovereignty when it delegates a degree of authority to a foreign commander, but this is dwarfed by the sovereignty the state exercises by placing limits on the degree of authority the foreign commander can exercise *plus* the higher degree of military efficiency achieved.

Moreover, ensuring a "piece of the action" or a "seat at the console" can also be considered an exercise of sovereignty. By taking an active role in continental air defence efforts with the United States by placing RCAF personnel in bilateral positions (notably, of course, with NORAD upon its establishment in 1957 – see Chapter 9), Canadians were able to directly monitor Canadian sovereignty and thereby reinforce it. This is best captured by Lackenbauer, who originated the "piece of the action" concept. "By actively participating in continental defence on and over Canadian soil, along lines determined by American (and joint) strategic interests," he argues, "Canada exerted her sovereignty and ensured that the burgeoning superpower to the south would not be obliged to take matters into its own hands."[45] Simply put, a nation's military can exercise sovereignty by conducting operations in defence of the state's territory or national interests (i.e., military actions), or, as this book posits, it can do so by other means, such as retaining command in command and control arrangements with other countries and ensuring that its forces have a functional role to play ("seat at the console") in bilateral military efforts.

Lastly, a key aspect of external sovereignty is its recognition by other states and its retention by the state exercising supreme authority.[46] The arrangements that Canadian and American officers, as agents of the state, negotiated (and that their respective governments approved) confirming that Canada would retain command of its forces in bilateral operations provided legally binding protections of sovereignty. These were symbolic and therefore had moral force. Not only did the United States recognize Canadian sovereignty over Canadian forces, but the accord was reciprocal: in agreeing to the arrangements, Canada also recognized American sovereignty over its forces.

Civil–Military Relations and the Profession of Arms

> *Canadian leaders were acutely sensitive to the need to protect Canada's sovereignty not only from Soviet attack, but from American encroachment as well. Canada was, therefore, constantly faced with the dilemma of how to help the United States while trying to establish clear lines beyond which the Americans were not to go.*
> – David Bercuson[47]

This study also addresses how the efforts to negotiate command and control arrangements that preserved Canadian command (and therefore sovereignty) relate to civil–military relations (CMR) theory and profession of arms doctrine.[48] In particular, emphasizing the military functional approach to protecting and asserting Canadian sovereignty in continental air defence addresses what it historically "means to be a Canadian military professional,"[49] as well as larger political, diplomatic, and civil-military issues.

The officers negotiating the command and control arrangements were representing their military service *and* their country – essentially, they were agents of the state. Thus they were political actors in addition to being military professionals. Moreover, as the preceding discussion of sovereignty shows, by negotiating limits on the command and control authorities in the arrangements, these officers were also exercising Canadian sovereignty. The great lengths to which the Canadian military went to assert command of its forces allocated to continental air defence, while allowing for operational control by US commanders to ensure that missions could be accomplished efficiently and effectively, reflected Canadian military leaders' robust understanding of broader political and diplomatic matters as well as the character of civil–military relations in Canada. For their part, US political and military leaders understood and appreciated this broader context.[50]

Perhaps the most influential CMR theory is Samuel B. Huntington's trailblazing concept, developed during the early Cold War period, of "objective control." Writing on democratic theory, Huntington believed in the complete separation between the government and its armed forces, stressing that a truly professional military is one that remains politically neutral and autonomous. He described the ideal CMR relationship as one of objective civilian control whereby the government recognized the expertise of a professional military and respected its autonomy, while the military recognized the legitimacy of the democratically elected government by remaining loyal and subordinate to it and by not interfering in politics. According to Huntington's theory, the autonomous military would be able to maintain and indeed maximize its professionalism once it had been freed to pursue its own professional agenda (including and especially technical military matters) independent of politics. This, he believed, would strengthen civilian control of the armed forces and maximize military security and strategic success. Huntington contrasts objective control with what he terms civilian "subjective control": the maximization of civilian power over the military. Here, the civilian authority breaches the military's autonomy and encroaches on its independence and professionalism, leading to its civilianization and politicization, which Huntington argues is undesirable.[51]

Political scientist Eliot Cohen describes Huntington's ideas on objective control as the commonly accepted "normal" theory of CMR.[52] They have also been referred to as part of the non-interventionist school of CMR literature, which differs with the interventionist school that argues that the military cannot stay disengaged from politics. Morris Janowitz, an advocate of the interventionist school, stressed that although the military should not be involved in partisan politics (i.e., "party politics"), it cannot be absent from the political affairs of the state. Due to the growing military involvement in the management and conduct of increasingly complex political and diplomatic affairs, officers must have an understanding of politics and interpersonal relations, in addition to appreciating the political implications and consequences of military affairs.[53] Such was the case with the Canadian officers who negotiated the bilateral continental air defence command and control arrangements with the United States from 1940 to 1957.

Canadian planners had expertise in military command and control to conduct the negotiations, as well as specific knowledge of military operations. They were keen to devise efficient command and control arrangements that would ensure effective bilateral continental air defence operations protecting North America from enemy attack. This effort to ensure military efficiency and effectiveness had to be balanced, however, with their responsibility to protect the Canadian government's interest in maintaining the country's sovereignty vis-à-vis the United States.[54] This was no easy task: as Daniel Gosselin has noted, "reconciling the demands of national sovereignty with the need for military efficiency is always a difficult challenge."[55]

Some officers might have viewed politicians as amateurs when it came to matters of command and control, but as Cohen has observed, the politicians exercise "supreme command" and hold the balance of power in the "unequal dialogue" between soldiers and politicians in liberal democracies like Canada.[56] Canadian military planners therefore had to have the concerns of their political masters about sovereignty in mind and also oversight from government officials during negotiations with the Americans in order to protect Canadian interests. Although desirous of devising command and control arrangements that were efficient and effective from a military functional point of view, these officers were also cognizant of political sensitivities regarding the implications and optics of these arrangements as they related to command.

The Canadian officers were therefore responding to what the modern Canadian profession of arms doctrine, *Duty with Honour*, describes as the two main imperatives of Canadian military professionals. Reflecting the Canadian military's historical adoption of the Janowitzian interventionist approach to CMR, there is "a societal imperative to ensure their ability to successfully fulfill their special responsibility to the community and a functional imperative to guarantee the necessary high quality and relevance of their systematically acquired body of knowledge," personified in an "obligation to be operationally proficient and effective."[57] Whereas the functional imperative emphasizes military effectiveness and efficiency, the Janowitzian societal/political imperative emphasizes an understanding of and appreciation for political issues. It was therefore essential to find a proper balance between the two imperatives. The officers, influenced by their functional approach, might have found what they saw as political sensitivities to the issue of command annoying at times, but they also empathized with their political masters, recognizing the importance they placed on such considerations and appreciating the need to reconcile them with efficient and effective command and control arrangements. Although the protection of Canadian sovereignty was more implicit in – and even in some cases a by-product of – the actions of the Canadian officers involved in command and control negotiations, by ensuring that Canadian forces did not come under American command these officers balanced the functional and Janowitzian societal/political imperatives.

Indeed, if the functional imperative was the officers' sole concern, then the most effective command and control arrangement would have been to place Canadian forces under American command, since the United States was the largest and most capable continental air defence partner in the relationship. However, to do so would have negatively impacted Canadian sovereignty – a risky situation that was politically unacceptable (not to mention an affront to the professional and national pride of Canadian officers). Command therefore became a key issue: by ensuring that Canada retained command over its forces, the Canadian military planners were able to protect political concerns about sovereignty.

It could therefore be argued that in negotiating command and control arrangements, the Canadian military planners, according to Peter Feaver's CMR theories, were "working with" the government rather than "shirking." Basing his thinking on microeconomic "agency theory," Feaver writes that the government is in a superior position as the "principal" controlling and overseeing the military, which is in a subordinate role as the "agent."[58] Focusing on the motivations of the two groups, he describes the tendency of the military to "shirk": not to disobey politicians but rather to interpret policies in a manner more consistent with their professional military expertise. He compared this to the tendency of the military to "work" with the government by adhering to its guidance and carrying out its stated government defence policies as faithful "armed servants."[59]

Maintaining Canadian sovereignty in its continental air defence arrangements with the United States was a shared objective of both the Canadian government (principal) and the military (agent). By negotiating command and control arrangements with the United States that were both efficient and effective (functional imperative concerns) *and* that allowed Canada to retain command over its forces (societal/political imperative concerns), Canadian officers were "working" with their government in a "good faith effort to represent the principal's interests" instead of "shirking," and thus they were doing their duty as faithful "armed servants."[60] Achieving military efficiency and protecting sovereignty were thus complementary.

Conclusion

The chapters that follow focus on the doctrinal command and control principles in which Canadian military planners were most strongly engaged and that evolved most dramatically in the time period under discussion. Nonetheless, it is important to keep in mind other theoretical and conceptual approaches to command and control; in particular, specific aspects of Pigeau–McCann's human-centred approach are relevant in highlighting the importance of the "human element" and personalities in fostering effective and efficient command and control arrangements.

Since military professionals practise command and control at the behest of the state, an understanding of civil–military relations theory and modern profession of arms doctrine also provides an important basis for historical analysis. Command and control systems reflected distinct service and national cultures, and this in turn influenced the development of Canadian–American continental air defence command and control arrangements.

2
Command and Control Culture and Systems

Command is a military matter, almost mystical.
– Brooke Claxton, Minister of National Defence, 1946–53

A CONCEPT FOCUSES on an idea or a notion. Command and control principles relate to the different degrees of command and control authority that can be granted to an operational commander. A command organization is a physical entity – "a command" – headed by an individual, the commander, who works in his or her own headquarters with his or her own staff. A commander is given a degree of command and control authority over the forces allocated to the organization, depending on the situation. That authority is embodied in the specific command and control principle assigned to the commander by a superior, usually a nation's chiefs of staff representing the government.

Systems of joint command and control are based on a particular country's military culture. These systems combine the *structure* of specific command organizations (and their interrelationships) with the specific command and control *authority* assigned to each commander for joint operations. They also reflect historical relationships. In this case, Canada–US *bilateral* air defence command and control arrangements evolved from Canadian, American, and British *joint* command and control culture. Five basic command and control systems influenced Canada–US bilateral command and control relationships for continental air defence: the British joint committee system; the American unity of command system; the Canadian joint committee cooperation system; the British operational control system; and the coalition supreme command system. Before examining these systems, it is first necessary to explore how Canada's wartime expeditionary command and control experience informed and also differed from its continental air defence relationship with the United States.

Canada's Second World War Expeditionary Command and Control Experience

When Canadian army and air forces were deployed overseas during the Second World War they came under the operational command of British officers. This was in stark contrast to Canadian military forces employed in the defence of

Canada, and those employed in continental defence in conjunction with the United States, as Chapters 3 and 4 will demonstrate.[1] Nonetheless, since the wartime expeditionary experiences informed Canada's postwar command and control arrangements and structures, they are worth discussing in some detail.

During the interwar period, Canadian Army leaders desired a degree of autonomy equal to or even greater than what the Canadian Corps had under the command of Lt.-Gen. Sir Arthur Currie in 1917–18.[2] Lt.-Gen. Andy McNaughton in particular was a champion of this cause. As C.P. Stacey has noted, while attending the British Imperial Defence College in 1926–27 as a colonel, McNaughton had "advanced the proposition that Canadian forces in a future war would, in effect, be in a position of allied forces operating in co-operation with British formations rather than being in a colonial and subordinate position."[3] Several Canadian army officers had chafed under British command during the First World War and developed a degree of animosity towards the British Army.[4] As we will see below, this phenomenon was similar to the feelings that RCAF officers had towards the RAF as a result of service overseas during the Second World War.

The 1931 Statute of Westminster was a watershed in Canadian history in terms of granting Canada greater foreign and defence policy independence.[5] The statute articulated that unlike during the First World War, any Canadian military forces deployed to England would be considered "foreign," just like the forces of any other country. Because British military law no longer applied to the Dominions, Britain and the Dominions would have to pass Visiting Forces Acts to legally address the administration of military discipline of any forces deployed to the United Kingdom.[6]

Canada and Britain passed the Visiting Forces Act into law in April 1933. It went a long way towards addressing Canadian concerns about autonomy in future allied expeditionary efforts.[7] The act specifically addressed the issue of discipline, which the Introduction indicates was a national prerogative.[8] The same act outlined what the Canadian forces' command and control relationship would be with Britain when deployed overseas. There were also important differences vis-à-vis Britain in how the act was to be applied to the Canadian Army and the RCAF.

The Visiting Forces Act made an important distinction between "serving together" and "acting in combination." Canadian Army forces in Britain were considered to be "serving together" with their British comrades when they were co-located but involved in non-operational tasks (i.e., training, or in garrison) in the United Kingdom. Since "serving together," as C.P. Stacey has noted, "implied equality and not subordination," British and Canadian forces came under independent national commands and simply cooperated when working in unison. When employed in operations, however – whether in the defence of England or in an overseas deployment – Canadian Army forces were considered to be "acting in combination" with British forces. In these circumstances, the Canadian forces were considered to be part of a British formation and came under the operational command of a British general.[9] However, the General Officer Commanding the Canadian Army overseas had the authority both to remove Canadian Army forces operating "in combination" with British forces any time he saw fit and to appeal to higher authority in Ottawa if he determined that the British theatre commander's orders were inconsistent with Canadian policy. In addition, since Canadian military forces placed "in combination" came under British operational command only for operations, administration and discipline remained a national prerogative.[10]

During the Second World War, the "serving together" / "in combination" expeditionary command and control arrangement worked well for the Canadian Army. Just as in the latter part of the First World War, Canada's overseas army forces were organized into a Canadian Corps. Canadian soldiers were largely "serving together" in Britain, where they undertook many months of training before being deployed on operations (with Dieppe and Hong Kong being the major exceptions). This was because there were no major Allied campaigns on the European continent until 1943 (Italy) and 1944 (Normandy).[11] And after Canadian soldiers were deployed to the continent to operate "in combination" with British forces, at no point did the Canadian General Officer Commanding have to resort to removing Canadian Army forces from British command or appeal to higher authority in Ottawa. Any difficulties the Canadian generals overseas had with their British superiors had more to do with personality conflicts than with command and control arrangements; in fact, Field Marshal Bernard Montgomery, the British commander of the 21st Army Group under which Canadian soldiers served, granted the First Canadian Army a great deal of freedom in Northwest Europe.[12] The situation for overseas RCAF forces was not as positive.

The difficulty facing the RCAF had to do with how it organized and operationally deployed its overseas forces during the Second World War, which was different from how the Canadian Army did these things. The RCAF was not organized into an identifiable Canadian organization such as the Canadian Corps or the First Canadian Army. This situation, however, was not a complete repetition of the First World War, when Canadian airmen were absorbed directly into British squadrons and there were no distinctly Canadian units.[13] During the Second World War, Canada's overseas air force was organized into a large number of RCAF squadrons, which were designated the 400 series to distinguish them from RAF squadrons.[14]

In terms of operational deployment, Canadian Army forces reflected traditional army geographical organization and were largely placed "in combination" with and under British operational command when deployed outside of Britain.[15] RCAF units, however, reflecting traditional air force functional organization, were incorporated directly into and became an integral part of the existing RAF functional command organizations in Britain such as Bomber Command, Fighter Command, Coastal Command, and so on. The result was that the air force conducted operations throughout the conflict (the Allies, having been ejected from the European continent, needed to stay active in the war against Germany). The Visiting Forces Act also applied to the RCAF, whose overseas forces were thereby considered to be "in combination" with, and thus an integral part of, the RAF commands operating from Britain; they accordingly came under RAF operational command for the duration of the war.[16] This command and control arrangement also applied to RCAF forces deployed outside of Britain as part of higher Allied formations (e.g., the Second Tactical Air Force in Normandy): these, too, were considered to be "in combination" and thus came under British operational command.[17]

Organizationally, some progress was made with regard to the status of RCAF forces deployed overseas, but this did not extend to greater operational command and control opportunities. Although Air Force Headquarters in Ottawa established an RCAF Overseas Headquarters in London under a Canadian Air Officer Commanding in January 1940, its main purpose was to administer the RCAF units in Britain, not to command and control them in air operations.[18] Since no separate RCAF operational command organization was established in Britain or elsewhere outside Canada's borders during the war,[19] overseas RCAF units remained a part of the RAF functional command organizations and deployed as part of the higher British formations mentioned above. A measure of organizational autonomy was achieved with the creation in 1943 of the distinctly Canadian No. 6 Group in Bomber Command, and with the successful campaign for "Canadianization" (consolidating Canadian personnel into Canadian units).[20] However, such organizational accomplishments were not matched by any kind of RCAF expeditionary command and control arrangements for forces deployed overseas or by operational command opportunities for RCAF officers. An RCAF air vice-marshal was Air Officer Commanding No. 6 Group, but he reported directly to the AOCinC Bomber Command, an RAF air marshal (later air chief marshal). Moreover, as Allan English and John Westrop have noted, the AOC No. 6 Group "played a very minor role in the planning and execution of the bomber campaign, and it could be argued that he was really only a high level tactical commander."[21] Therefore, even though the RCAF retained administration and discipline as part of its national command over its forces, Canadian wartime practice was to relinquish operational command of its forces to the RAF.

The subservient expeditionary command and control situation had a negative effect on the RCAF. Not only did it have minimal say in the employment of its overseas units, but as English and Westrop note, the command capabilities of its leaders were extremely restricted: "the absence of an operational-level RCAF command structure limited opportunities for senior RCAF officers to become exposed to strategic- and operational-level planning considerations."[22] English and Westrop conclude that the Canadian wartime policy of assigning RCAF units to RAF commands under British operational command "'broke the back' of the RCAF and prevented it from fielding a 'national air force' with the same higher command opportunities as those enjoyed by the Canadian Army."[23] As a consequence, Stacey has noted, the subservient expeditionary command and control situation "certainly produced feelings of resentment and frustration among RCAF officers, particularly in the senior ranks."[24] This was something the postwar Canadian air force brass did not forget.

Many senior RCAF officers of the early Cold War period had served in overseas units during the Second World War, and they remembered with distaste the subordinate position of the RCAF within the RAF commands. For instance, when speaking to the United Services Institute in London, Ontario, in 1954 about the possibility of the RCAF's NATO Air Division being assigned to the British sector, Air Commodore Keith Hodson, who had commanded a tactical fighter wing in 1943–44, expressed a Canadian "sentimental reluctance to joining the RAF to whom we surrendered our identity in the last war."[25] In his recent book on the formation of Canada's NATO Air Division in Europe in the 1950s, historian Ray Stouffer describes the belief among Canada's early Cold War air force leadership that the RAF had subjugated the RCAF and limited its command and control opportunities. "As a result of this wartime experience," Stouffer notes, "an important objective of the Canadian airmen in the establishment of the Air Division was to maximize RCAF autonomy within NATO's Central Region." Significantly, instead of assigning the Air Division to the British zone (which is where the Canadian Army sent its NATO brigade), the RCAF succeeded in placing it in the US zone to operate alongside the USAF.[26] This was part of the overall cultural movement of the RCAF away from the RAF and towards the USAF discussed in the Introduction. Although practical functional reasons (geography, similar equipment and doctrine, etc.) played an important part of this cultural development, the RCAF's expeditionary experience during the Second World War shows that its command and control situation vis-à-vis the RAF also was a vital factor.

Lastly, Canadian forces came under American operational command during the campaign in the Aleutians in 1942–43. As the Visiting Forces Act did not apply to Canadian operations with American forces, a special Canadian government order-in-council had to be issued placing Canadian army and air forces under the operational command of the American commander.[27] The effect was very similar to Canadian forces being placed "in combination" with British forces in that they came under the operational command of an overall American commander but administration and discipline remained a Canadian prerogative as a part of national command. There was also a provision for the commander of the Canadian contingent to refer to higher authority, and "in 'extraordinary circumstances' the right of withdrawal from the undertaking."[28] Canadian forces coming under American operational command for the Aleutians campaign set important precedents for Canada's expeditionary command and control arrangements with its southern neighbour and for greater Canadian–American military integration during the Cold War.

In summary, when it came to expeditionary arrangements during the Second World War, Canada granted its allies a great deal of operational command and control authority over Canadian overseas forces – especially the RCAF. Command and control over Canadian forces operating in Canadian sovereign territory, waters, and airspace was – and it could be argued, remains – quite a different story. Indeed, the expeditionary experience highlights the unique situation of continental defence command and control. The Canadian government was comfortable placing its forces under foreign operational command when they were operating overseas; this was not the case for Canadian forces operating in bilateral continental defence operations in Canadian territory, waters, and airspace. Canadian, American, and British military culture influenced Canadian–American continental air defence command and control arrangements, so an examination of specific national systems is in order.

The British Joint Committee System

The British joint committee system of command and control was derived from British constitutional and parliamentary government culture and practice, particularly the political custom of collective responsibility and dispute resolution through discussion and compromise. Political committees were formed to discuss important and potentially thorny issues. Over time, this cultural practice extended to the military sphere for joint command and control arrangements.[29]

This produced the British joint committee system, consisting of three service commanders-in-chief (CinCs), each jointly responsible in a geographical command area, whether domestic or overseas. The three officers organized a "body of equals," usually chaired by the army commander, to deal with joint issues and operations; this body in turn reported to the chiefs of staff.[30] In 1941, the British Joint Staff Mission to the United States explained:

> In an operational theatre, therefore, we adopt the system of co-equal Commanders unless one Service is clearly the predominating partner. In theory it may be argued that this system is not so effective as that of Unified Command. In practice, however, under modern conditions Command of any Service is a matter of such technical complexity that it can be undertaken effectively only by a Commander with the necessary background and experience. The handling by one Commander of Naval, Military and Air forces is, therefore, impracticable in our view as a normal procedure. In addition, as in the case of higher direction ... difficulties will arise in finding a Commander with the necessary standing and knowledge.[31]

The British joint committee system recognized service expertise and equality. Given the technical complexities of modern warfare and the resulting specialization of each individual service, no commander could have an absolute understanding of both his service and the other two. Since no one commander could be an expert in army, navy, and air force operations, to give him greater authority over the other three (as is done in a unified command) was impracticable in British eyes. In the British committee model, the term "co-equal Commanders" meant that no single-service operational-level commander had any greater command and control authority over his peers. The command and control relationship among them was one of simple cooperation and coordination of efforts. The phrase "unless one Service is clearly the predominating partner" allowed one service commander to have greater command and control authority over the other two if the military task in that area was clearly focused on that commander's area of service expertise. The British used this provision to form the maritime operational control system (see below).

The service chiefs of staff also had more oversight in the British joint committee system compared to American systems. "Each Service Commander in the field," the British Joint Staff Mission explained:

> is bound by his directive from the Chiefs of Staff at home, thus ensuring a common aim for all Commanders in one theatre. Further, the higher strategy for all

theatres is necessarily linked and must emanate from the central authority in London. Service Commanders in the field are concerned, therefore, only with the local strategy of their theatre, which in turn must be considered by them within the terms of their respective directives. Close collaboration is achieved by a local Joint Staff working in a Combined [i.e., joint] Headquarters."[32]

The directive that the British Chiefs of Staff gave to each of the three service commanders in an operational theatre ensured that there was a common aim for all regarding joint matters.

Issues of strategy remained purely the prerogative of the War Cabinet and Chiefs of Staff in London. The operational commanders received only the strategic information that London deemed it necessary to provide them in order for them to accomplish their task. The commanders' main concern was the operational situation in their own theatre – what the Joint Staff Mission called the "local strategy of their theatre" – and they had a minimal direct impact on the development of the strategic "big picture."[33] Their actions had to adhere to the chiefs' directives. "It was not sufficient for a Government to give a General a directive to beat the enemy and wait to see what happens," British prime minister Sir Winston Churchill explained. "The matter is much more complicated. The General may well be below the level of his task, and has often been found so. A definite measure of guidance and control is required from staffs and from the high Government authorities."[34]

According to General Dwight D. Eisenhower of the US Army, the British joint committee system's constant requirement to report back to London intruded on the authority of an operational commander. American military practice permitted commanders greater autonomy, with less oversight from higher authority. While the Americans "favored a broad delegation of responsibility and authority to a commander, on the principle that he should be assigned a job, given the means to do it, and held responsible for its fulfillment without scrutiny of the measures employed," the British "kept a vigilant check on [their] commanders, with little regard to channels of command."[35] The Canadian military followed the British practice of operational commanders referring to higher authority; this proved a constant irritant to the Americans.

In practice, the joint committee system served the British well. A joint staff consisting of officers from each service who would work together in a joint headquarters ensured operational efficiency for joint matters. Most of the decisions facing the commanders were service-specific, meaning they did not have to refer to the other service commanders in the theatre. When a decision arose that demanded a joint decision, the commanders dealt with it according to the policy laid down by the chiefs of staff; this normally brought ready agreement. In the rare instance when the three commanders could not agree on a joint decision, they referred it to higher authority in London to arbitrate. Although this seemed to be a complex and tedious undertaking on the surface, the British felt that modern communications technology permitted London to rule quickly on any controversial issue.[36] This provision to refer to higher authority became an important characteristic of the Allies' coalition supreme commands during the Second World War and NATO commands in the Cold War period.

The British Joint Staff Mission admitted that the British joint committee system was imperfect, but stressed that it reflected practical experience and English political culture and practices.[37] Although the system worked effectively in regions where only British and Commonwealth forces operated, this was not the case with combined operations involving American forces. Indeed, US officers felt that the inefficient joint committee system "disrupted a cardinal principle of war": unity of command.[38]

The American Unity of Command System

Unity of command places one person in charge of a theatre of operations, resulting in a clear and unbroken chain of command. There are differences, however, between unity of command/unified command (the two terms are used interchangeably) as a *concept* and as a command and control *principle*. Most of the literature on unity of command focuses on the concept, but this study also examines unity of command as a command and control principle as practised by the US military for joint operations. The American unity of command system combined the *concept* of unity of command/unified command of one overall theatre commander with command and control authority from the *principle* of unity of command.

Colonel Ian Hope provides an excellent definition of unity of command as a concept in American military culture: "unity of command requires the placement of *all* forces operating in a specific theatre to achieve a distinct objective under a single commander."[39] He emphasizes unity of effort in a theatre as one of the main objectives of an American commander exercising unity of command, describing it as a "long-standing and distinctly American practice." Unlike the British joint committee system, which evolved both from political and military culture and practice, the American system had more uniquely military origins. The first US definitions focused solely on unity of command as a concept; only later did it evolve into a command and control principle.

The origins of unity of command in US Army culture and practice go back to the American Revolutionary War. During that conflict, military authority was centralized under a single military commander, General George Washington, and this exercise of "unity of command over the Continental Army ... provided a model for the future."[40] The United States next applied the concept of having one overall commander in charge of a theatre during the American Civil War. By appointing Lt.-Gen. Ulysses S. Grant as the General-in-Chief of the US Army, President Abraham Lincoln united "all northern military efforts under one brain."[41] Shortly thereafter, the concept became entrenched in US Army doctrine.[42]

When the United States entered the First World War, the US Army championed the idea of implementing unity of command on the Western Front under France's Marshal Ferdinand Foch. In practice, however, the actual unity of command authority that Foch exercised over Allied forces in 1918 was limited. Nonetheless, the US Army leadership was convinced that the Allied victory in November 1918 proved the concept of one overall commander as "the sole means to achieve operational level cognition and cohesion essential to unity of effort."[43] Unity of command was popular among American soldiers for independent army operations; it was a different case for joint operations, however.

American military culture emphasized the primacy of service autonomy and restrictions on command and control authority that one service would have over another in joint operations. The US Army was an advocate of unity of command, but only for independent army operations in the field, not in joint efforts with the navy. The US Navy (USN) did not prefer unity of command. Instead, based on the tradition of absolute and unquestioned authority of the ship's captain (a legacy of pre-twentieth-century organizational practice, when USN ships sailed alone or only in very small groups), American sailors favoured more decentralized command and control that emphasized mutual cooperation over any kind of authority over forces that unity of command entailed.[44]

American service autonomy meant that US soldiers and sailors spurned unity of command as a basis for the command and control of joint operations. Instead, the US Army and Navy (later to include the air force) preferred what Kenneth Allard describes as "the *sine qua non* of interservice cooperation": "the doctrine of mutual cooperation – a descriptive and prescriptive term for the proper exercise of operations whenever both services were involved."[45] Because mutual cooperation emphasized service independence, it did not necessarily entail efficiency of joint operations. For instance, Allard described joint army–navy efforts during the US Civil War in particular as "examples of 'ad hockery [sic] writ large,' in which success often was due not so much to good will and common sense as to chance."[46] Yet the mutual cooperation approach was gospel and continued to dominate American joint culture (or lack thereof) well into the twentieth century.

After the First World War, the two US services established a Joint Board of the Army and the Navy. Consisting of the Army Chief of Staff and the Chief of Naval Operations, their deputies, and their planning staffs, the Joint Board's purpose was to conduct planning and to develop joint defence plans (e.g., War Plan Orange for a conflict with Japan). The Joint Board also devised doctrine to guide operational planners in joint operations and to clarify command and control arrangements. Unity of command was discussed during the 1920s, but soldiers and sailors still largely saw it as a challenge to traditional service autonomy.[47] Nonetheless, as the "US military edged towards joint operations in order to coordinate armies and navies to pursue common objectives," officers began to warm to the concept somewhat, and by 1935 unity of command was included as an option in the doctrine publication *Joint Action of the Army and the Navy*.[48] Significantly, in addressing unity of command in this document, the planners combined the concept of an overall theatre commander with proper command and control authority.

Joint Action of the Army and the Navy was groundbreaking doctrine in terms of joint command and control. It stipulated that the US Navy and Army (which also included land-based air forces) would coordinate their joint operations either by "mutual cooperation" or by "unity of command." Unity of command would only be implemented in the following cases: when the president ordered it; in specific joint agreements between the Secretaries of War and the Navy; and "when commanders of Army and Navy forces agree that the situation requires the exercise of unity of command and further agree as to the service that shall exercise such command."[49] The last provision meant that unity of command would only be implemented in the event of a major enemy attack, as an emergency measure; otherwise, the US services would coordinate their operations through "mutual cooperation."

The definition of unity of command in *Joint Action* emphasized the concentration of authority and responsibility in one overall commander but also restrictions on that person. It read:

> Subject to the provisions of subparagraph b. below, unity of command in an operation vests in one commander the responsibility and authority to coordinate the operations of the participating forces of both services by the organization of task forces, the assignment of missions, the designation of objectives, and the exercise of such coordinating control as he deems necessary to insure the success of the operation. Unity of command does not authorize the commander

exercising it to control the administration and discipline of the forces of the service to which he does not belong, nor to issue any instructions to such forces beyond those necessary for effective coordination.[50]

This definition of unity of command was very similar to the current definition of "operational command."[51] The clause that excluded authority over administration and discipline reflected American military culture's primacy of service autonomy and was also consistent with Canadian command and control culture and practice. These responsibilities remained with the service chiefs of staff as part of their national command authority.[52] Indeed, the exclusion of authority over administration and discipline would be a constant characteristic of all Canada–US command and control arrangements for continental air defence.

The American unity of command system thus vested operational command of joint forces in a theatre in a single commander. One commander from any service (usually the one with the most forces) commanded the air, ground, and naval forces in the theatre. This "single authority" would be able to choose among campaign plans, resolve conflicts over resources, and assign operational priorities. It was essential to avoid duplication of effort and competition for resources. Also, a clear chain of command would avoid delay in issuing orders.

Joint Action was groundbreaking in its articulation of the concept and principle of unity of command; even so, it must be remembered that the predominant American military culture of service autonomy meant that the normative American joint service command and control practice remained "mutual cooperation." Although technically, unity of command and mutual cooperation were accepted in *Joint Action* as "equally valid principles of joint operations to be used as the situation dictated," the fact was that service independence prevailed and "mutual cooperation was both the rule and the reality."[53] This was an important distinction, for mutual cooperation was also the preferred approach of the Canadian services entering the Second World War (see below). This similarity in Canadian and American command and control culture would be a vital consideration in continental air defence negotiations. As Chapter 3 demonstrates, Canadian officers were able to gain USN support to help ward off attempts by the US Army to impose unity of command.

The American unity of command system was an effective means of maintaining centralized command and control, unity of effort, and a clear chain of command. All were crucial to ensuring operational effectiveness and military efficiency. It was therefore not surprising that US Army officers sought to implement the unity of command system as the basis for the Canada–US command and control relationship for the defence of North America. The Canadians, of course, preferred their own system.

The Canadian Joint Committee Cooperation System

> *Canada, like any junior partner in a joint [i.e., combined] command, is hesitant to place its forces under any other than Canadian command, where direct responsibility to the government can be assumed.*
> – General Charles Foulkes, former Chairman, Chiefs of Staff Committee[54]

Canada's command and control system was a hybrid of the British joint committee system with unique Canadian characteristics. For one, whereas the British system was for both domestic and overseas military operational theatres, the Canadian one was only for domestic Canadian joint command and control, as Canadian forces served under the British in expeditionary endeavours (see above section on the Canadian Expeditionary Command and Control Experience). Other unique characteristics of the Canadian system epitomized the Canadian desire to minimize command and control authority of one service over another's forces in both joint and combined situations; so the main emphasis was on the principle of cooperation. That being the case, the Canadian joint committee cooperation system was based on Canada's British political and military cultural heritage while also reflecting the country's growing desire for greater autonomy after the First World War.

Canada emphasized the principle of cooperation in joint and combined operations at the outset of the Second World War. When Canada declared war on September 10, 1939, the British Admiralty immediately requested that Canada place the Royal Canadian Navy's (RCN) warships at the Royal Navy's (RN) disposal for operations in the Atlantic Ocean, just as it had in the First World War.[55] Having trained and served in the British naval service throughout most of their careers, Canadian senior naval officers had absorbed the Royal Navy's rich culture and heritage as their own and were therefore favourable to the Admiralty request. The Canadian government was not.

Prime Minister William Lyon Mackenzie King explained that on constitutional grounds Canada could not place its military forces under British operational command. He instructed Canadian ships to cooperate with the Royal Navy instead, and formalized this command and control arrangement in an order-in-council on November 17, 1939.[56] This was explicit government direction regarding military command and control, and Mackenzie King set an important precedent: Canada would resist having its forces commanded by an outside service; they would operate under the principle of cooperation.

Mackenzie King's decision was based on Canadian political and military thinking and practice since the First World War. The 1931 Statute of Westminster had granted Canada full autonomy to conduct its foreign relations independent of London. This achievement was not lost on Mackenzie King or on his Under-Secretary of State for External Affairs, O.D. Skelton. Protective of gains already made, Mackenzie King and Skelton were not willing to allow London to dominate Canadian foreign and defence policy, as it had in the past. In any conflict the Canadian Parliament would decide the degree of Canadian participation. The Second World War "put to the test the genuineness of the 'nationhood' [that Canada] had acquired," C.P. Stacey remarked, "and show[ed] just how significant 'constitutional progress' was when weighed in the power balance of world war."[57]

Canada's geostrategic situation and inter-service relationships also played a part in the military's utilization of the cooperation principle. The relative safety of North America from attack and hard feelings about Canadian experiences under British command during the First World War led Canadian political and military leaders to argue for a more independent stance, separate from the traditional British link.[58] Compared to the Canadian Militia (Canadian Army after 1940), the RCN and the RCAF were younger and smaller services (the RCN was created in 1910 and the RCAF only received full independence from the army in 1938). Wary of being dominated by the Army, the RCN and RCAF "jealously guarded their independence."[59]

All three of Canada's military services were opposed to any centralization of command and control: unity of command "was alien to Canadian doctrine and practice."[60] Some Canadian officers had studied unity of command during their time overseas at British staff schools, but these had been largely hypothetical explorations with minimal bearing on Canadian joint command and control practice.[61] Instead, the RCAF, RCN, and Canadian Army insisted on "mutual and voluntary cooperation as the only basis for joint planning and command."[62] Based on British custom, the three services coordinated their efforts through a Joint Staff Committee (later renamed Chiefs of Staff Committee) at the strategic level in Ottawa. This practice was repeated at the operational level, where the three commanders on each coast formed joint committees to oversee inter-service issues and operations and coordinated based on the principle of tri-service cooperation.[63]

Cooperation was the only accepted command and control arrangement for Canadian joint operations. As was the case with the USN and US Army's views regarding joint command and control, unity of command was a challenge to Canadian service autonomy. If Canada's air force, navy, and army were all reluctant to grant command and control authority to another Canadian service, then they were equally if not more reluctant to grant any similar authority to a foreign service. It was no wonder, then, that Canada even more strenuously insisted on cooperation for operations with armed forces of other nations; this was a way to guard Canadian sovereignty during the Second World War. It was this approach that Canadian planners used in their dealings with American planners for continental air defence command and control arrangements with the United States.

The British Operational Control System

In the British operational control system, a nation's chief of staff assigned operational control over the forces from one service or nation to a commander from another service or nation. The origins of this were in the British joint committee system's enactment of the "expertise" provision outlined above. This clause permitted one CinC to exercise greater authority over another service's forces in a theatre if the military task was clearly focused on the CinC's area of expertise.[64] In the case of the British operational control system, this was recognition of the navy's primacy in maritime trade defence operations during the Second World War.

In the late 1930s, poor joint maritime trade defence exercise performances prompted British military authorities to implement the "expertise" provision in the joint committee system. They established four joint headquarters called Area Combined Headquarters (ACHQs) for the geographical theatres on the British coast. Here the three service theatre commanders coordinated the maritime defence mission, ensuring greater unity of effort. In the summer of 1938, British military authorities concluded that the British Army's contribution to the maritime defence mission was negligible, so the ACHQs included only officers from the RN and the Royal Air Force's (RAF) Coastal Command. In each ACHQ, the RAF Air Officer Commanding the assigned Coastal Command group and his staff worked closely with the admiral commanding the naval commands and his staff.[65]

While sufficient for prewar RN–RAF operations, this arrangement proved unsatisfactory in wartime. When the campaign to protect convoys from German U-boat attacks intensified in late 1940, the Admiralty grew concerned about the inadequate performance of RAF maritime patrol aircraft. The Admiralty complained to the British War Cabinet that the RN had "no responsibility of the day-to-day operational control of Coastal Command aircraft which are carrying out what are essentially naval operations."[66] Feeling that the Air Ministry was not committing sufficient resources to Coastal Command to protect shipping properly, the RN leadership sought greater influence over maritime patrol operations.

The solution to this problem – to grant RN theatre commanders operational control over Coastal Command operations – was fraught with many difficulties. After much discussion on the status of Coastal Command, the British War

Cabinet announced that it "should remain an integral part of the Royal Air Force, but that for all operational purposes it should come under the control of the Admiralty."[67] The resulting Coastal Command Charter stated that "operational control of Coastal Command will be exercised by the Admiralty through the Air Officer, Commanding-in-Chief, Coastal Command."[68] This did not specify what command and control authority operational control actually entailed, causing the Air Ministry to describe the Admiralty's operational control over Coastal Command as a "polite myth."[69] The naval commander only had the authority to issue "general directives" to the air force commander as to the objectives to be obtained; it was the Coastal Command group AOC who actually exercised operational control, as he remained responsible for the "day-to-day detailed conduct of air operations."[70] Air Marshal Sir John "Jack" Slessor, a wartime Coastal Command CinC, perhaps put the command and control relationship best: "the sailor tells us the effect he wants achieved and leaves it entirely to us how that result is achieved."[71] The RN commander only had the authority to issue his requirements for air coverage to the RAF commander, who would then exercise operational control by assigning Coastal Command forces under his command to accomplish the mission. The RN commander's authority was therefore more akin to what came to be known as operational direction.[72] The 1941 Coastal Command Charter did not specifically define operational control. It did, however, establish an important precedent for the British operational control system: the detailed employment of forces to accomplish the task or complete the mission remained the prerogative of the service/component commander from which the forces came.

It was not until much later in the war that operational control was formally defined as a command and control principle. In 1944 the American military asked the British to explain the Coastal Command–RN operational control relationship in detail. Finally forced to sit down and articulate in writing what authority operational control actually entailed, the Admiralty and the Air Ministry jointly devised the following definition:

> Operational Control comprises those functions of Command involving composition of Task Forces or Groups or Units, assignment of Tasks, disignation [sic] of objectives and co-ordination necessary to accomplish the Mission. It shall always be exercised where possible by making use of normal organisation Units assigned, through the responsible Commanders. It does not include such matters as Administration, discipline, Internal Organisation and training of Units ... It is recognised that the Operational Authority may in emergency or unusual situations employ assigned Units on any task that he considers essential to effective execution of his operational responsibility.[73]

This 1944 British definition resembles very closely the modern definition of operational control.[74] Indeed, this study shows that operational control evolved significantly from the Second World War to the establishment of NORAD in 1957.

Some characteristics of operational control remained consistent throughout its evolution, and they appealed to military officials. Operational control did not include the original disposition of resources, which is an aspect of operational command. Nor did it include authority over administration, training, and discipline, which were service prerogatives as part of the national command that a service exercised over its forces. Most definitions of operational control also specifically did not include the term "command." These factors all made operational control attractive to services whose forces served under a commander from another service or country.

Unlike more offensive-focused and expeditionary systems, the operational control system was much more suitable for defensive efforts focused on specialized functional missions such as the defence of trade or air defence, as future chapters will demonstrate. Tailoring the operational control system to the bilateral North American air defence situation and synchronizing it with evolutionary changes to the US unity of command system proved to be an important consideration for Canada–US planners.

The Coalition Supreme Command System

The United States military's peacetime joint command and control practice of using mutual cooperation to coordinate army, navy, and air forces proved weak when tested by the friction of war. This was demonstrated most dramatically by the Japanese attack on Pearl Harbor. The lack of USN–US Army unity of command was cited as a major reason for the failure of American forces to coordinate Hawaiian defences and led to horrendous death and destruction. Although mutual cooperation preserved service autonomy in peacetime or when there was minimal enemy action, it was unsuitable for joint command and control in theatres of war where there was potential for frequent active engagement with adversaries.[75] It was therefore not surprising that the United States used its unity of command system for both American joint commands and Allied coalition commands during the Second World War.

The coalition supreme command system evolved during wartime to provide the basis for new joint American commands in the Unified Command Plan (UCP) and the coalition supreme command system. For American-only commands conducting joint operations, the unity of command system consisted of three service (army, navy, and air force) commanders organized into one geographic theatre. The service with the most forces was the one that usually provided the overall CinC (also called "supreme commander") for the theatre. The other service commanders were designated "component commanders" and made subordinate to the CinC. At first, it was one of the three service commanders who took on the role as CinC – a practice that today's militaries refer to as "double-hatting." He would thus also retain command over the component forces of his own service in the theatre in addition to his responsibilities as CinC.[76]

Operational experiences with Allied coalition "supreme commands" (see below) led the American military leadership to make changes to this arrangement. In 1943 the Joint Chiefs of Staff (JCS) released a new directive titled "Unified Command for US Joint Operations." It defined a unified command as an organization "in which a force composed of units of the Army and Navy operates as a single command unit under an officer specifically assigned by higher authority."[77] The JCS assigned the overall CinC based on what the theatre objectives were and which service had the dominant role. For instance, if the theatre consisted mostly of land operations, the JCS assigned an army officer; if maritime operations dominated, the CinC would be from the US Navy, and so forth.

The directive indicated that unless the JCS specifically directed otherwise, the CinC would not be double-hatted. He would not assume command of the component forces from his service in the theatre; instead, a separate component commander from his service would be assigned for this purpose. A joint staff consisting of representatives from the component commands supported the CinC "to insure an understanding of their several capabilities, needs, and limitations, together with the knowledge essential to efficiency in integration of their efforts." The supreme commander was therefore responsible only to the Joint Chiefs of Staff and exercised his command and control authority through the component commanders. This authority "normally" consisted of the assignment of missions to the forces; however, the JCS directive specified that in "carrying out its mission the tactics and technique of the force concerned" was the responsibility of the individual component commanders.[78] This provision left the component commanders much leeway for the detailed operational employment of their forces and thus adhered to the principles of mission command and centralized command and decentralized execution.[79] As with the definition of unity of command in *Joint Action*, the JCS directive specified that administration and discipline were excluded from the authority of the CinC and remained a service prerogative. These changes to the American unity of command system were an important precedent for component commanders in the unified commands that the United States established after the war.

The Second World War showed the American military leadership that successful joint operations and unity of effort in a theatre required a single CinC with component commanders under his authority. In 1946 the JCS repeated the success of its wartime unity of command system by institutionalizing it in the new Unified Command Plan. Approved by President Harry Truman in December 1946, the UCP established seven unified commands in specially designated "strategic areas." They were: Far East Command (established in 1946), Pacific Command (1946), Alaska Command (1946), Atlantic Command (1947), Caribbean Command (1947), European Command (1947), and Northeast Command (1950).[80] Each of these unified commands had a single CinC who was responsible directly to the Joint Chiefs of Staff. He exercised unity of command over all forces in the command, including those from other services, as provided for in *Joint Action*. The "general principles" of each unified command were:

> Unified command in each command will be established in accordance, in so far as practicable, with Chapter 2, paragraph 12, of *Joint Action of the Army and the Navy*, [with] component forces consisting of Army, Army Air, and Naval forces. Forces assigned to a command will normally consist of two or more components and each will be commanded directly by an officer of that component. Each commander will have a joint staff with appropriate members from the various components of the Services under his command in key positions of responsibility. Commanders of component forces will communicate directly with appropriate headquarters on matters such as administration, training, and supply, expenditure of appropriate funds, and authorization of construction, which are not a responsibility of unified command. The assignment of forces and the significant changes therein will be as determined by the Joint Chiefs of Staff.[81]

These principles ensured greater centralization of command and control, military efficiency, and also maintenance of the maxim that operational command/unity of command did not include administration and discipline.

The new unified commands were purely operational-level commands, leaving the strategic-level issues to higher authority. Strategic direction and national command authority remained the purview of the JCS, which was responsible for assigning the forces and dictating the mission and tasks of each unified command. Whereas the unified commanders had responsibility to conduct joint operations, "the sustainment of these operations remained squarely in the hands of the JCS, because only they could manage overall global war strategy and set priorities between unified commands."[82]

Because the mission the JCS assigned to a unified command was usually oriented towards a specific service, UCP practice was to affiliate the unified commands with a parent service chief of staff as its "executive agent." This custom, which the RCAF later adopted for its Air Defence Command (see Chapter 5), guaranteed that specific services dominated certain UCP commands. For instance, the US Navy dominated Pacific Command, the US Army dominated European Command (later double-hatted as NATO's Supreme Allied Commander Europe [SACEUR]), and the USAF dominated Continental Air Defense Command (CONAD) when it was established in the mid-1950s.[83]

In addition to regional/geographic unified commands, the UCP provided for the establishment of what became known as "specified commands." These were unified commands tasked with a specific mission and consisting of forces from only one service. Strategic Air Command (SAC) was the first "specified command" to be established (1946). It was a purely air force organization responsible for the US nuclear deterrent.[84] As future chapters will demonstrate, the unity of command system embodied in the UCP had an important impact on how the RCAF organized its early Cold War air defences; it also heavily influenced Canada-US continental air defence command and control relations. The American unity of command system also shaped the development of Allied and NATO multilateral command arrangements of the 1940s and 1950s.

During the Second World War, the Allies adopted a hybrid version of the American unity of command system – the coalition supreme command system – to suit the needs of coalition warfare. Although the Americans desired to implement their own unity of command system, the complexities of alliance warfare and in particular national concerns of sovereignty necessitated certain changes to the system as concessions to the countries involved. Meeting with British officials in Washington shortly after the United States' entry into the war, US Army Chief of Staff General George Marshall was adamant that in all theatres where the Allies would be fighting together the American unity of command system would have to be implemented. In his opinion, "the most important consideration" for successful joint and combined operations was "the question of unity of command." Marshall stressed: "With differences between groups and between services, there must be one man in command of the entire theater – air, ground and ships."[85]

The British were hesitant to accept this American practice, instead preferring their joint committee system. To make their proposal more palatable to the British, the Americans compromised by suggesting that an officer from Britain be the first Allied supreme commander. British opposition melted away, and the American–British–Dutch–Australian Command (ABDA) was created in late December 1941 with the British Army's Lt.-Gen. Sir Archibald Wavell at its head. As a Supreme Allied Command, ABDA was responsible to the British–American Combined Chiefs of Staff (CCS). Though ABDA had a short life (Japanese advances in Southeast Asia and the Pacific led to its disestablishment on February 25, 1942), it set an important precedent, and the CCS subsequently established additional coalition supreme commands. They included the Southeast Asia Supreme Command, established in August 1942 under RN Admiral Lord Louis Mountbatten; the Mediterranean Supreme Command, established in November 1942 under US Army General Dwight D. Eisenhower; and the Northwest Europe Supreme Command, established in late 1943 under Eisenhower.[86]

The Allied coalition supreme commands were geographically structured organizations with component air, land, and sea forces and staffs from contributing nations. The supreme commander was always either an American or British officer. He was responsible to the CCS and, "as the ranking representative of his own nation in the area," also to the political authority in his own country. An important hallmark of the supreme command system was that the supreme commander did not have authority over the administration and discipline of the various national forces under his command.[87] Once again, this remained a national service prerogative.

There were a number of differences and similarities between the American unity of command system and the Allied coalition supreme command system. One was that the definition of unity of command was not included in the supreme command system, so the actual command and control authority assigned to the supreme commander depended on the specific CCS directive. One of the uniform characteristics was that the supreme commander was not allowed to directly command the component forces of his service in the theatre; he was only to exercise his command authority through his component commanders, who were responsible for the actual conduct of operations. This gave the supreme commander the overall "big picture" of the situation in his theatre and meant that he would not get bogged down with detailed operational and tactical matters. The supreme commander's role was as a coordinator in a headquarters whose authority was to assign missions, oversee deployment of forces in the theatre, and have "the final responsibility of choosing among various possible courses of action."[88]

There were also specific restrictions on the authority of the supreme commander. One focused on avoiding interference with the composition of the forces of varying nationalities under his command: he was not allowed "to alter the 'major tactical organization' of national forces or to disperse them among multinational task forces."[89] Again, this authority fell under national command and was a national service prerogative. The right of national commanders to appeal to higher authority was another important restriction on the supreme commander. This principle had its origins in Marshal Foch's exercise of unity of command during the last year of the First World War. Foch was the overall commander on the Western Front, but each subordinate national commander – notably, British Field Marshal Sir Douglas Haig – had the right to appeal to his government if he received any orders from Foch which he believed put his soldiers needlessly at risk. During the Second World War, superiors at home expected their in-theatre national component commanders to be "watchdog[s] for national interests" and to keep them fully informed. As a result, the national commanders had the right to appeal to their own chiefs of staff if they received an order that they felt jeopardized the interests of their country or put their own forces unnecessarily in peril. The difference between the First World War provision and the Second World War right to appeal was that the component commander had to first inform the supreme commander of his intention to appeal to higher authority and give him the reasons for doing so. The purpose here centred on the issues of professional courtesy and time; it provided an opportunity to discuss the matter in the hope that the two officers could resolve the dispute before taking it to a higher authority.[90]

This right to appeal to higher authority proved to be a major irritant to some supreme commanders. American officers in particular saw the provision as "the symbol and the crux of the problem of command in a coalition." However, it was necessary in order to ensure civilian control over the military and the maintenance of national interests. Since a supreme command consisted of multinational forces, there was always the possibility that the advice of a supreme commander would differ from that of a nation's chiefs of staff and that national leaders would fear "los[ing] control over how their troops were used and to what ends."[91] As Douglas Bland has noted, the national component commander's right to appeal to higher authority

> was the concept that allowed generals and politicians to subordinate their forces to an allied commander but at the same time to retain control of those forces. This concept is a confidence-building measure that is rarely, if ever, exercised but, like deposit insurance in a bank, it is the device that allows trust to develop where none might exist otherwise. Without it little trust would prevail, but once in place the idea unlocks almost boundless potential for alliance cooperation.[92]

The right to appeal was therefore an effective measure to safeguard a country's sovereignty and enhance Allied cooperation.

The Second World War supreme commands set an important precedent for coalition command arrangements and formed the basis for the NATO commands established in the early 1950s. Formed in 1949 to counter the growing Soviet threat, the North Atlantic Treaty Organization soon established five Regional Planning Groups (RPGs) to oversee planning in the most strategically important areas: Western European RPG, North Atlantic Ocean RPG, Northern Europe RPG, the Southern Europe/Western Mediterranean RPG, and the Canada–US RPG. Consisting of military officers, each RPG was responsible for developing defence plans for its area and providing advice to NATO's political standing group to resolve differences among members.[93]

The detonation of the first Soviet atomic bomb in 1949 and the outbreak of the Korean War in 1950 convinced NATO's leadership to stand up four of the RPGs as NATO command organizations. They were based on the Second World War coalition supreme commands. Each supreme commander was responsible to the NATO Military Committee (consisting of the member nations' chiefs of staff) for developing defence plans, determining the type and number of forces required, and for "the deployment and exercise of the forces under their command." The restrictions on the supreme commander's authority over the multinational forces under his command were similar to those in place during the Second World War.[94]

The only RPG that NATO did not stand up as a supreme command was the Canada–US Regional Planning Group (CUSRPG). CUSRPG originally consisted of the Canadian Chiefs of Staff Committee and the US Joint Chiefs of Staff, plus a Regional Planning Committee working group staffed by their representatives. Both the Canadian and American chiefs of staff took part in CUSRPG's activities until 1951, but thereafter delegated most of the CUSRPG's duties to the Planning Committee. The officers of the Planning Committee were also the members of the bilateral Canada–US Military Cooperation Committee (MCC). When the MCC met, these individuals simply put on their additional CUSRPG "hat" to discuss NATO matters. The only distinction between the MCC and CUSRPG was that the chair for each organization was different.[95]

There were a number of important reasons why Canada and the United States decided against standing up CUSRPG as a NATO supreme command. The most significant was that the two nations – especially the United States – saw the task of defending North America as an exceptional one that was separate from the mostly European-focused NATO. It was a bilateral responsibility, not a multilateral one, that formed a "special relationship" between the two countries, and the Americans felt that including other allies in a North American

NATO command would be too "cumbersome."⁹⁶ NATO was not leak-proof from a security perspective, and the Americans did not want other countries to be privy to the intelligence and research and development information on North American defence matters that the United States shared with Canada. Furthermore, because the Canada–US air defence system was so important in protecting the SAC deterrent, the Americans were fearful of a multilateral NATO command that would have undue influence over the SAC.⁹⁷ From an American perspective, it was essential to keep the SAC deterrent in American hands to ensure that the protection it offered, in the form of continental air defence arrangements with Canada, remained a purely bilateral arrangement. American strategic culture, not to mention the US Congress, would never agree to any arrangements that would have restricted US freedom of action in the direct defence of the United States. Cooperation with Canada, which was a strategic, operational, and tactical necessity, was as far as Washington would go. The United States therefore consistently opposed any proposals to establish CUSRPG as a NATO command.

The Canadians, too, questioned whether a NATO supreme command for North America was in their best interests. William Willoughby and General Charles Foulkes have both described the NATO supreme commands as the "European pattern command system" and not a good option for Canada. If Canada and the United States set up a NATO command for North America, a US officer would be the supreme commander. As Willougby remarked, when Canadian authorities realized this probability, "their enthusiasm quickly subsided."⁹⁸ Other NATO members also had no interest in the establishment of a Canada–US NATO command, fearing that it would draw resources away from Europe. In any event, European NATO members felt that the defence of North America was purely a Canada–US matter. The multilateral aspect of NATO, however, still appealed to a number of Canadian political and military officials. NATO's coalition supreme command system thus continued to have an impact on Canada–US continental air defence command and control arrangements during the 1950s.

Conclusion

Canada and the United States at varying points considered all of these different command and control systems for their bilateral continental air defence relationship. Each country's command and control system was based on its own unique experiences and culture, and the two cultures often conflicted as Canada and the United States began to work together to defend the continent from aerial attack. The key issue after 1940 was reconciling each country's own *joint* system of command and control for effective *bilateral* continental air defence. It is how this relationship evolved from a compromise cooperation–unity of command system during the Second World War to a hybrid operational control–unified command system by 1957 that will be the focus of the remainder of this book.

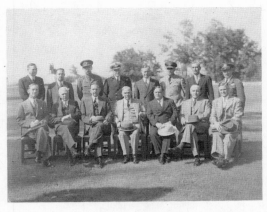

The Permanent Joint Board on Defence, August 26, 1940. Prime Minister Mackenzie King sits at the centre flanked by Canadian PJBD Chairman O.M. Biggar on his right and American PJBD Chairman Fiorello La Guardia on his left. *LAC photo C-005767.*

Throughout the Second World War, RCAF aircraft such as this Lockheed Hudson from 11 Bomber Reconnaissance Squadron provided essential protection to Allied shipping off Canada's east coast. *DND photo PL 1185.*

RCAF Eastern Air Command Badge. *DND Directorate of History and Heritage.*

Centralized command and control of maritime air power made aircraft a very effective weapon against U-boats targeting Allied convoys in the North Atlantic. The most common U-boat the Germans employed was the type VII, such as *U-970* seen here, upon which the author's grandfather served during the Second World War. *Author's collection.*

In April 1943, Air Vice-Marshal G.O. Johnson became Air Officer Commanding-in-Chief Eastern Air Command and was granted operational control over all Canadian and American maritime air power in the Canadian Northwest Atlantic. *DND photo PL-12772.*

Group Captain F.V. Heakes kept a close eye on American maritime air power developments in Newfoundland and later served as the Air Officer Commanding No. 1 Group in St. John's as an Air Vice-Marshal. *DND photo PL 1309.*

After the Second World War, the RCAF Auxiliary employed P-51 Mustangs, largely flown by wartime pilots. The Auxiliary's squadrons were located in large cities across the country. *DND PL-54769.*

In addition to Mustangs, the RCAF Auxiliary's DH.100 Vampire jet fighters were the first line of Canadian air defence following the Second World War. *DND PL-135220.*

DH.100 Vampire and F-86 Sabre jet fighters of the RCAF Auxiliary played a vital role in Canadian air defence during the early Cold War. *DND photo PL-55012.*

Armed with the atomic bomb the Soviet Tu-4 Bull constituted the main aerial threat to North America during the early 1950s. *Open source Wikipedia.*

A group of RCAF Staff College students stand in front of the RCAF's standard air defence interceptor of the mid-1950s, the CF-100 Canuck, at the Avro Malton plant in Toronto. *Canadian Forces College Image 00120067.*

RCAF Air Defence Command Badge.
DND Directorate of History and Heritage.

Air Marshal Wilfred Curtis, Chief of the Air Staff from 1947 to 1953, was instrumental in bringing the RCAF closer to the USAF during the early Cold War. He is seen here laying the first bricks for the Curtis Building being constructed at the RCAF Staff College in Toronto. *Canadian Forces College Image 00120013.*

A USAF Airman Second Class tracks the movement of an aircraft on a glass plotting board in 1953. *Courtesy of NORAD History Office.*

RCAF Station Parent, Quebec, was one of several radar stations as part of the Pinetree Line established in the early-to-mid-1950s to detect enemy aircraft and control friendly interceptors. *DND photo PL-122400.*

3
Wartime Planning for Command and Control

> *Canadians were just as reluctant as Americans*
> *to place their troops under foreign command.*
> – Stetson Conn and Byron Fairchild,
> US Army official historians

In bilateral Canada–US defence discussions in 1940–41 the issue of command and control arrangements was a delicate one. Canadian authorities, reluctant to surrender command of their forces to the Americans, championed the Canadian joint committee cooperation system, while the United States desired to implement its unity of command system. Opinion as to the suitability of these unique command and control cultures often conflicted, but in the end a compromise was struck. Canada successfully resisted American pressure to place Canadian forces under US command in the resulting defence plan, ABC-22, thereby retaining national command and ensuring Canada's sovereignty.

Early Coordination and Planning

It was by no means preordained that Canada and the United States would cooperate in the defence of North America. Although these countries both fought against the Central Powers during the First World War, there was no formal Canadian-American alliance. Aside from some mutual assistance in air training and the USN's deployment of flying boats to Nova Scotia in the summer of 1918 to counter German U-boats operating in the Western Atlantic,[1] there was minimal bilateral collaboration between the militaries of the two North American countries. In fact, in the years following the First World War, Canadian military officials continued to consider the United States a potential adversary and planned accordingly. Brig.-Gen. James Sutherland "Buster" Brown's resulting famous "Defence Scheme Number 1 (The United States)" stipulated the Canadian military defending an American attack through a slow and stubborn retirement towards the Canadian vital points of Edmonton, Quebec City, and Halifax while waiting for British reinforcements from overseas, and launching offensive spoiling attacks into US territory to disrupt their invasion. Although the strategic basis of the plan could be considered dubious – C.P. Stacey has called Brown "one of the greatest optimists of Canadian history" – and the likelihood of an American invasion very remote, "Defence

Scheme Number 1" demonstrated that Canada–US cooperation in the defence of North America that all Canadians and Americans currently enjoy was not automatic after the First World War.[2] Continental and international developments during the interwar period soon changed this situation.

Growing commercial and cultural ties between Canada and the United States, in addition to the deteriorating international situation, pushed Canada closer into the American embrace during the 1930s. Both countries trumpeted the more than 100 years of peace between them since the War of 1812 and took pride in sharing the longest undefended border in the world as "a symbol of outstanding friendliness and an ideal for other nations to follow."[3] Yet as international tensions continued to increase with the rise of fascism in Germany and the militarization of Japanese government and society, the security of the continent became a growing concern in both Ottawa and Washington. Although the defence of the approaches to Canada was a remote consideration in the vast British Empire, it was not so remote in American strategic thinking because they were also the routes that any potential enemy would take to attack the United States. The defence of Canada thus became a growing American concern and led to a greater desire for continental defence cooperation. The Monroe Doctrine, which pledged that the United States would respond to any external aggression in the Western Hemisphere, extended north as well as south, and this caused concern in Ottawa. In the case of war, Canadian officials knew that US security interests would lead the Americans to take their own steps to prevent an enemy from being able to attack the United States through Canada – with or without Canadian consent. It was therefore important for Canada to have a credible military presence in its territory to ensure that such steps would not be necessary.[4]

This situation was made apparent in August 1938 in Kingston, Ontario, during the Munich Crisis. Addressing the crowd after receiving an honorary doctorate from Queen's University, President Franklin D. Roosevelt affirmed that American security was tied to that of Canada and pledged that the United States would "not stand idly by if domination of Canadian soil is threatened by any other empire." In response, Prime Minister William Lyon Mackenzie King stated that Canada also had its obligations as a friendly neighbour to ensure that "enemy forces should not be able to pursue their way either by land, sea or air to the United States across Canadian territory."[5] These reciprocal pledges became known as the Kingston dispensation and are often referred to as the roots of continental defence cooperation between the United States and Canada.[6] This exchange, its close, friendly, and cooperative nature notwithstanding, captured the essence of the bilateral security relationship: Canada could not become a security liability for the United

States. It was a classic "defence against help" situation: it was not that the Americans would help defend Canada but that the United States, if necessary, would have to help Canada defend the United States. The challenge facing Canada, as Mackenzie King noted, was how to recognize this reality while minimizing American "help" so as to protect Canadian sovereignty.[7] The Kingston dispensation set the stage for North American defence cooperation in the near future.

The United States did not enter the war in September 1939 like Canada; even so, President Roosevelt and American military planners kept a close eye on Adolf Hitler's actions in Europe from the beginning. The German *Wehrmacht*'s defeat of Poland in 1939 was not surprising, but France's collapse in the late spring of 1940 came as a great shock to Ottawa and Washington, and indeed to the entire world. France's defeat caused a re-evaluation of Canada's defensive situation in North America. Prewar assessments had predicted possible attacks consisting of bombardment by two eight-inch-gunned heavy cruisers or one battleship and landings by small raiding parties. The scale of such an attack grew with the Nazi victory in continental Europe; now only Britain stood in the way between the Germans and North America, and the chances of British survival were rather bleak. Air attacks alone against Canada would result in material damage, but they would also strike fear in Canadians – a negative psychological effect that could have a destructive impact on the country's war effort. Senator Raoul Dandurand's famous claim during the interwar period that Canada was a "fireproof house, far from inflammable materials," was becoming increasingly dubious.[8]

Having dispatched the bulk of its forces overseas to aid the British, Canada had few means to defend itself from an enemy air attack. Only four RCAF Bomber Reconnaissance (maritime patrol) squadrons and one Coast Artillery Co-operation squadron were left to defend Canada. Just 23 of these aircraft were considered modern (they were all on the east coast), enough for only a paltry 1.5 full-strength squadrons. To help make up for the lack of aircraft, the RCAF organized a civilian Aircraft Detection Corps to keep an eye on Canadian skies.[9]

With meagre resources, Canada's armed forces reorganized as best they could to face the enemy threat. On August 1 the Canadian Army created Atlantic Command (headquarters in Halifax), which encompassed eastern Quebec, the Maritimes, and Newfoundland. Pacific Command (headquarters in Victoria) was formed in October, covering British Columbia, Alberta, Yukon, and the District of Mackenzie. The army completed this reorganization to correspond to the respective RCAF commands that had been established in 1938–39 – Eastern Air Command and Western Air Command – desiring to achieve the greatest degree of cooperation between the two services.[10] The Joint Services Committee Atlantic Coast and the Joint Services Committee Pacific Coast, which were based on the Canadian joint committee cooperation system, would coordinate operations. Each committee consisted of the three operational commanders from each service and was chaired by the army commander. Unlike in Britain, where the three operational commanders operated in an Area Combined Headquarters, no joint headquarters was established in Halifax or Victoria at this time.[11] This oversight proved to be a detriment to the effective execution of operations.

Although the Chiefs of Staff Committee established a system of "Joint Command" in its August 1940 Defence of Canada Plan, this sounded more impressive than it really was. The plan did not authorize Canadian army, air force, or naval officers to exercise any degree of operational command or control over other services. Instead, each of the service commanders on both coasts was to have "a collective as well as an individual responsibility. Collectively they are responsible for the success of the enterprise as a whole. Individually each is responsible for the control and employment of his own forces."[12] This was just a simple reiteration of the joint committee cooperation system.

The Canadian military tried their best to shore up the country's defences, but the Chiefs of Staff Committee understood that Canada could not defend itself against a major enemy attack. Canada's only hope hinged on American assistance, and the Canadian military leadership pressed the government to begin defence conversations with the United States. The timing was ideal: though neutral at the beginning of the Second World War, the United States had become alarmed by the growing German threat to North America after France's shocking defeat in 1940, and it desired continental defence arrangements with Canada. In July, the Canadian government very discreetly made provisions for the deputy chiefs of the army, navy and air force to travel to Washington to commence secret talks with their American counterparts. Because the United States was still neutral, secrecy was of utmost importance. The Canadian officers, in the words of Deputy Chief of the Naval Staff Captain L.W. Murray, "went and put on plain clothes and disappeared across the border to Washington, where we [were] asked to dinner by the American Chiefs of Staff."[13] At the meeting, the Canadian officers "put all our cards on the table," presenting Canada's need for "assistance from the U.S. in the way of provision of equipment and material supplies" and proposing to discuss "facilities for the common action of U.S. and Canadian forces in Eastern Canada and in our Eastern Waters subsequent to possible U.S. intervention in the war." The American officers sympathized with the Canadians and had them speak with top officials in the Navy and War Departments. However, these were only very preliminary and hypothetical discussions, for little action could be taken unless "the Canadian Government [approached] the U.S. Government and made them appreciate the common nature of our problems of direct defence."[14] Before the Canadians left, US Army Chief of Staff General George Marshall and USN Chief of Naval Operations Admiral Harold Stark reiterated the need "for maintenance of the strictest secrecy" regarding the conversations, stressing that any leak to the media "would have the effect of curbing any further preparatory co-operative effort."[15] The next meeting between Canadian and American officials was not so secretive.

On August 17, 1940, President Franklin D. Roosevelt met with Prime Minister William Lyon Mackenzie King at the border town of Ogdensburg, New York, to discuss continental defence. The result of this meeting was the Ogdensburg Agreement, which created the Canada–US defensive alliance that continues to this day. The agreement also established the Permanent Joint Board on Defence. Consisting of civilian and military members, the PJBD was an advisory body (it had no executive authority) that met in secret and made recommendations on defensive measures requiring governmental approval before implementation (the PJBD remains in existence today).[16] Formal defence discussions began shortly thereafter. With Canada reliant on the United States for continental defence, the American military expected to dominate any bilateral command and control defence relationship.[17]

The first Canada–US defence plan, dated October 10, 1940, painted a bleak picture. It was officially called the Joint Canadian–United States Bas ic Defence Plan–1940, but it was commonly known as the "Black Plan" because it was based on the strategic scenario of British defeat and a major Axis assault on North America. It provided for defence responsibilities in territorial waters and land areas on the lines of national sovereignty: Canada would protect Canadian territory and the United States would protect American territory.[18] There was no specific section on command and control in the Black Plan, but this is deceptive. Though the US Army official history of Canadian–American military relations states that no command provisions appeared in any drafts of the plan,[19] subsequent examination of archival documents has revealed that this is not the case.

The command section in the original drafts of the Black Plan proposed a relationship based on a compromise between the Canadian joint committee cooperation system and the American unity of command system. Canadian and American forces would "normally" practise mutual cooperation, and unity of command would only be required when provided for in joint agreements between the two governments or when commanders of Canadian and US forces agreed that the immediate operational situation required unity of command and concurred on who would exercise it.[20]

The command and control arrangements in the draft Black Plan closely mirrored those in the American 1935 doctrine publication *Joint Action of the Army and the Navy*, and were actually copied right from it. This meant that early on, Canadian planners had access to American command doctrine and in particular could observe that mutual cooperation was the primary command and control principle on which US joint operations were based. They would use this knowledge to great effect in subsequent defence planning discussions. Nonetheless, the PJBD decided to leave out the section on command from the final version of the Black Plan, largely at the behest of the Canadian officers. Demonstrating an appreciation of political matters, they astutely predicted that the command and control provisions would be controversial with authorities in Ottawa and that this would cause unnecessary delay in securing official approval of the entire plan.[21] This did not mean the end of discussion on the issue, however.

Officers in the US Army War Plans Division (WPD) were disappointed with the lack of command and control provisions in the Black Plan, feeling that "the whole issue of command [had] been side-stepped."[22] They were adamant that mutual cooperation was not sufficient and warned that it would present "a most difficult problem."[23] Nor were the responsibilities for the defence of Newfoundland entirely clear. In August 1940, Canada and the Newfoundland government signed an agreement placing the British colony's tiny military forces under Canadian command.[24] The following month, the United States and Britain concluded the "Destroyers-for-Bases Deal" where in return for 50 old US Navy destroyers the Americans would receive lengthy leases on British bases. The sites in Newfoundland included in this agreement were a naval base at Argentia, 131 miles west of St. John's, an air base at Stephenville on the island's west coast, and an army base named Fort Pepperrell on Quidi Vidi Lake, just outside St. John's. These American bases, which opened at various points throughout 1941, were considered US territory and were thus to be defended by US military forces.[25]

These developments led to an awkward situation whereby both American and Canadian forces were stationed in Newfoundland to defend the colony. Although the Black Plan indicated that "initial" defensive responsibility for Newfoundland remained with Canada, the Americans felt that it would eventually fall to the United States. Such an assumption came from the belief that Canada, as a minor military power, "possessed little if any mandate beyond its own territories." Since Canada had approached the United States for defensive assistance in its time of need, the Canadians had therefore "accepted the need for American strategic direction in the western Atlantic."[26] Although the Black Plan had no formal command provisions, American officials felt that the territorial responsibilities outlined in it "presumably included command responsibility."[27] These American presumptions proved to be a major challenge for the Canadian military in subsequent command and control discussions.

By 1941 it was becoming increasingly probable that Britain would survive and continue to fight. At the PJBD meeting of January 20, the Canadian members proclaimed the Black Plan out of date and stated that a new one was required.[28] The resulting Plan 2, later known as ABC-22 (both terms will be used), was a much different document. Nevertheless, because there was still a possibility that Britain could fall, the Americans insisted on finalizing the command arrangements for this contingency.

The Black Plan was only an outline plan that listed certain responsibilities for future Canada–US continental defence collaboration; it left the detailed arrangements to the army–air force operational implementation plan, called Plan 1. Because of the dire strategic situation envisioned in the Black Plan, Canada was willing to grant the United States greater command and control authority over Canadian forces. Based on recommendations from the Canadian Section PJBD and the Chiefs of Staff Committee, the Canadian government agreed that the US Army Chief of Staff could, with Canadian consent and consultation, exercise "strategic direction" (the term "strategical direction" was also used) over Canadian land and air forces in Plan 1.[29]

The Canadian planners pressed their American counterparts to explain precisely what strategic direction entailed. Their response was: "By the term strategical [sic] direction is meant the assignment of missions and the allocation of the means to accomplish them."[30] This description included authority that was inherent in *Joint Action*'s definition of unity of command, which was a predecessor of the modern definition of operational command. Based on modern command and control terminology, strategic direction entailed the exercise of operational command by the United States over all military forces in Canada and Newfoundland, subject to consultation with Canadian authorities. It was therefore not surprising that American planners began using the terms "strategic direction" and "unity of command" interchangeably.

The similarities among strategic direction, unity of command, and the modern definition of operational command highlight the importance of language and terminology during this time period. They illustrate the evolutionary state of most command and control principles during the Second World War. The use of the word "strategic" to describe what today is known as operational command demonstrates how the term "operational" was still entering into the lexicon of military terminology. It also reveals how the relationship between the operational level and other levels of conflict was also still evolving during the war.

There was a provision in Plan 1 that gave the individual Canadian service chiefs or a Canadian operational commander "the right to appeal to the Canadian government if, in his opinion, Canadian national interests are imperiled by the strategic directives received from the United States."[31] This was consistent with the right of national commanders serving under Marshal Ferdinand Foch in 1918 to appeal to higher authority, and would be repeated later in the coalition supreme command system (see Chapter 2). In Plan 1, Canada recognized that in the event of an Axis assault on North America, the United States would be the pre-eminent power and should have strategic direction over its forces. By making strategic direction subject to Canadian consultation and having a provision for the referral to higher authority, however, the Canadian government ensured that it would have a say in the deployment and movement of its forces. The situation was not ideal, but it did give at least a small degree of protection to Canadian sovereignty given the dire strategic situation envisaged in the Black Plan. This protection would be strengthened in ABC-22.

Negotiating ABC-22

Unlike Plan 1, Plan 2 envisaged British survival and American entry into the war as an ally – a scenario that was becoming much more likely by the spring of 1941. Since no large-scale enemy attack was anticipated, Canadian planners felt that the more strategically favourable climate gave them leverage in their discussions with the Americans.[32] In particular, they believed that Canada did not have to accept US strategic direction. It was therefore with a much more optimistic attitude that the Canadians commenced negotiations on command and control arrangements for ABC-22 with the American planners. This Canadian steadfastness led to several disagreements, however.

Ongoing American–British defence talks in Washington resulted in significant delays in the negotiation of ABC-22, for the Americans did not want to disclose their arrangements with the British to the Canadians before finalizing them. Begun in January, these were purely British–American service discussions: the Dominions, including Canada, did not take part, although they were each allowed to have an "observer," whom the British delegates would brief after each session.[33] The resulting ABC-1 Staff Agreement assigned responsibility for the defence of the Western Hemisphere to the United States. Although this brought forth the possibility that Canadian forces might come under American command in North America, Canadian military authorities fought hard to ensure that the British–American planners acknowledged Canada's special concerns regarding continental defence. The planners expected that any Canada–US defence plan would "conform generally" to ABC-1, but they also recognized that the British–American agreement should not prejudice any Canada–US arrangements for North American defence.[34] ABC-1 therefore contained a special provision stating that the United States would not automatically have strategic direction over forces operating in Canadian "waters and territories"; instead, that direction would be "defined in United States–Canada joint agreements."[35] This was a notable achievement for the Canadians, and it would prove to be a crucial asset in the ongoing ABC-22 discussions.

ABC-1's recognition of American strategic responsibility for the defence of the Western Hemisphere encouraged the American PJBD section to insist on US unity of command for ABC-22. Canadian military officials were not intimidated, however. The improved strategic situation upon which both ABC-1 and ABC-22 were based actually strengthened the Canadians' position in North America and fortified their resolve to resist American unity of command efforts. The two sides were at loggerheads, and as C.P. Stacey noted, "the provisions of ABC-22 concerning command led to the most serious difference of opinion between the two national Sections of the Board that took place during the war."[36]

The military members of the PJBD met in Montreal on April 14, 1941, to begin preliminary negotiations on Plan 2. The Americans advocated having US strategic direction over all Canadian and American forces, but the Canadians, arguing that the improved strategic situation did not warrant it, insisted that the command and control relationship be based on cooperation. While admitting that cooperation was "possible," the Americans insisted that it "is definitely wrong, inconsistent with Plan No. 1 and invites confliction [sic], delay and uncertainty."[37] Unable to find a middle ground, the two sides agreed to disagree for the moment and to seek further advice from their military and government authorities. Nonetheless, the American planners' admission that cooperation was at least a possibility encouraged the Canadians to push even harder to make it a reality.

The main concern of the Chiefs of Staff Committee was that American strategic direction would potentially allow the United States to redistribute Canadian forces to other areas of North America, which was a national command responsibility. The Chiefs particularly feared that American commanders would be able to remove forces from Canadian territory to provide added protection to their own vital points without Canada's consent. This would have been a significant blow to Canadian sovereignty, besides leaving Canadian territory undefended.[38]

The Chiefs of Staff Committee instead endorsed Capt. H.E. Reid's suggestion. "Rastus" Reid, who was the RCN member of the PJBD (he would become Chief of the Naval Staff after the war), recommended that the United States retain strategic direction over its forces in its territory and that Canada retain strategic direction over its forces in Canada and Newfoundland. When Canadian and American forces operated together, the command and control relationship should be "the same mutual co-operation which has been so evident between the U.K. and Canadian forces now operating in the Atlantic area."[39] The Cabinet War Committee concurred and gave government sanction to the Canadian PJBD Section's position on resisting US strategic direction.[40]

Although now armed with government endorsement, Canadian PJBD chairman O.M. Biggar acted far too aggressively in his subsequent correspondence with his American counterpart, New York mayor and US PJBD chairman Fiorello La Guardia. Biggar did not consult with the Canadian military members of the PJBD before writing to La Guardia, and this only made the negotiations for ABC-22 more difficult. Biggar wrote that it was "very unpropitious" for Canada to "surrender to the United States what she has consistently asserted vis-à-vis Great Britain," especially since the United States was not yet even an active belligerent in the war.[41] The problem was that Biggar did not make any distinction between Plan 1 and Plan 2, which was – an important consideration since the Americans still felt there was a real possibility of the Plan 1 scenario – that is, a British defeat and an Axis assault on North America. The Americans were thus quite upset with Biggar's position: Canada, they believed, was reneging on its agreement to place its forces under US strategic direction for Plan 1. In his replies to Biggar's letters, La Guardia only muddied the waters further by also failing to make the distinction between Plan 1 and Plan 2. In one letter to Biggar, LaGuardia wrote: "I fear that we are getting dangerously apart." Reiterating the US position for strategic direction, he remarked "it is far better to trust in the honor of the United States, than to the mercy of the enemy."[42] Unfortunately, this correspondence only made the confusion over the ABC-22 command and control situation worse.

Members of the US Army War Plans Division were "amazed and shocked" with what they viewed as Canadian backtracking and concluded that this military issue had now become a political one.[43] When La Guardia asked President Roosevelt to intervene directly by approaching Prime Minister Mackenzie King, the president refused and insisted that the PJBD had to sort out this issue on its own. Roosevelt did, however, offer the following as an official US government position as a basis for further discussions with the Canadians:

> The United States not being an active belligerent is, nevertheless, virtually ready to undertake the defense of the Canadian eastern coast, including the land and waters of Newfoundland and Labrador. Canada has not either the men or the

> material to undertake this except as a participant on a much smaller scale than the United States. Canada is really devoting its war effort to sending as much in the way of men and materials across the ocean as possible. In the active carrying out of war plans the strategic responsibility ought to rest with the United States, in view of the fact that in actual defense nine-tenths of the total effort will fall on the United States.[44]

Once again, there was no differentiation between the two defence plans or between the strategic scenarios upon which they were based. As Stacey has noted, Roosevelt's arguments were "more relevant to Plan 1 than to Plan 2"[45] and thus did little to solve the debate on ABC-22. Brig.-Gen. Maurice Pope, the Canadian Army PJBD member, feared that the recent confusion over the two plans was leading to growing apprehension between the two sides as they sought a solution.[46] These concerns were confirmed at the next PJBD meeting at the end of May 1941.

"Don't let the bastards grind you down!"

The 19th Meeting of the PJBD was the most strained in the organization's history. The Americans were intent on securing strategic direction over Canadian forces for Plan 2, and the Canadians were equally intent on resisting the Americans. This was demonstrated vividly by Pope's recollection of the advice the Minister of National Defence gave him before the Canadian Section departed for the meeting in the American capital. It was a clear political directive to the planners to resist any proposals that Canadian forces come under US command: "As far as I can today recall, Mr. Ralston's admonition to me as I left Ottawa could be summed up by a new-fangled dog-Latin tag current in Washington a year or two later, *Non illegitimus carborundum* [sic], which was said to mean, 'Don't let the b … s grind you down!'"[47] The official record contained in the PJBD journals does not reveal the level of friction between the two sides.[48] Brig.-Gen. Pope's official report and his recollections of the meeting thus provide important insights.

Pope described the negotiations as "a nightmare" in which the Americans launched "an all-out offensive, and to us fell the less comfortable role of defending our positions." The meeting started at 9 a.m. and lasted well into the evening, with only a short break for lunch. It was a hot day, but after 6 p.m. the building's air conditioning system was shut off, and the Canadians, in Pope's words, "wondered if our American friends were subjecting us to a special form of heat treatment." Finally, at about 10:30 p.m., "after prolonged wrangling," the two civilian Secretaries of the PJBD, John Hickerson for the United States and

Hugh Keenleyside for Canada, presented a draft command and control proposal for Plan 2.[49]

This proposal has never before been discussed in the historical literature, and its contents are examined here for the first time. It gave Canada "strategic direction and command of its forces operating in Canada and Newfoundland." This arrangement dealt only with forces for continental defence; the one for maritime forces was quite different. The proposal granted Canada strategic direction and command over its naval and maritime air forces only for coast defence and the protection of coastal shipping; Canadian forces engaged in the defence of ocean-bound maritime trade were to be placed under United States authority.[50]

Showing a keen awareness of political concerns, Pope knew that Canada's military and political leadership would find this arrangement unacceptable, so he spoke out against it. The Americans stressed that strategic direction/unity of command (they again used both terms interchangeably) was "essential in war," citing the precedent of Marshal Foch's successful exercise of unity of command over Allied armies in 1918 as proof. Pope disagreed with this line of reasoning, arguing that the huge number of forces (350 divisions) on the Western Front and the strategic setting in 1918 were completely different than the situation in North America, where the enemy threat consisted only of minor raids.[51] ABC-22 was based on the assumption that the British would survive and that there would be no major Axis assault on North America – circumstances that did not warrant US strategic direction over Canadian forces. Hitler's invasion of the Soviet Union just three weeks later, on June 22, 1941, would lend further credence to Pope's assessment that North America was not going to be a target of a major German effort. Prime Minister Mackenzie King recorded in his diary that day his "immense relief" that the Soviet Union was now fighting Germany and that this development in the war "must appreciably and greatly improve Britain's chances for ultimate victory."[52]

Pope's stance had the desired effect, and the proposal was withdrawn. The official record contained in the PJBD journals stated that "since no mutually acceptable solution of the problem of command relationship[s] was found after a full discussion of the subject," American and Canadian planners agreed to proceed with their ABC-22 talks "on the basis of command by cooperation."[53] The first battle had been won: the Canadian planners understood that they had secured a great concession from the Americans, but they also recognized that their US colleagues were not happy with cooperation. The Americans felt it "had the effect of whittling away the power to such an extent as to leave what remains of little value."[54] Thereafter, the ABC-22 negotiations switched focus to the Canadian preference for cooperation and the US desire to have at least some kind of provision for unity of command.

Finalizing the ABC-22 Arrangements

The service members of the PJBD met at various points throughout the early summer of 1941 and released a final draft of ABC-22 on July 28. Its command and control arrangements mirrored those originally proposed for the Black Plan, which were, in turn, very similar to those in *Joint Action of the Army and the Navy*.[55] Overall, the final ABC-22 arrangements represented a significant victory for the Canadian planners.

Each nation would retain strategic direction and command of its forces, which effectively recognized national sovereignty. The combined Canada–US military effort was to be based on "mutual cooperation," and each country's forces were required "to their utmost capacity, [to] support the appropriate forces of the other nation." Like *Joint Action*, ABC-22 included a provision for the exercise of unity of command over Canadian and American forces in an emergency situation. Unity of command could be implemented if both countries' chiefs of staff ordered it or if the local operational commanders agreed that the military situation required it and concurred as to which of them should exercise it (subject to later confirmation by the Canadian and American chiefs of staff). The final definition of unity of command in ABC-22 was very similar to the one in *Joint Action* – which of course closely resembles the current definition of operational command – with only minor differences in wording to reflect bilateral operations:

> Unity of command, when established, vests in one commander the responsibility and authority to co-ordinate the operations of the participating forces of both nations by the setting up of task forces, the assignment of tasks, the designation of objectives, and the exercise of such co-ordinating control as the commander deems necessary to ensure the success of the operations. Unity of command does not authorize a commander exercising it to control the administration and discipline of the forces of the nation of which he is not an officer, nor to issue any instructions to such forces beyond those necessary for effective co-ordination.[56]

Unity of command thus entailed vesting operational command of joint *and* combined forces in one single operational commander. By specifying that unity of command did not include authority over administration and discipline, ABC-22 ensured that these remained a service prerogative under the national command.

A clause appended to the definition of unity of command regarding authority over the movement of forces from one area to another also safeguarded national command. It read: "In no case shall a commander of a unified force move naval forces of the other nation from the North Atlantic or the North

Pacific Oceans, nor move land or air forces under his command from the adjacent land area, without authorization by the Chief of Staff concerned."[57] In an emergency situation ABC-22 thus allowed the temporary movement of forces if the operational situation required it. Lastly, the plan permitted the exchange of liaison officers between commanders at all levels.

ABC-22 combined the Canadian joint committee cooperation system with the American unity of command system. It also maintained the chain of command by ensuring that each nation's chiefs of staff would retain national command over their forces. Although there was a provision for unity of command in an emergency, ABC-22 clearly established cooperation as the main principle for the Canada–US operational-level continental defence command and control relationship. The new plan was therefore consistent with Canadian joint command and control culture and practice, and protected Canadian sovereignty.

The American planners were never fully satisfied with ABC-22 arrangements. The War Plans Division in particular felt that that "mutual cooperation is an ineffective method of coordination of military forces" and that the plan was therefore "defective in its provisions relative to command arrangements."[58] The senior US Army PJBD member, Lt.-Gen. Stanley Embick, agreed with this assessment but concluded that the difficulties already experienced in negotiations with the Canadians meant that the ABC-22 arrangements "probably represent the best compromise possible under present conditions."[59] Besides, as the American planners all too easily forgot, cooperation was also the main command and control principle for US joint operations according to *Joint Action*. The ABC-22 arrangements were therefore also consistent with American joint command and control culture and practice, and safeguarded American sovereignty.

Both governments formally approved ABC-22 by the early autumn of 1941.[60] It was appended to existing American and Canadian defence plans and distributed to the relevant operational commanders to be put into effect only upon the entry of the United States into the war.[61] The service members of the PJBD took satisfaction that they, not the civilian members, had completed the final negotiations on the command and control provisions for ABC-22, as they felt that the issue was a military one. In their opinion it was best to avoid political interference, such as the unfortunate Biggar–La Guardia correspondence, which had caused so much consternation and confusion, and leave the service members of the PJBD to come to a mutually agreeable solution.[62] In the end, ABC-22 was a functional military plan to handle a military situation. This practice of having military planners complete the detailed command and control arrangements set an important functional precedent for future Canada–US air defence planning.

Conclusion

When Canada and the United States began collaborating on bilateral continental defence plans in the summer of 1940, the issue of command and control over Canadian and American forces became very important and controversial. Canada agreed to US strategic direction over its forces in the event of a British defeat and a major Axis assault on North America but was also able to secure important concessions. By early 1941, the improved strategic situation put Canadian planners in a much stronger position to secure a more favourable command and control arrangement in a new plan. Though the negotiations for ABC-22 were difficult and often heated, the Canadians stood their ground and remained firm in their convictions. They prevailed in the end by securing cooperation as the basis for Canada–US operations. ABC-22 was consistent with Canadian command and control practice and culture, and by maintaining national command authority over its forces Canada was able to protect its sovereignty. The provisions for mutual cooperation and unity of command in ABC-22 closely mirrored those in *Joint Action of the Army and the Navy* and were therefore also consistent with American command and control practice and culture.

By and large, the command and control provisions in ABC-22 were favourable to Canada. Nonetheless, even though the US desire for unity of command had hit a roadblock, it did not dissuade the Americans from trying again. The debate over mutual cooperation and unity of command would now spill over into the relations between Canadian and US operational commanders in the field.

4
Wartime Operational Level Command and Control

DEFENDING NEWFOUNDLAND IN 1941 was a new experience for Canadian and American military forces, and significant differences of opinion soon surfaced on what their command and control relationship should be. American commanders wanted to implement the US unity of command system, but this was contrary to the established Canadian joint committee system. Though both countries had agreed to the ABC-22 compromise in late July and the United States was still neutral, American commanders continued to press for unity of command, and Canadian commanders successfully resisted.

The Japanese attack on Pearl Harbor on December 7, 1941, put ABC-22 immediately into effect and with it the command and control arrangement for mutual cooperation. Disagreements persisted about whether the enemy threat to the continent warranted the activation of the unity of command clause in ABC-22. The Americans were still in shock from Pearl Harbor and immediately pressed Canada to implement unity of command under a US officer, not only in Newfoundland but also on the Pacific coast. In both theatres, however, Canadian military officials successfully defended their position on mutual cooperation, stressing that the enemy threat was still minimal and did not require unity of command. In maintaining this stance in the face of intense American pressure, Canada safeguarded its sovereignty by retaining national command of its forces and ensuring that its forces did not come under American command.

The air forces that Canada and the United States used in a continental defence role performed a vital function protecting Allied shipping, notably off the east coast. Since it was in this maritime trade defence role that the two nations' air forces conducted most of their combined operations, the efficient coordination of maritime air power became the most important aspect of the overall Canada–US debate over command and control. When mutual cooperation proved incapable of achieving this coordination, Canada and the United States implemented the British operational control system in the Northwest Atlantic.

The Period of American Neutrality

By the summer of 1940, there was a growing Canadian military presence in Newfoundland. The RCAF had established one flight of Bomber Reconnaissance (maritime patrol) aircraft there to patrol the British colony's sea approaches, and in August the small Newfoundland Militia was placed under Canadian command.¹ The Canadian Army posted Brig.-Gen Philip Earnshaw to St. John's as the commander of the Canadian Army's Force "W" (two divisions) and tasked him to oversee the military buildup. In November the CSC gave Earnshaw the title "Commander Combined Newfoundland and Canadian Military Forces, Newfoundland." This was not an independent command, but subordinate to the Army's Atlantic Command under Maj.-Gen W.H.P. Elkins in Halifax.²

In May 1941, the Royal Canadian Navy established the Newfoundland Escort Force (NEF) at St. John's under the command of Cmdre. Leonard W. Murray. This organization of navy escort ships was responsible for protecting North Atlantic shipping from German U-boat attack. That July, the RCAF created Number One (No. 1) Group Headquarters in St. John's under the command of former First World War fighter ace Group Captain (later Air Commodore) C.M. "Black Mike" McEwen. No. 1 Group was responsible for all RCAF forces in Newfoundland and, most importantly, for the control of air operations in support of the NEF, flying Stranraer, Digby, Catalina, Canso, and Hudson maritime patrol aircraft out of RCAF Stations Botwood, Gander, Goose Bay, and Torbay. Operational command of Canadian air forces in Newfoundland, however, remained with Eastern Air Command in Halifax, where its Air Officer Commanding (AOC), Air Vice-Marshal N.R. Anderson, passed orders on to McEwen. Coordination of the Canadian services in Newfoundland was based on the Canadian joint committee system with the establishment of a Joint Services Sub-Committee, Newfoundland, consisting of the senior officers of the three Canadian services in St. John's. It was subordinate to the Joint Services Committee Atlantic Coast in Halifax.³

The first US forces began to arrive in Newfoundland in January 1941. Army forces came under Col. Maurice D. Welty's Newfoundland Base Command at Fort Pepperrell and consisted of 1,000 troops and a squadron of US Army Air Corps (US Army Air Forces or USAAF after June 1941) B-17 bombers at the Newfoundland Airport in Gander. Starting in July 1941, Rear Adm. A.A.L. Bristol commanded the USN's Task Force 4 and all other naval forces in Newfoundland from his base in Argentia on the west side of the Avalon Peninsula. Canadian service personnel accepted the American presence but were cautious in their dealings with their US counterparts.⁴ This was prudent, as the Americans fully intended to implement their unity of command system. Nonetheless, in early 1941 Canadian military forces remained predominant in Newfoundland.

Canadian and American officers had different ideas on how Newfoundland fit into the overall defence of North America, and this only amplified the command and control debate. The Americans saw Newfoundland as a separate entity outside of the continental United States and Canada; as the larger partner in the Canada–US alliance, they naturally desired that one of their commanders be allowed to exercise unity of command. The Canadians viewed Newfoundland as an integral part of the overall defence of eastern Canada. Having deployed forces in the colony first, at the behest of the British and Newfoundland governments, Canada felt that its special interests in Newfoundland were paramount. Responsibility for the defence of Newfoundland – and choosing the command and control system – should therefore be Canada's prerogative.[5]

From the beginning, the Canada–US operational-level command and control relationship in Newfoundland was ambiguous because the PJBD was still drafting bilateral defence plans (as discussed in the previous chapter). In accordance with *Joint Action of the Army and the Navy*, Col. Welty's superiors had ordered him to cooperate with the USN to defend the US bases in Newfoundland. Welty's instructions, however, were silent regarding what his command and control relationship with Canadian commanders would be; they stated only that he was to participate with them in defending Newfoundland from attack.[6] Brig.-Gen. Earnshaw's situation was no better and despite several requests to superiors to clarify the situation, he was also unable to secure concrete policy guidelines.[7] The situation should have been clarified with the approval of ABC-22 by the autumn of 1941, but ambiguities persisted throughout the year. Part of the problem was that ABC-22 was not in effect because the United States was not yet officially at war. Another was the arrival of a new US Army officer in St. John's.

With the numbers of US forces in Newfoundland growing and USAAF maritime air power operations increasing by the autumn of 1941, the US Army decided it was time to appoint a general-rank officer. In October 1941, sixty-one-year-old Maj.-Gen. Gerald Clark Brant arrived in St. John's as the new Commanding General (CG), Newfoundland Base Command.[8] Brant's personality did not make him amenable to accepting mutual cooperation as a basis for command and control in Newfoundland. A former West Point classmate of General Douglas MacArthur, Brant was, in the words of historian W.A.B. Douglas, "a soldier moulded in the tradition of the old army."[9] Although Brant was a USAAF officer, unlike most American airmen at the time he had not attended the Air Corps Tactical School and was not an advocate for an independent air force. Brant was also a staunch believer in unity of command, having lamented its absence in the 1924 joint US Army–US Navy exercises when he testified at the court martial of American air power advocate Brig.-Gen. William "Billy" Mitchell in 1925.[10] It was therefore not surprising that immediately upon his arrival in Newfoundland, Brant began to pressure the Canadians to establish unity of command, which was to be exercised by himself. This attitude only alienated him from the Canadian operational-level commanders.[11]

Maj.-Gen. Brant overestimated the strategic threat to Newfoundland. With German forces heavily engaged with the Red Army on the Eastern Front, it was highly unlikely that Germany would attempt an assault on North America. This development in the war in Europe had a direct impact on how Washington regarded the strategic threat to Newfoundland. No longer did American planners anticipate a heavy attack on the colony; instead they had begun to adopt a view similar to the Canadian one that the main enemy threat was hit-and-run raids.[12] Nevertheless, since American commanders had greater autonomy than Canadian commanders, Brant was able to disregard the most recent strategic assessments and continue to advocate a more dire appreciation of the threat to Newfoundland. Unfortunately for those who had to deal with Brant, the lack of oversight from Washington permitted him to continue his unity of command crusade in Newfoundland.[13] The Japanese attack on Pearl Harbor only made Brant more apprehensive.

Further American Attempts to Impose Unity of Command

Fulfilling its alliance commitments, Canada quickly declared war on Japan on December 7, 1941 – one day before the United States. ABC-22 immediately came into effect against Japan, and then also against Germany when Hitler declared war on the United States a few days later.[14] With the US at war, all confusion about the command and control relationship between Canadian and American forces was apparently eliminated: mutual cooperation was the means by which Canadian and American commanders would coordinate their forces according to ABC-22. Nevertheless, Canadian–American disagreement persisted regarding whether to enact the plan's unity of command provision.

The main point of contention was conflicting interpretations of the seriousness of the threat to North America. Devastated by the Japanese offensive, the Americans were anxious to take all the defensive precautions they felt were necessary, and this meant implementing unity of command in a number of theatres.[15] This included Newfoundland, where the potential for enemy attack grew in the mind of Maj.-Gen. Brant, and also now British Columbia, where early Japanese successes in the Pacific brought forth a new seriousness to Canadian–American defensive arrangements on the west coast.

Pearl Harbor was a shock to the population of BC, and frightened citizens made calls to their parliamentary representatives in Ottawa demanding that substantial reinforcements be sent to the province immediately. Formal

American requests for greater defence centralization on the western seaboard followed shortly thereafter. According to one War Plans Division study, the entire west coast from the Mexican border to Alaska constituted a theatre of operations that should come under one operational headquarters.[16] On the east coast, an alarmed Maj.-Gen. Brant "anticipated the fall of Great Britain, German victory in Africa, and 'devastating air raids' against all his installations as a prelude to a 'probable attempt in later stages to capture and hold Newfoundland.'"[17] Unsurprisingly, Brant immediately pressed for unity of command of all Canadian and American forces under himself.

The Canadian Chiefs of Staff Committee felt that the strategic situation still did not merit the activation of ABC-22's emergency unity of command clause. The Canadian brass correctly predicted that in the Pacific the Japanese would focus their offensive to the southeast, not towards the North American west coast, and that Germany, still heavily engaged in the Soviet Union (by December 1941 the German assault was stalled outside Moscow), would not risk an attack on the east coast. Thus, mutual cooperation was satisfactory for both the Pacific coast and Newfoundland.[18] The next task was to convince the Americans.

Regarding the Pacific coast situation, it took a direct approach with the American military leadership to secure concurrence that the enemy threat to the area did not require unity of command. After Pearl Harbor, the United States PJBD Section immediately pressed Canada to place all its military forces in BC under US unity of command. Chairman Fiorello La Guardia argued that BC "geographically was an enclave" within the newly formed US Western Theatre of Operations, which stretched from southern California up to and including Alaska. He stressed that the Puget Sound–southern BC area in particular was of crucial strategic importance and therefore "cannot be most effectively defended under the control of several commanders." La Guardia formally requested that the Canadian army and air forces (except for those involved in maritime trade defence operations) in BC be placed under the "Supreme Command" of the US Army commander, Lt.-Gen. John L. Dewitt, according to ABC-22's unity of command clause. He emphasized that this was "a wise precautionary measure, in advance of the occurrence of an actual attack."[19]

To appease the Canadians, La Guardia reassured them Dewitt's powers over Canadian forces would be limited to giving "strategical and tactical directives" to Canadian forces: DeWitt would not be permitted to move Canadian forces from Canada without permission from either the Canadian operational commander or the Canadian government, nor would he be allowed to alter the composition of these forces or have any power over their administration or discipline.[20]

After meeting on January 3, 1942, the Canadian PJBD Section unanimously concluded that the United States had not made its case for unity of command. "As usual," they noted, the Americans did not include any updated forms or scales of attack to justify their position. The strategic threat to the west coast was actually exactly the same as it had been when ABC-22 was drafted the previous spring. Since the Canadians had then held out for mutual cooperation, there was no reason why they should not do so again.[21]

In early January 1942, the Canadian Chiefs of Staff took advantage of their presence in Washington for discussions with the Americans and the British to resolve the command and control issue by speaking directly to their American counterparts, USN Chief of Naval Operations (CNO) Admiral Harold Stark and US Army Chief of Staff General George Marshall.[22] Unbeknownst to the Canadian brass at the time, Stark had been advocating to Marshall that *Joint Action of the Army and the Navy*'s mutual cooperation should be the primary command and control principle on which to base American defensive joint operations. Although Stark felt that unity of command might be necessary for mixed task forces for offensive operations, he believed that mutual cooperation was better suited for defensive ones.[23]

This opinion aided the Canadian position. The Canadian Chiefs asked Stark and Marshall the specific reasons why, in their view, a change from mutual cooperation to unity of command was necessary. Stark and Marshall agreed the threat was no more than sporadic hit-and-run raids and "stated unequivocally that they did not subscribe to the necessity of Unified Command either in Newfoundland or on the West Coast."[24] This official US military opinion satisfied the Canadian military leadership and convinced them of the effectiveness of a direct approach with the US Chiefs of Staff. It is a good example of Canada's historical ability to press its case in Washington sometimes being strengthened by American political culture. While unified in national purpose in times of crisis, the American political structure still retains the "checks and balances" inherent in its overall republican system of government, which often also characterize relations between various parts of the executive branch, especially the armed services. In these circumstances, Ottawa can often find "allies" within the US government who, for their own bureaucratic reasons, agree with the Canadian position.[25] Few countries have managed to play Washington politics better than Canada.

After the CSC's success with the direct approach with the US Chiefs of Staff, henceforth deliberation and decisions regarding the Canada–US command and control relationship would no longer come under the purview of the PJBD; instead they would be the prerogative of American and Canadian military leaders. Only *if* the US Chiefs of Staff directly approached the

Canadian Chiefs of Staff about implementing the unity of command provision in ABC-22 – which appeared unlikely – would the CSC consider the matter again. Although members of the US PJBD Section tried at various times during the winter and spring of 1942 to convince their Canadian counterparts of the need for unity of command on the west coast, because they gave no new proof that mutual cooperation had broken down and there were no formal requests from the US Chiefs of Staff (by March 1942 the Joint Chiefs of Staff, or JCS), the Canadian military leadership declined to contemplate the requests.[26]

Public pressure in BC did, however, affect Canadian military organization at the operational level. In March 1942, over the objections of the Chiefs of Staff Committee, the Cabinet War Committee approved a system of Canadian unified command on the east and west coasts. In the new scheme, the two senior members of each Joint Service Committee in Halifax and Vancouver respectively became "commanders-in-chief, East and West Coast Defences," and were granted the authority to oversee all operations in their areas of responsibility in addition to retaining operational command over their own service's forces. In Newfoundland, the senior member of the Joint Service Sub-Committee in St. John's, Maj.-Gen. L.F. Page, was designated as "commanding Newfoundland defences," although he remained responsible to the Commander-in-Chief, East Coast Defences, in Halifax.[27] This was purely a Canadian arrangement, not a bilateral one with the United States. It was also a façade: "in practice ... all that changed were the titles. The commanders-in-chief did not interfere in the operations of the other services."[28] Although there was the appearance of the *concept* of unity of command in that all services on each coast came under one officer, in command and control *practice* the joint committee system of cooperation remained.

On the east coast, where the enemy threat remained a key consideration, the professional relationship between the Canadian and American commanders proved vital. One measure the Canadian military took was to reassert Canadian authority through promotions. The RCN commanders respectively in Newfoundland and Halifax, Commodores Leonard Murray and George C. Jones, were both promoted to Rear-Admiral. The Canadian Army replaced Brig.-Gen. Earnshaw in December 1941 with Maj.-Gen. L.R. Page as the General Officer Commanding Canadian-Newfoundland troops. Page was two days senior to Maj.-Gen. Brant, but his attitude was much more conciliatory towards the American officer than Earnshaw's had been.[29] In a series of meetings in January 1942, Page carefully outlined the Chief of Staff Committee's conclusions regarding the enemy threat to North America to Brant as the reasoning for the continuation of mutual cooperation. This personal approach went a long way towards smoothing things over. Page acknowledged there was not enough cooperation between the Newfoundland Base Command and the Canadian services and admitted that "the fault was largely ours." Brant agreed the cooperation with the Canadian commanders was not as close as it should be, and he reassured Page of his "desire to co-operate and get on with the war."[30] This appeared to put the unity of command issue to rest – for the moment.

Brant's improved relations with the Canadian commanders in Newfoundland proved to be fleeting. He had a particularly rocky relationship with the RCAF commander in Newfoundland, Air Commodore (A/C) McEwen. Brant was a US Army "traditionalist," and his opposition to independent air forces meant that he was not enamoured with the upstart younger Canadian airman.[31] Unhappily, the resulting difficult working relationship between Brant and McEwen had a negative effect on RCAF–USAAF maritime trade defence cooperation (see below). It also led to an unfortunate incident between Brant and his Canadian opposite numbers.

Friction between Brant and the Canadian commanders in Newfoundland spilled over during a meeting at St. John's in June 1942. Unlike Rear Adm. R.M. Brainard, the new USN commander at Argentia, Brant refused to adhere to RCAF air traffic control and identification regulations. To do so, in Brant's opinion, would mean RCAF control over USAAF aircraft, which was unacceptable since it was none of McEwen's business what American aircraft did and where they went.[32] Brant concluded that "the whole situation is that the tail (RCAF) is trying to wag the dog (US Army)."[33] After Brant proceeded to ridicule the RCAF "at considerable length," Page and McEwen issued him a thinly veiled threat that they would shoot down any USAAF aircraft not adhering to the regulations. Brant retorted that he was willing to take that risk and stormed out of the meeting. This clash of personalities had a direct impact on the command and control situation in Newfoundland. As Rear Adm. Murray noted shortly after the meeting with Brant, "the system known as cooperation, which we favour, is very difficult to achieve with an officer of this type commanding one of the five services involved."[34]

Brant's superiors reprimanded him for the incident, and he later apologized to the Canadians for his behaviour. The Chiefs of Staff Committee accepted the apology and considered the matter closed.[35] Thereafter relations between Brant and his Canadian counterparts were "extremely good," and mutual cooperation was deemed effective for continental defence coordination. Brant did not publicly bring forth the unity of command issue again; however, he never did feel comfortable with mutual cooperation. In October 1942, Brant officially agreed that in an emergency he would place himself under the unity of command of Maj.-Gen. Page, who was the senior officer in Newfoundland, as per ABC-22.³⁶ Privately, however, Brant revealed to his superior Lt.-Gen. Hugh A. Drum his opinion that mutual cooperation was defective and that the Canadian forces in Newfoundland were insufficient and unreliable. He was convinced that in the event of an actual attack, the US Army would carry out the main defensive effort, and he stated that he had "no misgivings [as to] who the actual Commander would be [i.e., himself]."³⁷ Fortunately, there were no enemy attacks on Newfoundland to test whether Brant would have imposed his personal views on unity of command. Although the Canadians were apprehensive that the US Army might promote Brant to lieutenant-general, this fear proved unfounded. When Brant was reassigned in December 1942, his replacement was an officer of lower rank, Brig.-Gen. John B. Brooks, although he was shortly promoted to major-general.³⁸

The account of Brant's relationship with Canadian officers in Newfoundland is an example of the important role that the "human element" – notably, personalities and their effect on relations between commanders – plays in operational-level command and control relationships. Brant, a hardnosed old-fashioned type of US Army officer who was set in his ways, made the relationship between Canadian and American commanders a divisive and strained one. Fortunately, as future chapters will demonstrate, in the Cold War period, RCAF and USAF officers would draw on their common bonds as airmen and air power advocates to ensure a much better rapport.

In the end, there were no major enemy attacks on North America during the Second World War to determine whether or not mutual cooperation was effective for Canada–US continental defence operations. Nonetheless, mutual cooperation was put to the test in the Northwest Atlantic as Canadians and Americans coordinated their maritime air power efforts, and the results of this are worthy of discussion.

Early Maritime Air Power Coordination

Although continental defence remained an important concern for the Canadian and American air forces in Newfoundland, their main focus by 1941 was on coordinating maritime patrol operations to protect Allied shipping from German U-boat attacks in the western Atlantic, which were increasing in number.³⁹ That year, the United States agreed to provide destroyer escorts for American ships in Commonwealth convoys, and by July the USN had assumed responsibility for the defence of all American and Icelandic merchant vessels travelling between North America and Iceland. When Franklin Roosevelt and Sir Winston Churchill met at Argentia for the Atlantic Conference in August, they agreed to implement the USN's Hemisphere Defence Plan No. 4 (commonly known as WPL-51). Canada was not consulted in this decision, and the

Canadian naval leaders were perturbed to discover that the plan gave the Americans responsibility for the western Atlantic and placed the operations of the RCN under American direction.[40] Specifically, Rear Adm. A.L.L. Bristol, Commander of USN Task Force 4 at Argentia, was given "coordinating supervision of the operations of Canadian escort units, which latter will be effected through and with" RCN Commodores Murray in St. John's and Jones in Halifax.[41] Murray and Jones, however, still retained operational command of their forces, which included authority over the availability of RCN escorts.

WPL-51 did not specifically address command and control authority over Eastern Air Command's maritime patrol aircraft in the Northwest Atlantic. Nonetheless, as the RCAF official history notes, "since the American doctrine of unity of command assumed naval control and direction of maritime air operations far from shore, the US Navy was inclined to exercise command over the RCAF for these purposes as well." Believing they could dictate which forces could be assigned to the defence of Allied shipping, the USN informed the RCAF in September 1941 that American air forces (naval and USAAF) would conduct all long-range cover for convoys, while Eastern Air Command, including No. 1 Group in Newfoundland, would be relegated to the coverage of Canadian and Newfoundland coastal waters.[42] Such responsibilities, however, fell under the operational command that Eastern Air Command exercised over its squadrons: the USN had no authority to dictate RCAF operations.

Senior Canadian air force officers immediately began to build a case against having their maritime units come under American command. The Air Officer Commanding Eastern Air Command, Air Vice-Marshal N.R. Anderson, pointed out that the RCAF had more experience in maritime patrol operations, having conducted sorties from distances of 600 to 800 miles out to sea since the war began. It was not logical, nor was it conducive to the maintenance of morale in the squadrons, to relegate Eastern Air Command coverage to coastal zones. Group Captain F.V. Heakes, an officer on the staff of No. 1 Group, concurred. Heakes had a first-hand view of the military situation in Newfoundland, and he could see that the Americans did not have sufficient aircraft to take on full-scale convoy coverage duties without RCAF help.[43]

In early October 1941, Admiral Stark formally requested that the RCAF place Eastern Air Command under USN unity of command. He noted that ABC-22 allowed for unity of command if the other country's chiefs of staff agreed to its implementation, and he stressed that this change was necessary to ensure "maximum efficiency" of the Canadian-American effort to provide protection for the convoys.[44] The Air Staff in Ottawa disagreed. Regardless of the fact that ABC-22 was not even in effect because the United States was not yet in the war (a reality that both sides for some reason ignored), the RCAF leadership placed greater emphasis on the clause in ABC-22 that provided for unity of command only "in case of extreme emergency" and subject to confirmation by both nations' chiefs of staff. As no such emergency existed at the time, USN unity of command was not necessary. Since the beginning of the war, cooperation had been a satisfactory command and control arrangement between the RCAF and British forces, and so it also should be with American forces. Besides, the USAAF was only required to *support* the USN's maritime patrol operations (see below), yet "strangely enough," Stark was requesting that the RCAF place its maritime air power forces under USN *command*.[45] These arguments satisfied the Canadian Minister of National Defence for Air, C.G. "Chubby" Power. At the Cabinet War Committee meeting of October 9, he convinced his colleagues there was "no reason to give to a foreign neutral power more than had been given to the Canadian and British Navies."[46]

With the full support of his government, the RCAF Chief of the Air Staff, Air Marshal L.S. Breadner, informed Admiral Stark that even though the closest possible coordination between all forces in the Northwest Atlantic was desirable, the situation did not necessitate unity of command at that time. Breadner instead advocated continuing the system of cooperation between Eastern Air Command and American air forces, and notified Stark that the RCAF was setting up a system of coordination in the form of liaison officers – a system provided for in ABC-22.[47] This was not the response that Stark was hoping for, but he respected the RCAF's decision and issued orders for the USN air forces in Newfoundland to cooperate with Eastern Air Command. Upon hearing of Stark's acceptance of the RCAF position, Breadner informed Power and added a message of relief: "We have held them off, so far!"[48]

Renewed American Pressure for Unity of Command

After Pearl Harbor, American officers began a fresh attempt to impose unity of command over Eastern Air Command's maritime patrol operations. In Newfoundland, their main focus was on gaining control over the RCAF's No. 1 Group. The primary American concern was A/C McEwen's apparent lack of independence from Eastern Air Command Headquarters in Halifax to take "immediate action" in support of USN trade defence efforts.[49] The Americans took their concern to the PJBD, which released its 22nd Recommendation on December 20, 1941. It called for the decentralization of command and control to permit easier local cooperation and to give commanders in Newfoundland greater freedom of action to deal immediately with the operational situation.[50] This measure was consistent with the military principles of mission command and centralized command and decentralized execution. However, the Americans felt that the wording of the 22nd Recommendation was not very precise and could be interpreted very broadly in terms of implementation measures.[51] The RCAF leadership was also uncomfortable with the measure. Because the recommendation decreased Eastern Air Command's authority over RCAF forces in Newfoundland, the Air Staff immediately suspected that the Americans would try to capitalize on this situation by implementing US unity of command over No. 1 Group. The RCAF was even more suspicious of American intentions now that the United States was formally at war.[52]

The division of responsibility for maritime air power between the USN and the USAAF was also a concern. During the interwar period, the United States had not established a separate air force like Canada; instead, it had permitted the USN and US Army to each have its own air arm. *Joint Action of the Army and the Navy* granted the navy responsibility for all inshore and offshore patrol for the protection of shipping and for defending the coastal frontiers. The US Army Air Corps was responsible for the defence of the coast itself through the tasking of long-range bombardment aircraft to destroy any approaching hostile forces. *Joint Action* also included a provision that US Army aircraft could temporarily execute USN maritime patrol functions in support of or "in lieu" of the navy if the latter did not have sufficient aerial resources.[53] This is precisely what happened during the Second World War: a shortage of maritime patrol aircraft forced the USN to rely on the long-range bombers of the USAAF's Number I Bomber Command to fill the gap. The problem was significant disagreement between the two services about to how carry out this defensive role.

USN maritime air power doctrine emphasized the more defensive function of convoy escort and patrol of fixed sectors of coastal waters, which mirrored British and Canadian practice, while USAAF doctrine had more of an offensive focus. Heavily influenced by strategic bombing theories developed at the Air Corps Tactical School, USAAF doctrine emphasized the concept of "forward air defence," an offensive form of defence that used aircraft to seek out and destroy attacking enemy forces. The USAAF thus adopted a "seek and sink" doctrine for its anti-submarine operations instead of the proven defensive one that the Royal Air Force's Coastal Command had developed and the RCAF used. *Joint Action* stipulated that USAAF support to the USN was only a "temporary" expedient, and this gave the USAAF little incentive to dedicate time and resources away from its offensive strategic bombing role and towards the development of more defensive anti-submarine doctrine.[54]

Because of the division of responsibility for maritime air power between the USN and the USAAF and their differing doctrinal focuses, it is easy to understand why the RCAF was so opposed to the idea of placing its maritime patrol forces in Newfoundland under a US commander. The Air Staff feared that

American unity of command over No. 1 Group aircraft would lead to the splitting of RCAF forces in Newfoundland into two parts, one under the USAAF and the other under the USN. Such a division of RCAF squadrons along US service lines would have been disastrous for Canada's air force. It would have contravened the fundamental concept of the indivisibility of air power;[55] it would also have meant a complete loss of the RCAF's responsibility for the air defence of Newfoundland, even though Canada had more military aircraft there than the Americans. Moreover, these factors all promised to have a negative effect on the RCAF's morale and on the efficiency of its maritime patrol operations in support of navy convoy escorts – something that also concerned Rear Adm. Murray, the RCN commander in Newfoundland.[56]

Group Captain Heakes made the important observation that the strength of the American position would decline exponentially with the decreasing number of US forces in Newfoundland, as US commanders would no longer be able to argue that the country with the largest forces should exercise unity of command. He predicted – accurately as it turned out – that the Americans would soon transfer many of their forces from Newfoundland to the Pacific to face the Japanese threat and that this would only solidify the RCAF's position on mutual cooperation.[57] Air Vice-Marshal Anderson agreed with these conclusions, adding that "Canadian commanders cannot relinquish their command or responsibility to their government."[58] Mutual cooperation would have to suffice for the time being.

Anderson also understood that the responsibilities of both A/C McEwen's headquarters in St. John's and his own headquarters in Halifax were growing every week. To ensure greater efficiency and to placate American concerns about McEwen's lack of local authority over air operations, Anderson agreed to the decentralization of command and control of No. 1 Group aircraft in Newfoundland in fulfillment of the PJBD's 22nd Recommendation. However, because the RCAF still viewed Newfoundland as an integral part of the overall defence of the Canadian Atlantic coast, Anderson decided that it was imperative for Eastern Air Command to retain operational command over all RCAF forces in eastern Canada so that he would have access to all of the command's aircraft; this would allow him to reinforce units in any area that came under enemy attack. Consequently, Anderson only granted McEwen "local operational control" (the term "tactical control" was also used, though neither was defined) over No. 1 Group forces in Newfoundland. The Americans concurred, and the measure came into effect on January 20, 1942. Thereafter, the AOC Eastern Air Command in Halifax gave only "general directives" to the AOC No. 1 Group in St. John's. This arrangement ensured that McEwen's maritime patrol aircraft would be available, through cooperation, to support USN forces based in Newfoundland.⁵⁹

The US Army official history states that with Anderson's implementation of the PJBD's 22nd Recommendation, "the US Navy task force commander at Argentia finally achieved the unified operational control of all the air and naval resources of the two countries available."⁶⁰ This is an incorrect assessment. Actual operational control of RCAF aircraft in Newfoundland did not rest with the American commander at Argentia, Rear Adm. Bristol; rather, it was exercised by A/C McEwen, the Canadian AOC No. 1 Group in St. John's. The arrangement between Canadian and American air forces in Newfoundland therefore continued to be based on the principle of mutual cooperation, as provided for in ABC-22. It took the form of a system of air coverage for convoys whereby "cooperation of the RCAF with the US Navy [was] accomplished by means of proposals mutually acceptable."⁶¹ In Argentia, Rear Adm. Bristol sent "proposals" for air coverage to McEwen in St. John's, who then had the right to decide whether he would employ his resources to meet the USN's request. To ensure that proper air coverage was provided to shipping, the Americans and No. 1 Group also developed a system of mutual assistance for operations: if one maritime patrol force (e.g., USN air forces at Argentia) was unable for any reason (e.g., weather conditions) to provide aircraft to protect shipping in any given area, it could request that another force (e.g., No. 1 Group RCAF or the USAAF at Stephenville or Gander) provide aircraft to perform the task in its place.⁶² This system worked well for most of 1942; at no time did the No. 1 Group commander refuse any of the US admiral's requests for air coverage.

The Failure of Mutual Cooperation and the Implementation of Operational Control

In March 1942, Canada and the United States faced a massive U-boat campaign against shipping in the western Atlantic. Before Pearl Harbor, Hitler had sharply restricted U-boat operations in waters south of Newfoundland to avoid infringing American neutrality and bringing the United States into the war. After Germany's declaration of war on the United States in December 1941, Hitler gave his U-boat commanders a free hand to attack all shipping in the western Atlantic just as the United States was beginning to move the bulk of its maritime forces to the Pacific to make up for losses to the Japanese. Only in the spring of 1942, after U-boats had ravaged unescorted shipping off the American coast, did the United States finally agree to implement the proven convoy system in its waters.⁶³ The RCAF used the weak American anti-submarine performance off its eastern seaboard as further evidence that Eastern Air Command aircraft should not come under American unity of command.

By April 1942, with fewer easy pickings off the US east coast, the U-boats had once again begun focusing their operations against convoys on the main North Atlantic shipping lanes. This put greater pressure on Canadian and American maritime patrol forces in eastern Canada and Newfoundland. By the middle of the year, escorts were complaining of poor communication and scanty air coverage and it was becoming clear that efforts to coordinate operations based on mutual cooperation were insufficient.[64] To help the Canadians and Americans remedy this problem, the British sent experts from RAF Coastal Command to North America to assess the situation and make recommendations for improvement.

Coastal Command had been conducting maritime patrol operations in the North Atlantic since the beginning of the war. Its personnel worked side by side with the Royal Navy in joint Area Combined Headquarters according to the British operational control system (see Chapter 2). This proved to be an efficient means of coordinating maritime trade defence operations, and combined with the development of the best anti-submarine weapons and doctrine, Coastal Command became the premier maritime air power organization in the world. The British officers who visited North America were therefore well placed to comment on the effectiveness of Canadian and American maritime air power forces.[65]

One of the main criticisms concerned the command and control arrangement in the Northwest Atlantic. The British officers deplored the existence of several command organizations in the area and the reliance on mutual cooperation to coordinate air forces for multiple tasks (i.e., the defence of shipping *and* continental defence). As a solution, they recommended command and control centralization: Canada and the United States should implement the British operational control system and place all of their anti-submarine forces, naval and air, under a single authority.[66]

The disadvantages of the mutual cooperation system were starting to show at the operational level in Newfoundland. By autumn 1942 the RCAF had begun to adhere to the RAF Coastal Command practice of providing air coverage only to convoys that intelligence indicated were actually threatened by U-boat attack. As a result, No. 1 Group frequently did not provide all the air patrols the USN admiral at Argentia (by now Rear Adm. Brainard) proposed because intelligence from Ottawa indicated there were no U-boats in the area.[67]

An incident in late November provided further proof of the inadequacy of mutual cooperation. When fog at Argentia prevented USN aircraft from providing coverage to a nearby convoy, the RCAF attempted to compensate by tasking aircraft from Sydney. A/C McEwen requested further assistance from USAAF B-17 aircraft at Gander, but Maj.-Gen. Brant's headquarters failed to respond. The resulting absence of air protection allowed a German U-boat to sink one ship and damage two others. The failure of mutual cooperation greatly frustrated the RCAF. Its commanders on the east coast were fed up with both Brant's lack of cooperation and Rear Adm. Brainard's constant "proposals" for air cover of convoys that intelligence indicated were not threatened. For these reasons, the Air Staff decided to join the RCN in its campaign during the winter of 1942–43 to have all air and sea trade defence forces in the Northwest Atlantic brought under one Canadian authority.[68] With Canadian aircraft and escort vessels outnumbering American forces in the Northwest Atlantic after Pearl Harbor, there should have been greater Canadian command and control influence in the area. However, there was a holdover effect that allowed the United States to retain its superior command and control position in the Northwest Atlantic for several months despite the redeployment of most of its forces to the Pacific. The situation was not lost on the leaders of Canada's navy and air force, and it only added impetus to their push for a Canadian theatre and commander-in-chief.

This idea made sense for Eastern Air Command both in terms of operational efficiency (underscored by the demonstrable failure of mutual cooperation), and because the RCAF had the majority of maritime air forces in the area. By late November 1942, the USN had only two flying boat squadrons conducting patrols from Argentia, and just two USAAF long-range B-17 squadrons were conducting anti-submarine patrols interchangeably out of Gander and Stephenville. By contrast, Eastern Air Command had ten maritime patrol squadrons, four of which were in Newfoundland under No. 1 Group. By late 1942 the RCAF was also providing all of the convoy coverage to the north and east of Newfoundland, which is where the bulk of U-boat operations against Allied shipping were taking place.[69] It was time for a change.

During negotiations with their American counterparts in early 1943, Canadian officers soon discovered how influential the RAF Coastal Command experts' visit to North America had been on American thinking about the defence of shipping in support of the war effort. Constructive criticism from the British had convinced the USAAF to transform I Bomber Command into the Army Air Forces Anti-Submarine Command, eliminate its continental defence bombardment role, and task it solely with anti-submarine operations. The most important effect of the consultation with Coastal Command officers, however, was that the USN and USAAF became amenable to the idea of centralizing command and control in the Northwest Atlantic under Canadian authority.[70] To the Canadians' pleasant surprise, the USN and the USAAF stated that they were willing to surrender operational control over all their anti-submarine forces in the Northwest Atlantic to Canada provided that a Canadian commander-in-chief was made responsible for the entire anti-submarine effort in the area. This proposal appealed to the RCAF leadership since it would give the AOC Eastern Air Command operational control over American maritime patrol aircraft.[71] Discussions continued in Washington during the winter of 1943 and culminated at the Atlantic Convoy Conference in March.

The major outcome of this meeting of Canadian, American, and British experts on the anti–U-boat campaign was the creation of a new theatre of operations in the Northwest Atlantic. Stood up on April 30, 1943, the Canadian Northwest Atlantic Command was based on the British system of operational control. Canada was granted operational control over all air and surface escorts in the area west of 47°W and north of 40°N, including Newfoundland. Rear Adm. Murray became the theatre commander-in-chief, with operational control over all naval forces and operational direction over all maritime patrol aircraft. Murray exercised this operational direction through his Eastern Air Command deputy, Air Vice-Marshal G.O. Johnson, who oversaw maritime patrol operations.[72]

Johnson, whose title was now Air Officer Commanding-in-Chief (AOCinC) Eastern Air Command, exercised "general operational control" over all Allied air forces employed in the defence of shipping in the Canadian Northwest Atlantic Command. In Newfoundland, Johnson delegated the "local operational control" of all maritime patrol operations to the new AOC No. 1 Group, Air Vice-Marshal (formerly Group Captain) Heakes.[73] "General" operational control meant that the AOCinC Eastern Air Command continued to pass on "general directives" to the AOC No. 1 Group, although now they included instructions for the use of USN and USAAF aircraft in the defence of convoys. "Local" operational control meant that Heakes retained operational control over RCAF forces in Newfoundland and now also had operational control over all American maritime patrol forces.

On March 30, 1943, a team from the USAAF's new 25th Anti-submarine Wing joined Heakes's staff at No. 1 Group Headquarters in St. John's to facilitate the new arrangement. Although the Americans technically did not have to make this move until the Northwest Atlantic Command was officially stood up on April 30, they did so in good faith, desiring to get a head start on learning the RCAF's methods so that they would be "ready to operate under the new scheme immediately the word go is given."[74] Finally, the USN liaison staff from Argentia arrived at Heakes's headquarters in May, shortly after he had assumed local operational control of all Newfoundland maritime patrol aircraft.[75] Heakes appreciated these measures, which pointed to a future positive

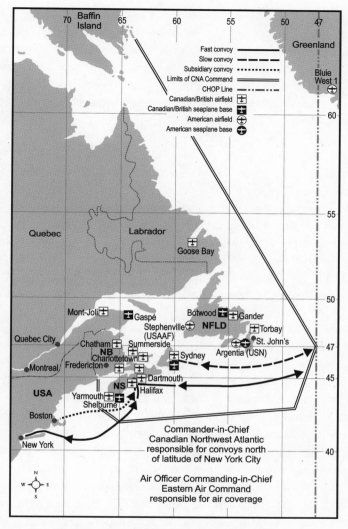

MAP 1   The Canadian Northwest Atlantic Command.
Map drawn by Mike Bechthold.

working relationship between Canadian and American officers for Cold War continental air defence coordination.

The creation of the Canadian Northwest Atlantic Command was a significant accomplishment in Canada's military history. It marked the first time that Canadian naval and air force officers had commander-in-chief status in an active theatre of war.[76] Canada and its allies had recognized the necessity of command and control centralization to ensure efficient defensive operations.

The US Army official history states that the Canadian Northwest Atlantic Command was "the only instance of unified command under ABC-22."[77] To make this claim, however, is to confuse the issue. Although the *concept* of unity of command – the centralization of command and control under one commander – was achieved, it was done through the implementation of the British operational control system, not the command and control *principle* of unity of command in ABC-22.

Conclusion

Canada was able to protect its sovereignty during the Second World War because at no time did its forces operating in North America come under American command. Although Canadian commanders found themselves under frequent pressure from their US counterparts on both coasts to implement unity of command (especially after Pearl Harbor), they were able to resist. By consistently relying on the most recent strategic estimates of the enemy threat, Canadian officers successfully argued that the danger did not require unity of command and that mutual cooperation, as provided for under ABC-22, was sufficient.

By deciding to deal directly with the American Chiefs of Staff, the Canadian Chiefs of Staff Committee reasserted that it would be the Canadian service chiefs, acting on behalf of the Canadian government, who would decide on all command and control arrangements with the United States. The Newfoundland experience demonstrated that ambiguities and disagreements regarding command and control could lead to a difficult situation for commanders. Moreover, the encounters with Maj.-Gen. Brant showed the importance of the "human element" and of maintaining a cordial and professional relationship to ensure effective command and control and coordination of combined operations.

For organizational, operational, and doctrinal reasons, the RCAF leadership insisted that its maritime air power forces could not come under American command; instead, coordination of Canadian–American maritime air power operations should be based on mutual cooperation. Although mutual cooperation was an effective means to coordinate continental defence in times of minimal enemy action, the Canada–US air effort against German U-boats demonstrated its weakness. The implementation of the British operational control system in the Canadian Northwest Atlantic Command after April 1943 provided the required command and control centralization. It also safeguarded national sovereignty because operational control did not include authority inherent in national command. The resulting improved coordination of maritime patrol operations was an important factor in the eventual Allied victory over the U-boats in mid-1943. Canada and the United States had set important precedents for operational control and the centralization of command and control during the war. The two nations' air forces would revisit and enhance them during the Cold War.

5
Replacing ABC-22

> *I think it safe to say, therefore, that if we do not push the Canadians too rapidly we may expect slow but steady progress toward that integration of our defense systems which seems so essential to the defense of the North American continent.*
> – Ray Atherton, US Ambassador to Canada, August 1946

> *... the United States wishes in every way to respect Canadian sovereignty.*
> – Maj.-Gen. Guy Henry, US Army Member of the Canada–US Military Cooperation Committee, August 1946

As the end of the Second World War drew near, Canadian and American planners began pondering the future continental defence relationship. Maj.-Gen. Maurice Pope, chairman of the Canadian Joint Staff Mission in Washington and the Canadian Army member of the PJBD, accurately predicted that the two North American nations would continue their "intimate defence relationship" in the postwar period. Sensing the growing animosity between the United States and the Soviet Union, he anticipated that the Americans would approach the Canadians about a new defence plan sooner rather than later "so as to have it ready to be put into effect when the next war comes." Pope harboured no doubt that Canada would become directly involved in any such conflict. "To the American," he wrote, "the defence of the United States is continental defence, which includes us, and nothing that I can think of will ever drive that idea out of their heads."[1]

Given this situation, Pope warned that Canada had more to fear in "a lack of confidence in the United States as our security, rather than enemy action." It was crucial for Canada to maintain sufficient military forces, not to defend against any potential enemy (though this would also be a requirement) but "to ensure that there was no apprehension as to our security in the American public mind."[2] In Pope's professional opinion, Canada should continue its close defence relationship with the United States to avoid a "defence against help" situation. Replacing the wartime Canada–US defence plan, ABC-22, was the first and most important step. Although American officials sought greater command and control authority, the two nations ultimately agreed to maintain the status quo – the mutual cooperation–unity of command compromise arrangement – in the June 1946 Canada–US Basic Security Plan (BSP). Nevertheless, with the development of an Air Interceptor and Air Warning appendix to the BSP, the foundation was laid for greater Canada–US air defence integration.

Air Power, Strategic Culture, and Air Defence

> *If there is another war it will come against America by way of Canada from Russia.*
> – William Lyon Mackenzie King, Diary Entry, September 11, 1945[3]

The early Cold War period was uncharted waters for Canada–US continental defence collaboration. The two nations had come together to defend North America in 1940 because of wartime expediency. Now they initiated peacetime discussions to plan how best to defend their continent during the next war. In September 1945 the Soviet threat remained distant – at least for the moment. Canadian–American planners had the luxury of time. Nevertheless, officers were aware of potential threats to the continent. Until 1953, the danger to North America was still a more traditional one from all three environments: sea, land, and air. As such, the focus of planners largely remained on overall continental defence. The growing Soviet strategic bomber threat to North America, however, required attentiveness to continental air defence needs. This aerial danger was a relatively new development, but it grew to dominate the Canada–US command and control defence relationship.

At the end of the Second World War, American airmen were convinced that strategic bombing had been the "winning weapon" in their victories over Germany and Japan. Even though most of these officers did not acknowledge that an enemy could use offensive air power against North America (given their own messianic belief in US offensive air power and their internal struggle to form an independent air force), they placed the building of a strong Strategic Air Command and the production of an atomic bomb stockpile at the forefront of USAAF (after 1947 USAF) doctrine and force development.[4]

Indeed, the dropping of the atomic bombs on Japan had a significant effect on American strategic thinking. It was quite widely appreciated that the new weapon would change the nature of war and future threats to the United States.[5] This was best captured in Bernard Brodie's 1946 book, *The Absolute Weapon: Atomic*

Power and World Order, in which the noted nuclear theorist astutely predicted the emergence of strategic deterrence and mutual retaliation:

> If the atomic bomb can be used without fear of substantial retaliation in kind, it will clearly encourage aggression. So much more the reason, therefore, to take all possible steps to assure that multilateral possession of the bomb, should that prove inevitable, be attended by arrangements to make as nearly certain as possible that the aggressor who uses the bomb will have it used against him. If such arrangements are made, the bomb cannot but prove in the net a powerful inhibition to aggression ... Thus, the first and most vital step in any American security program for the age of atomic bombs is to take measures to guarantee ourselves in the case of attack the possibility of retaliation in kind.[6]

The dropping of the atomic bombs on Japan also had a significant impact on Canadian strategic thinking. As a minor participant in the Manhattan Project (including supplying uranium), Canada had some knowledge of this new weapon; and in fact, as James Eayrs has noted, "Mackenzie King, C.D. Howe, and C.J. Mackenzie were the only persons outside similar inner circles in the United States and United Kingdom who knew beforehand when and where the bombs would fall."[7] Still, the revolutionary power of the atomic bombs to cause such death and destruction shocked the Canadian prime minister. Mackenzie King recorded in his diary his "mixed feelings" that although "we were now within sight of the end of the war with Japan ... we now see what might have come to the British people had German scientists won the race."[8]

There was minimal discussion of Canada possessing nuclear weapons at that time. Public statements instead emphasized the peaceful civilian applications of atomic energy. As a main producer of uranium, Canada was keenly interested in subsequent efforts to control the atom, and was therefore instrumental in the establishment of and participation in the UN Atomic Energy Commission.[9] Nonetheless, the danger posed by the new nuclear age unleashed by Hiroshima and Nagasaki was not lost on Canadian officials. Hume Wrong, the Associate Under-Secretary of State for External Affairs, articulated the DEA's views in a memo to the prime minister, expressing that "the successful development of the bomb is of such far-reaching importance that it may profoundly affect international affairs even to the point of altering the whole balance of international forces overnight."[10] Nor was the fact overlooked that it was air power that had delivered the atomic bombs. As Professor J.I. Jackson from the RCAF Staff College remarked:

The surrender of Japan is a significant date in the history of air power. For one thing, the surrender was brought about without invasion, mainly by air attack, and with 2,000,000 Japanese soldiers still under arms in the home islands. For another, the dropping of the first atomic bombs, even as it marked the war's end, marked the opening of a new era in man's use of the air ... With the first atomic bomb, the destructive capacity of a single heavy bomber was increased three thousand times, which is to say that a single bomber could effect the same damage as had previously been done by three thousand bombers carrying conventional high explosive weapons.[11]

The Soviet Union's devious actions, exposed by revelations about Soviet espionage in Canada (the infamous Gouzenko Affair) only one month after Hiroshima and Nagasaki, also shook Mackenzie King.[12] As the quotation at the outset of this section indicates, the prime minister became concerned that the Soviet Union would become the next enemy and that it might try to use Canadian airspace to attack the United States.

During the Second World War the Soviet Union had focused mainly on tactical air power in support of Red Army ground operations. After the war, however, Josef Stalin was intent on developing an atomic bomb, along with a strategic bomber force to deliver it. He knew that the United States' monopoly on atomic weapons gave the Western democracies a major advantage, and he wanted to close the atomic gap – quickly. Therefore, when a few USAAF B-29 Superfortress bombers were forced to make emergency landings on Soviet territory in 1944–45, Stalin seized them as a template to build his own strategic bomber force capable of reaching the North American industrial heartland. Soviet technicians meticulously took the American B-29s apart piece by piece and by 1949 had re-engineered the aircraft to produce the Tu-4 Bull long-range strategic bomber.[13] Canada – indeed, the entire continent – was now vulnerable to aerial attack, and thoughts about continental air defence in the Cold War era did not escape the attention of Canadian–American planners.

In January 1945, the Working Committee on Post-Hostilities Problems final report concluded that any future threat to North America would be significantly different from the one faced during the Second World War. Canada was situated between the Soviet Union and the United States, which might well go to war with each other. The Atlantic, Pacific, and Arctic Oceans no longer provided protection to the continent, given advances in air power technology and doctrine. The most likely route for the new Soviet strategic bomber force to the industrial heartland of North America was the shortest one: over the North Pole and Canada.[14]

Despite this potential Soviet threat, American strategic culture in this new nuclear age was more offensively oriented. It was based on maintaining a monopoly over the atomic bomb (before 1949) and on Strategic Air Command's deterrence capability – that is, its capacity to launch pre-emptive or retaliatory nuclear attacks. Strategic defence, even of the American homeland, was not a priority for the United States, so investments in air defence were rather modest. The USAF's focus on offensive air power meant that many of its leaders, with support in the Department of Defence, opposed improved strategic air defences out of concern that funding, personnel, and equipment resources for such expensive projects might take away from the building up of the SAC deterrent.[15]

The Americans' approach to continental air defence would affect discussions on the future of air defence and Canada's role and relative importance. American strategic culture placed little priority on strategic defence, whereas Canadian strategic culture placed greater emphasis on continental air defence for the related goals of protecting both North America and the SAC deterrent from attack. Although the RCAF was rightly proud of the role that No. 6 Group had played in Bomber Command during the Second World War, the growing size and expense of Cold War heavy bombers and Canada's early decision not to possess nuclear weapons led Canadian airmen to abandon strategic bombing and to focus instead on other air power roles – especially fighters.[16] Strategic defence thus became a high priority for Canada in terms of political and material commitment. Ottawa therefore devoted greater relative political and military attention to continental air defence arrangements than Washington during the early Cold War. Indeed, this early focus on strategic defence helps explain how and why Canada was able to use its functional approach to protect Canadian sovereignty.[17]

Immediately after the Second World War, Canada placed a growing priority on continued defence collaboration with the United States. A revised Canada–US defence plan that addressed the growing strategic bomber threat in addition to traditional maritime and ground threats became a top priority for both Canadian and American defence planners. The command and control relationship between the two nations' armed forces was a pivotal aspect of this new scheme.

Formulating a New Plan

At the end of the Second World War, the Canadian and American sections of the PJBD formally agreed that bilateral defence collaboration should continue and that the PJBD should remain the primary forum for discussions. At its meeting on January 16–17, 1946, the PJBD drafted the "Canada–United States

Security Plan" (short title CA-1), which was designed to protect both countries from attack. The command and control provisions of CA-1 were similar to those of ABC-22: it provided that "a unified command" could be established for any Canadian–American force working in a particular area (geographical) or for any common purpose (functional). Those authorized to establish a unified command included a proposed combined Canada–US Chiefs of Staff organization (discussed below) or the local operational commanders in an emergency situation, subject to later confirmation from higher military authority. This latter provision was consistent with ABC-22, as was the definition of unity of command, which remained the same. There was one important difference between CA-1 and ABC-22: the former did not contain the provision that the principal command and control relationship between the two nations' armed forces would be based on mutual cooperation. This may have been a simple oversight, but it foreshadowed American thinking on the ideal Canada–US command and control relationship.

The most groundbreaking aspect of the CA-1 draft plan was the proposal for a combined Canada–United States Chief of Staff (CANUSA). This organization would be established in an emergency situation to assume responsibility for "the implementation and strategic direction" of the plan.[18] This proposal was the brainchild of the US Section, and in particular the senior US Army PJBD member, Maj.-Gen. Guy Henry. What the Canadians did not know was that Henry, by devising the CANUSA proposal without first consulting his superiors, had "far exceeded the mandate" the JCS had given the US PJBD members.[19]

Throughout the Second World War, the United States had gone to great lengths to avoid encirclement by smaller allies when it came to the higher direction of the conflict. Even though Canada was the United States' close partner in continental defence, it was too small a power to justify sharing strategic command and control authority (as the US had done with Britain in the wartime Combined Chiefs of Staff organization). During the war, the Canadian military leadership had pondered the idea of requesting membership on the Combined Chiefs of Staff, but it did not pursue the matter when preliminary investigation indicated that the Americans would surely refuse.[20] It was doubtful that the JCS would now accept CANUSA or any similar proposal that would give Canadians some control over the strategic direction of a future war, especially one that included not only defensive bilateral operations but also offensive multilateral ones. Canada–US defence collaboration – and any command and control arrangements therein – would be bilateral only.[21]

Nevertheless, Canadian-American planners avoided these considerations and continued planning based on the assumption that Maj.-Gen. Henry's CANUSA proposal reflected formal US defence policy. Only after the CA-1 document came to the attention of the JCS did the American military leadership set the record straight. Fearing that CANUSA would set a dangerous precedent for similar bilateral and multilateral arrangements with other smaller countries such as Mexico or Brazil, the JCS instructed Henry to withdraw the CANUSA proposal.[22] Although some Canadian critics continue to believe that the United States is anxious to draw Canada into entangling, sovereignty-threatening defence arrangements, the reality is that the United States is wary of any alliances or security arrangements that would in any way restrict its freedom of manoeuvre when it comes to its national security interests.[23] NATO, as the title of Robert Osgood's 1962 book noted, was *The Entangling Alliance*, but American military and political dominance of that alliance, and the US nuclear deterrent on which it rested, would ensure that multilateralism would be an instrument of, and not a restraint on, American actions.[24]

The only Canadian commentary on CA-1's command and control provisions came from Canada's Ambassador in Washington (and future prime minister), Lester B. Pearson, and it was pessimistic. He interpreted the unity of command provision to mean that the single designated commander "could, I suppose, do pretty well what he wished."[25] This was a fundamental misunderstanding of the power and authority – and, specifically, the limitations on both – that unity of command entailed. Pearson's alarmist reaction, however, was understandable. Because CA-1 did not contain ABC-22's mutual cooperation provision, Pearson – and his Department of External Affairs associates – remained concerned that Canadian forces might come under American command. However, CA-1 was only a preliminary draft. If nothing else, the document demonstrated the need for a new bilateral organization to undertake detailed planning.

In keeping with the wartime tradition of having the military members of the PJBD do the detailed bilateral defence planning, the Canada–United States Military Cooperation Committee (MCC) was established in 1946 to formulate the particulars of the new plan. The MCC consisted of Canadian and American PJBD service members and also included officers from the two nations' chiefs of staff organizations: the JCS's Joint Planning Staff; the CSC's Joint Planning Committee (JPC); plus one civilian representative each from the Canadian Department of External Affairs and the US State Department. Two of the RCAF members of the MCC were Air Vice-Marshal Wilfred Curtis and A/C C.R. "Larry" Dunlap.[26] Both would later rise to the rank of air marshal and become Chief of the Air Staff. This meant that the future senior leadership of the RCAF had an intimate understanding of Canada–US postwar air defence planning in its earliest stages. Such awareness would help foster the development in the early Cold War period of a close bond between Canadian and

American airmen regarding air power, integrating Canada–US continental air defence efforts, and centralizing command and control.

Meeting alternately in Ottawa and Washington, the MCC was responsible for drafting Canada–US defence plans and reporting back to the PJBD. The draft plans would go to the Canadian and American chiefs of staff for authorization and then proceed for formal approval to the appropriate government officials (i.e., the Secretary of Defense and the president for the US, and the Cabinet Defence Committee for Canada). If approved, the defence plan and recommendations would be implemented.[27]

Having established the process for detailed bilateral defence planning, during the spring of 1946 the Canadian and US sections of the MCC began working separately on drafts of the new Canada–US defence plan. On May 6, American planners completed a preliminary continental defence plan titled "Matchpoint." Reflecting US strategic culture, the document emphasized that the best defence was a good offence against the enemy. Any defensive plan devised in conjunction with the Canadians, therefore, would be "prepared for execution in *conjunction with* and *not as a substitute for* the offensive campaign."[28]

Defending bases, especially those from which offensive operations might be launched, was of critical importance to the Americans. Matchpoint also stressed the importance of an appreciation of enemy capabilities and the requirement to provide for bilateral defence measures in the areas of air defence, reconnaissance, trade defence, and defence against enemy lodgements, plus a system of command capable of effectively coordinating combined forces in an emergency situation.[29] The American planners devised two separate command and control appendices, one for US eyes only (hereafter "US version") and one for Canadian consumption (hereafter "Canadian version") – and there were significant differences between the two.

The US version advocated establishing combined Canada–US commands exercising unity of command in peacetime. The American planners felt that since it was probable Canadian and US forces would be co-located in strategically important (and vulnerable) locations, unity of command would be more efficient in times of peace as well as war. It was also imperative that all air defences (including aircraft, anti-aircraft artillery, and air warning services) be placed under unified command during wartime.[30] There was no mention of mutual cooperation.

This proposal went further than any previous Canadian–American arrangements. ABC-22 had been a *wartime* defence plan that provided for unity of command only if approved by the two nations' chiefs of staff or by operational-level commanders in an emergency, subject to later confirmation. It did not presuppose unity of command upon the outbreak of hostilities. In addition, the United States and Canada had never enacted the unity of command provision in ABC-22 during the Second World War. Therefore, to propose establishing it in peacetime would mark a clear shift in the bilateral command and control relationship. The American planners thus kept this proposal away from Canadian eyes.

A significant clause in the US version specified that any principles and provisions it contained "are necessarily dependent upon the views of the Canadians as well as the United States toward combined or unified command."[31] Having learned from the experiences of the American PJBD Section in 1941, the US planners understood that they could not bully the Canadians into accepting their preferred ideas on command and control. Instead they would have to respect Canadian opinion and reach a compromise that satisfied both parties. The United States would not impose a defence-against-help situation on Canada.

The Canadian version settled on the status quo from ABC-22 for command and control. Mutual cooperation would remain the primary Canada–US command and control relationship, with a provision for unity of command in an emergency.[32] This provision was very close to the one that would appear in the final version of the new bilateral defence plan.

The Basic Security Plan

At their meeting of May 20–23, 1946, in Washington, the MCC produced two documents: a Joint Appreciation of the threat to the continent, and a Canada–US Basic Security Plan to replace ABC-22. The Joint Appreciation outlined the prospective threat to the continent during the early postwar period. Although the immediate danger was slight, amounting to only minor nuisance raids, the geographic barriers of the Atlantic and Pacific Oceans and the frozen Arctic no longer protected North America. The Joint Appreciation took a page right out of the USAAF's strategic doctrine, emphasizing that by 1950 the growth of enemy capabilities would permit attacks over the North Pole on the political, military, economic, industrial, and population "nerve centres" of North America. The "potential enemy" (no country was named, although it was obvious it would be the Soviet Union) could be expected to capture objectives in the Arctic regions of Canada, while using strategic bombers armed with conventional and atomic bombs, guided missiles, rockets, and even aircraft launched from submarines.[33]

Canada and the United States needed to devise a new defence plan to counter this threat – one that included a comprehensive air defence system, with radar providing warning as well as interceptor aircraft and other anti-aircraft measures capable of destroying incoming enemy bombers. Since it was expected that this threat would develop within a relatively short time (about four years), the Joint Appreciation concluded that work on these air defences should "be initiated immediately."[34]

The Basic Security Plan echoed some of the assessments in the Joint Appreciation, outlining the main tasks the two countries were required to undertake in order to protect North America. Joint Task One was a combined effort between Canada and the United States to defend the continent from aerial attack by setting up and operating "an effective common air defence." Canada had the primary responsibility for Joint Task Two – defence of northern Canada and Labrador – with US assistance if needed. Joint Task Three assigned responsibility for the defence of Alaska to the American military, with help from Canadian forces as required. The defence of Newfoundland (Joint Task Four) remained the primary responsibility of Canada. The principal role of the United States in Newfoundland was to defend its bases, and the secondary one was to cooperate with Canadian forces in the British colony's overall defence. Joint Tasks Five and Six, the defence of Eastern Canada and the northeastern portion of the United States, and the protection of western Canada and the American northwest, fell along national lines, with provisions for each nation to provide assistance to the other as required. Finally, for Joint Tasks Seven and Eight, the defence of shipping in the Northwest Atlantic and North Pacific respectively, responsibility was to be a combined effort, although the wording suggested that the United States would be the primary actor in this endeavour.[35]

The BSP was a war plan; however, it also called for the implementation of peacetime measures to ensure readiness in the event of a wartime emergency. The most important of these was the establishment of a bilateral air defence system.[36] In addition to command and control provisions (explained below), coordination in the BSP included the provision of liaison personnel and cooperation between commanders; the establishment and use of common base facilities in each country for combined training, tests, and exercises (which included the peacetime stationing of forces from one country on the territory of the other); and military missions in each nation's capital to ensure greater coordination at the strategic level.[37] Each nation's chiefs of staff would enact the BSP only when their respective governments directed them to do so. The plan would also be subject to annual revision.

Like ABC-22, the Basic Security Plan established mutual cooperation as the primary means by which Canada–US forces would coordinate their actions and provided for the establishment of unity of command by the two countries' chiefs of staff in an emergency. Unlike ABC-22, the BSP included a clause permitting the commander exercising unity of command to move forces from one operational theatre to another.[38] This had been specifically forbidden in ABC-22 without the commander first receiving authorization from the chiefs of staff; this protected the chiefs' operational command responsibility and authority for the movement of forces. The authors offered no explanation for this new clause. The change, however, addressed the frustrations that American and British supreme commanders had experienced during the Second World War. In several instances – most notably the short-lived American–British–Dutch–Australia unified command established in late 1941 – the supreme commanders had been hamstrung by national regional jurisdictional restrictions.[39] Given the need to act quickly in an emergency in the atomic and jet age, securing national approval from a country's service chief to move forces from one operational area to another could result in a dangerous delay. The new clause would facilitate rapid reinforcement in threatened areas. This led to the establishment of new PJBD Recommendations in the 1950s for cross-border interception and mutual reinforcement (discussed in the next chapter).

The process for approving and implementing the provisions of the BSP proved to be quite complex. In July 1946 the US Joint Chiefs of Staff and the Canadian Chiefs of Staff Committee approved the plan.[40] Prime Minister Mackenzie King and the Department of External Affairs remained concerned about proposed US installations on Canadian soil, the high expenditures involved (especially for the Air Interceptor and Air Warning Appendix), and threat assessments, so they insisted that specific "implementation programs" be approved each fiscal year. The MCC developed these programs, which then went to the Chiefs of Staff Committee for review and, if all three service chiefs agreed to them, were submitted to the government for decision.[41] In Canada, it fell to the Cabinet Defence Committee to consider the implementation programs, which were then subject to government examination "on the basis of what is necessary if and when the security of the North American continent is threatened."[42]

In the end, the Basic Security Plan was a watershed in the Canada–US defence relationship. It accomplished in peacetime what the two countries had previously done only in times of war. Although the BSP was a war plan, the Canadian section of the MCC made it clear to their superiors that peacetime preparations were essential to ensure that the plan could be quickly and efficiently implemented in wartime.[43] Thus, as Peter Haydon observed, the BSP became the "umbrella for a series of interrelated military agreements for defending the continent and its ocean frontiers."[44] The BSP appendices outlined detailed force requirements to implement the plan. Completing these appendices occupied the attention of MCC planners for the next few months,[45] with the most significant provisions relating to air defence.

The Air Interceptor and Air Warning Appendix

The appendix was a bombshell, calling for, in essence, Fortress North America.
— Joseph Jockel, *No Boundaries Upstairs*[46]

Throughout the mid- to late 1940s, Canadian and American planners focused on completing the appendices to the BSP. In August 1947 the RCAF member of the JPC tabled a study titled "Command Relations – Canada–U.S.,"[47] which suggested that North America be separated into four theatres of operations: Air Defence, Northwest, Northeast, and Southern. These would consist of joint and combined Canada–US unified commands, each headed by a supreme commander. The 1946 US Unified Command Plan (discussed in the next chapter) clearly influenced this study, but it remained in preliminary form, for the JPC decided it was unsuitable for tabling as a BSP appendix at the next MCC meeting.[48] Not until the drafting of the Canada–US Emergency Defence Plan and its Command Relations appendix near the end of the decade would concrete command and control provisions for continental air defence during the Cold War be established. One appendix did contain certain command and control provisions, however: it was titled "Air Interceptor and Air Warning" (the AIAW plan).

Completed in December 1946, the AIAW plan had two distinct phases. The first, to be placed in effect by 1950, was designed to counter the enemy threat of bombers flying at subsonic speeds. Planners anticipated that these aircraft would be armed with conventional bombs and, after 1950, with atomic bombs. This phase also considered the possibility of a threat posed by shorter-range guided missiles similar to the German V-1 and V-2 rockets from the Second World War. To defend against this threat, the plan called for Canada and the United States to construct an integrated early warning and ground control intercept system consisting of radar installations and fighter-interceptor aircraft defences. Radar chains would be established across North America, located on the Atlantic and Pacific coasts and along the Arctic Ocean seaboard from north Alaska to northern Labrador and Greenland; as well, there would be radar stations around the vital economic and populated areas of the Great Lakes and the US east coast. An extensive fighter-interceptor force of 975 American and 828 Canadian aircraft, distributed throughout the continent, would operate alongside the radar system.[49]

The second phase, to begin in 1955, was designed to deal with an augmented and more dangerous enemy threat: large numbers of supersonic jet bombers armed with atomic bombs and also long-range rockets and guided missiles. It called for air defences to be moved farther north; these would include a distant early warning radar network in the Arctic augmented by an increased fighter-interceptor aircraft presence as well as advanced ground-to-air missile batteries. The planners stressed that it was essential to complete both phases of the entire program and that any omissions would negatively impact the effectiveness of this air defence system, which was designed to complement the overall offensive forward strategy.[50]

An effective system of command and control remained essential to integrating this air defence scheme. Each country would establish its own Air Defence Command headquarters to coordinate its respective early warning and air interceptor measures. Each national command organization would be responsible for the air defence of its country's territory, and each headquarters would have liaison personnel from the other nation's armed forces. The AIAW plan also called for a combined Canada–US Air Defence Command to be established "at the proper time to control, coordinate and operate the common air defense system in time of an emergency." Its headquarters would include staff representatives from both nations' armed forces, and this new combined command would exercise "operational control" (not defined) of all Canadian and American air defence units deployed in North America.[51] The plan also called for air defence units to be grouped into geographic zones within Canada and the United States. Each zone was to have its own headquarters to control the forces, and each headquarters was to be under the command of its respective national air defence command organization.[52] Operational command of Canadian zones would remain in Canadian hands, and operational command of US zones in American hands.

There was no mention of the nationality of the person who would be in command of the new overall combined Canada–US Air Defence Command, although it likely would be an American air force officer. Exactly what command and control authority "operational control" entailed was unspecified. The inclusion of this term in the appendix, however, made it clear that Canadian and American airmen were beginning to think of operational control in a continental air defence context. As subsequent chapters demonstrate, the meaning of operational control would continue to evolve as the air defence relationship between the two nations' air forces tightened. Nevertheless, despite the lack of precise definition, the role of the commander of the combined Canada–US Air Defence Command would be limited to coordination: operational command remained with the component commanders, who in this case were the commanders of the (still only proposed) Canadian Air Defence Command and US Air Defense Command. National command would continue to remain the prerogative of the national chiefs of staff, thereby safeguarding national sovereignty.

The seeds of North American air defence integration were clearly planted in the Air Interceptor and Air Warning appendix. Nevertheless, the Canada–US effort to protect North America from the Soviet aerial threat remained at the planning phase. Because of the complex process for approving BSP implementation, the appendix (including the provision for the combined Canada–US Air Defence Command exercising operational control) was tentative. The MCC described the situation best: "it must be borne in mind that this appendix can serve as a planning document only. It is not intended that it should stipulate precise commitments for future years."[53]

In the end, both nations' chiefs of staff approved the AIAW Appendix in principle. However, since it was only a planning document, there was no formal obligation for Canada or the United States to implement the appendix's vast and expensive scheme.[54] The key was how much each nation was willing to invest and which parts of the appendix to implement. The focus of the AIAW plan thus switched to the annual implementation programs that the chiefs of staff submitted to the government for approval. Although many of the most expensive aspects of the appendix were never implemented, each nation did form its own Air Defence Command organization for the command and control of air defences.[55]

The implementation program for 1948–49 was notable for assigning the US responsibility for the "initiation and formulation of a standardized air defence doctrine and operating procedure."[56] This represented a paradigm shift for Canadian air power and doctrinal culture: whereas the RCAF had before mirrored the practices of its parent organization, Britain's Royal Air Force, now it would adhere to the US Air Force's way of defending the continent from aerial attack. Continental air defence integration was inevitable, and the rough basis for its command and control was established by 1947.

Conclusion

During the early Cold War, Canada and the United States maintained the command and control status quo established during the Second World War by retaining the mutual cooperation–unity of command compromise from ABC-22 because there was no urgency to change this arrangement. The BSP included a new clause allowing a commander exercising unity of command the freedom to move forces within his theatre of operations. In the age of modern aerial warfare, promptly moving forces from one area to another was an important practical consideration. The new clause in the BSP began to address this concern, and laid the foundation for cross-border interception and mutual reinforcement arrangements implemented in the 1950s.

Canadian and American planners anticipated that the period of relative global calm would not last, and heightened worries in the mid-1940s prompted Canada and the United States to reassess their air defence command and control relationship. The Basic Security Plan and especially its Air Interceptor and Air Warning appendix acknowledged that North America was no longer immune from attack. Accordingly, Canadian and American planners sought better ways to counter the growing Soviet aerial threat. They planted the seeds for integrated Canadian–American air defences, including enhanced command and control. The command and control "status quo" of the BSP and even the proposed operational control air defence system in the AIAW appendix ensured that Canada would retain command over its air forces. The United States thus did not force a "defence against help" dilemma on its northern neighbour, and this allowed Canada to protect its sovereignty.

Canadian and American air force officers meet in Ottawa in February 1954 to discuss continental air defence. The meeting included the top Canadian and American national air defence operational commanders at the time, Air Vice-Marshal A.L. James (fourth from right), Air Officer Commanding RCAF Air Defence Command, and General Benjamin W. Chidlaw, Commanding General USAF Air Defense Command (fifth from right). *Courtesy of NORAD History Office.*

Canada's deployment of much-needed F-86 Sabres to the RCAF's NATO Air Division in Europe and air defence requirements elsewhere in Canada meant there were no RCAF interceptor aircraft available to defend Newfoundland. *DND photo MV5829-1.*

F-94B Starfire interceptors similar to this one were based at Goose Bay in November 1952 as part of the USAF's 59th Fighter-Interceptor Squadron. They were later replaced with F-89 Scorpions and, subsequently, F-102 Daggers. *Courtesy of NORAD History Office.*

F-89 Scorpions from the USAF's 59th Fighter-Interceptor Squadron operated from RCAF Station Goose Bay in Labrador from the mid-1950s. *USAF Museum.*

Head-on view of two CF-100 aircraft. The Canuck, affectionately remembered by some as the "Clunk," became the RCAF's frontline interceptor by the mid-1950s. *DND photo PC-1011.*

The Soviet Union's revealing of the Myasishchev M-4 Bison on May Day, 1954, combined with its explosion of a thermonuclear bomb eight months prior caused considerable consternation in North America that eventually led to the centralization of Canadian and American continental air defences under NORAD operational control in 1957. *Open source Wikipedia.*

Group Captain Keith Hodson was one of the key RCAF staff officers who assisted in the development of the operational control arrangements for US Northeast Command in the early 1950s. As an Air Vice-Marshal he became NORAD's first Deputy Chief of Staff for Operations, a position that has been reserved for an RCAF officer to this day. *Courtesy of the Canadian Forces College.*

The RCAF's most prolific thinker and writer on air power during the early Cold War, Air Commodore Clare Annis, gave the crucial briefing to the Canadian Chiefs of Staff Committee on April 6, 1955, that convinced Canada's air force leadership of the need to integrate Canadian–American air defences. *Courtesy of the Canadian Forces Aerospace Warfare Centre.*

Air Vice-Marshal L.E. Wray, Air Officer Commanding Air Defence Command. Air Defence Command was "plugged in" to NORAD when the latter was established in 1957, and Wray became the RCAF component commander. *DND photo PL 108476.*

With the establishment of NORAD in 1957, Air Marshal Roy Slemon moved from RCAF Chief of the Air Staff in Ottawa to NORAD Deputy Commander-in-Chief in Colorado Springs. NORAD's operational control arrangement was Slemon's brainchild.
DND photo PL 103213.

Air Marshal Roy Slemon and General Earle Partridge's clear understanding of the close partnership between USAF CinC NORAD and the Canadian Deputy CinC produced an effective professional working relationship that has continued with successive NORAD commanders and their deputies to this day. Here we see Slemon (l) and Partridge (r) chatting with Chairman of the US Joint Chiefs of Staff General Nathan Twinning at NORAD Headquarters in Colorado Springs, October 1957. *DND photo PL-107207.*

NORAD Emblem. *Courtesy of NORAD History Office.*

"NORAD Formation over Lake Ontario": an RCAF CF-100 Canuck leads a USAF F-89 Scorpion, F-102 Dagger, and a F-104A Starfighter on their way to the Canadian International Airshow in Toronto, September 1958. *DND photo PL-95876.*

RCAF and USAF personnel each with a seat at the console. *Photo NP69-622 courtesy of NORAD History Office.*

A CF-100 Canuck from RCAF 410 "Cougar" Squadron and a USAF F-102 Dagger from Ernest Harmon Field in flight over Newfoundland during air defence training. This picture was taken on October 1, 1957, just weeks after the establishment of NORAD. *DND photo PC-261.*

6
Organizing and Coordinating Canada–US Air Defences

CANADIAN AND AMERICAN PLANNERS devised the 1946 Basic Security Plan at a time of reconstruction and of hope for lasting peace. However, a period of global calm never really materialized. From the end of the Second World War onward, tensions between the United States and the Soviet Union continued to mount in Europe, while war and unrest plagued other parts of the world. By 1948, the possibility of an armed conflict erupting between the West and the Soviet Union was more real. As the Soviet Union built up its strategic bomber force and developed its own atomic bomb, the need to protect North America from aerial attack became increasingly apparent. An air defence "crisis," precipitated by the outbreak of the Korean War in 1950, made the issue a key planning priority, forcing a re-examination of the Canada–US air defence command and control relationship with an eye to further coordination and integration of the two nations' efforts.

Significant developments occurred in air force command organization in North America from 1949 to 1953 – the years between the Soviet Union's possession of the atomic bomb and its detonation of a thermonuclear weapon. These included the move towards functional commands and the arrangement of air defence assets under centralized command and control. There was also a major shift in Canada–US planning away from long-term projections towards short-term capabilities-based emergency plans. The growing importance of RCAF and USAF fighter-interceptor resources to the air defence mission, and the resulting requirement for emergency mutual reinforcement and cross-border interception provisions, generated questions about sovereignty and under whose authority these forces would fall. Most importantly, the Canada–US command and control relationship changed from the mutual cooperation–unity of command paradigm to one centred on the principle of operational control.

The Canada–US Emergency Defence Plan
After the MCC began work on the BSP appendices, it remained conscious of the need for a short-term implementation plan based on military resources available at the time – what is termed "capabilities-based" in modern military terminology – to be put into effect in an emergency. The communist

takeover of Czechoslovakia and the Berlin Airlift Crisis in 1948 made the requirement for such a plan even more apparent. These events caused a major confrontation with the Soviet Union, and the leadership of the USAF grew increasingly concerned that its air defence capabilities were insufficient to adequately protect the United States from the growing possibility of Soviet Tu-4 bomber attacks.[1]

In the summer of 1948, the MCC revised the bilateral planning process to address new concerns. It separated Canada–US defence planning into three distinct parts: an outline long-range BSP (MCC 100/4); a Basic Security Program (MCC 200, later MCC 122) that dealt with implementation issues; and a capabilities-based emergency plan (MCC 300) to be put into effect immediately during a crisis.[2] The latter document eventually became known as the Canada–US Emergency Defence Plan (EDP), and it was the only one that included command and control provisions.

The MCC released the first version of the EDP, MCC 300/1, on March 25, 1949. Based on "forces currently available," it addressed three main defence tasks that Canada and the United States would undertake in the event of an emergency: the defence of vital areas against air attack, the protection of sea lines of communication, and the employment of mobile striking forces to reduce Soviet amphibious/airborne lodgements. This document would be revised annually to keep it consistent with other plans and up-to-date with Canadian and American defence capabilities. The EDP would be brought into effect when so ordered by each country's chiefs of staff.[3]

Although planners expected the Soviets to mount an attack on shipping in wartime, the Soviet Navy of this era was relatively weak. With its submarines and land-based naval air power, it could attack shipping close to Europe, but it lacked true high-seas capability.[4] MCC 300/1 anticipated minor amphibious operations and a small air threat to the continent. The latter was projected to be limited attacks by Soviet Tu-4 strategic bombers flying one-way missions and carrying conventional bombs or a small number of chemical and biological weapons; the EDP estimated that the Soviet Union would not be in possession of an atomic bomb in 1949. The EDP's main focus was on maintaining high military production in North America to sustain offensive capabilities in the event of total war against an enemy; any Soviet strike on North America would only be a minor one in support of a main assault in Eurasia. As such, defensive measures to protect the continent from attack were to be minimal so as not to distract from and interfere with the principal military effort overseas. In this way, the EDP remained consistent with previous Canadian-American plans and American strategic culture by ensuring the maintenance of an overall offensive wartime strategy during a total war. To achieve the limited defensive strategy, the plan called for a combined Canada–US emergency air defence system, which included an effective command and control setup to ensure its efficiency. Each country was to be responsible for defending its own territory and sea and air approaches, and specific arrangements for the operational command and control of combined Canada–US forces were to be spelled out in a separate appendix, which would be developed.[5]

A draft EDP Command Appendix was completed in November 1948. It considered the establishment of a combined "Canada–US Defense Command" that would oversee all forces engaged in the defence of the continent. The planners justified the proposed new command by noting that a "well-integrated and unified defense effort" was required in order to counter a multifaceted and wide-ranging enemy attack. It would be, in the words of the MCC, a "combined command" based on the concept of unity of command: it would place all forces under a single commander to ensure the successful prosecution of the continental defence mission. The Canada–US Defence Command was to be headed by a US officer, who would oversee and coordinate the overall continental defence effort. Under him would be the various component commands, the commanders of which would have "operational control" (not defined) of Canada–US forces and would be responsible for detailed planning and execution of the actual EDP missions and tasks as per their operational command authority.[6] This was consistent with the British operational control system from the Second World War.

The Canadian and American planners forwarded the draft Command Appendix to their superiors – who subsequently rejected it. The documents that may explain why remain restricted and therefore unavailable to researchers. It appears, though, that it was opposition by the US Joint Chiefs of Staff to the combined command idea that led to this rejection. Continental air defence remained a low priority in American strategic culture, so a combined Canada–US command to protect North America (and the SAC deterrent) from a still-remote Soviet strategic bomber threat was not necessary – especially given the continuing American monopoly of the atomic bomb.[7] Canadian military leaders were in favour of a combined Canada–US command as late as the summer of 1949. When informed of JCS opposition to the concept, however, the Chiefs of Staff Committee decided to let the matter drop, believing that it would be dealt with during the talks surrounding the formation of NATO commands.[8] The combined command proposal was rejected; even so, it showed that planners were *thinking* along the lines of greater continental air defence command and control centralization during the late 1940s. This would prove to be a valuable asset for air defence planning later in the mid-1950s and helps explain why

Canada was ultimately able to implement a functional approach to air defence and thus protect its sovereignty.

When the EDP was released in March 1949, it did not mention a combined Canada–US command, nor did it contain the customary mutual cooperation–unity of command provision that had appeared in previous Canada–US defence plans. Instead, command and control authority fell to the individual Canadian and American commanders in each area; the EDP did not mention explicit command and control arrangements. Specifically, the appendix indicated that "any force located in Canada, including Newfoundland and Labrador, and employed in execution of the tasks set forth in this plan, will operate under a commander designated by Canada." The same was true for any forces located in the United States, which were to operate under a commander designated by that country.[9] This arrangement meant that operational control usually would be based on national lines depending on where the forces were located: if they were in Canada, they would come under a Canadian commander; if in the United States, under an American officer.

The words "designated by" were deliberate and significant. Although this usually meant that American forces would operate under the operational control of a Canadian commander in Canadian territory and that Canadian units would be under the operational control of an American commander in the United States, this would not always be the case. Each country had the authority to assign the commander, but this did not necessarily mean it had to choose a commander from its *own* nation's armed forces. For instance, if American forces were located in Canada for EDP operations, Canadian authorities could designate either a Canadian *or* an American officer to exercise operational control. This was an important provision because it meant that American forces could operate under US operational control in Canada. Alternatively, Canadian forces could technically operate under Canadian operational control in American territory, though this was less likely.

The purpose of this provision was not to erode the sovereignty of the host nation. It was a functional arrangement to permit the nearest and most capable commander (i.e., in terms of the number of forces available to him), regardless of nationality, to ensure efficient military operations in an emergency situation. Because the host nation remained the "designator," this was a functional exercise of sovereignty. As the discussion of the RCAF–USAF mutual reinforcement and cross-border interception agreements will demonstrate, this "designated by" arrangement in the EDP also provided a solution to the problem of finding a way to defend certain areas of each country because it permitted the deployment of forces from either nation into threatened regions. For example, if Canadian forces were absent in northern BC but American forces were present in nearby Alaska, the EDP would allow the Americans, operating under their own commander, to react to an attack on the Canadian province. The provision guaranteed that local commanders would have the authority they needed to deal with an immediate threat to an area, which ensured operational flexibility and military efficiency.

Lastly, the EDP Command Appendix contained two provisions that ensured the autonomy – and sovereignty – of forces engaged in operations. The first stipulated that once a force entered the other nation's territory, even though it might come under the operational control of the other country's commander, it would remain under the operational command of its own officers. This maintained national command by the country from which the forces originated; it also ensured that units would not be broken up. There would be a clear national chain of command, and the unit commanders would still be able to give detailed operational instructions to their forces, thereby avoiding confusion. Second, a clause stated that "regardless of the area in which [they are] operating, the internal administration of forces furnished by either country shall be the prerogative of the country furnishing the force."[10] This reasserted a long-held Canadian and US practice in joint and combined operations: administration and discipline remained a service prerogative as part of national command authority.

In summary, the EDP Command Appendix had finally abandoned the mutual cooperation–unity of command provisions that had formed the principal Canada–US command and control relationship since 1941. The EDP confirmed the shift to the operational control paradigm. As the next two chapters demonstrate, the definition of operational control and its application to the Canada–US air defence relationship would continue to evolve. Because operational control still ensured that Canada would retain command over its air forces, it also protected sovereignty. Again, no "defence against help" dilemma was being forced upon the country by its southern neighbour.

The Joint Chiefs of Staff approved the EDP on April 21, 1949; the Chiefs of Staff Committee did the same in late October.[11] Because the EDP was based on current force capabilities, it required annual MCC revision (usually by July) and reapproval by each nation's chiefs of staff. This gave each country the opportunity to include updated force levels and other capabilities and to make changes to any part of the plan they deemed necessary, including the Command Appendix. The first revision of the EDP (MCC 300/2) was completed in May 1950 in response to the air defence "crisis" caused by the Soviet detonation of an atomic bomb the previous year. With the Cold War threatening to turn "hot" (North Korea would invade South Korea only one month after this revision), military leaders became convinced of the need for enhanced continental air defences. This led the United States and Canada in the early 1950s to establish an extensive system of radar installations similar to the one proposed in the 1947 Air Interceptor and Air Warning Appendix to the BSP.[12] The revised EDP reflected the new air defence priorities.

MCC 300/2 coldly reassessed how North American air defence forces would be employed in the event of a sudden Soviet attack. It was based on new intelligence that the Soviet Union possessed anywhere from 25 to 45 atomic bombs – and was prepared to use them. The MCC agreed that it would be impossible to defend every mile of the two countries, so they had to make hard choices about priorities. "After exhaustive study," the combined planners decided that the only way to utilize the two countries' limited air defence resources was to concentrate them in five "critical areas" and accept risks elsewhere. The priority areas were (1) Montreal–Boston–Norfolk–Chicago (i.e., the Northeast and Great Lakes); (2) Vancouver–Spokane–Portland; (3) Fairbanks–Anchorage–Kodiak; (4) San Francisco–San Diego; and (5) central New Mexico. These considerations reflected the Soviets' most probable course of action; the vulnerability of the various areas, particularly to air attack; the density of essential elements of North American war-making capacity within certain areas; the extent of Canada–US defensive capabilities; and the extent to which the defence of selected areas would indirectly contribute to the protection of other areas by creating defence in depth. The two most pressing priorities were to extend and integrate Canadian and American early warning radar systems and to strengthen anti-aircraft defences. The latter priority included providing more modern weapons and more all-weather fighter squadrons and bases. Planners also reiterated the long-held stance that investment in North American defence should not take away from the main focus, which was Europe and the strategic offensive.[13]

The CSC approved MCC 300/2 on July 11, 1950, and the JCS did so one month later.[14] The Command Appendix remained essentially the same throughout the early 1950s, with only minor revisions.[15] During those years, the RCAF and the USAF each focused on reorganizing their air defence forces and negotiating specific bilateral arrangements between them, moving ever closer to the operational control paradigm.

Air Defence Command Organization

Post-1945, Canadian and American airmen endeavoured – with varying degrees of success – to centralize command and control of all national air defence resources under a functionally based unified command organization headed by an air force officer. Based on the success of its wartime expeditionary command organization, the United States established new regionally/geographically based unified commands under the JCS's Unified Command Plan in 1946. This did not include unified command for continental defence, however.

In stark contrast to the US Army's traditional geographical/regional command organization and based on wartime air power culture and doctrinal lessons, the USAAF leadership established three new functional commands after the Second World War: Strategic Air Command, Tactical Air Command (TAC), and Air Defense Command (ADC). The latter was stood up at Mitchell Field, New York, in March 1946.[16] After a lengthy struggle, American airmen finally achieved their long-held goal of a separate and independent air force in 1947. The United States Air Force maintained its predecessor's new functional command organizations and assigned ADC the mission to defend the continental United States against air attack.[17]

In the first few years of the USAF ADC's existence, its Commanding General was limited in his ability to carry out the air defence mission by both insufficient command and control authority and lack of resources. Although authorized to exercise "direct control of all active and coordinate all passive means of air defense," this power was limited in practice, for much of the authority for planning and command and control of operations fell to the subordinate air force component and unit commanders (i.e., the USAF's First and Fourth Air Forces).[18] The USAF's strategic culture was offensive, and US war plans for this period envisioned an immediate knockout blow against the Soviet Union at the start of any war. Due to this overwhelming strategic and doctrinal focus on the offensive – and the development and expansion of Strategic Air Command in particular – strategic defence was not a US priority, so Air Defense Command had minimal resources in personnel, aircraft, and equipment in the early postwar years. When created, ADC consisted of just 2 percent of the entire air force's manpower, boasting only two fighter squadrons, few radar resources, and six component commands (two active) for the defence of the entire United States.[19] As one USAF Historical Division narrative put it, "an Air Defense Command existed but not an air defense."[20]

American airmen believed strongly in the concept of the indivisibility of air power and felt that all the nation's air defence resources, regardless of service, should be centralized under the operational control of one USAF commander, who would be responsible for the air defence of the continental United States.[21] ADC partly achieved this goal through the Key West Agreement of April 1948, when the Secretary of Defense assigned primary functional (i.e., as opposed to regional) responsibility for air defence to the USAF. At the same time, the US Army was assigned responsibility to maintain anti-aircraft artillery (AAA) capabilities for continental air defence "in accordance with joint doctrines and procedures approved by the Joint Chiefs of Staff."[22] Unfortunately, the JCS never issued any policies or procedures, and inter-service disagreements plagued efforts to consolidate American air defences into an effective and centralized command organization. Although the heads of both the US Army and the USAF at different times suggested the formation of a unified air defence command, the JCS members were unable to come to any kind of unanimous decision, and thus had to rely on less efficient joint inter-service cooperation agreements.[23]

Air Defense Command continued to face challenges during the late 1940s and early 1950s. At one point in 1948, ADC was disbanded altogether, its resources combined with Tactical Air Command under a new USAF command, Continental Air Command (ConAC), which took over the air defence mission.[24] The air defence "crisis" of 1950 finally made American leaders pay more attention to protecting the country from aerial attack. Although SAC remained the main priority, Congress began allocating more funds to the USAF to build up active continental air defence capabilities. Greater centralization was also achieved when the US Army formed the Anti-Aircraft Command in July 1950 and moved its headquarters to Mitchell Field to be closer to ConAC HQ.[25] In August 1950, the JCS assigned the Commanding General of ConAC limited authority over Anti-Aircraft Command forces, including the designation of states of alert and Rules of Engagement (ROE) for AAA, and "operational control [not defined], insofar as engagement and disengagement was concerned." Finally, at the beginning of 1951, the USAF disbanded ConAC and re-established Air Defense Command as the primary organization for the defence of the continental United States. Its new headquarters was established at a more isolated and less vulnerable spot in Colorado Springs, Colorado; shortly after, it was joined there by Anti-Aircraft Command.[26] Colorado Springs thus became the Combat Operations Center of the US air defence effort; as a future chapter will demonstrate, six and a half years later it would become the headquarters for the newly established NORAD.

The issue of establishing some kind of tri-service air defence command organization, however, remained unsolved in the United States during the early 1950s. Until American military leaders could set aside their inter-service differences and agree on the need for a unified joint air defence command, nothing would be accomplished. Fortunately, the situation in Canada was marginally better.

Despite the success of the Canadian Northwest Atlantic Command, which was based on the British system of operational control, the Canadian armed forces reverted to the joint committee cooperation system immediately after the Second World War. Nevertheless, the CSC remained open to British and American ideas and concepts, including unified commands, throughout the early Cold War era.[27] Its Joint Planning Committee undertook several studies in the late 1940s and early 1950s to examine command organization for the Canadian armed forces. Some of these studies looked into establishing unified commands in Canada, notably the May 1951 proposal titled "A Proposed System of Unified Operational Commands."[28] Nevertheless, against the advice of CSC chairman General Charles Foulkes, Canada's service chiefs decided to retain the joint committee cooperation system for the time being and to take a wait-and-see approach regarding the American continental defence command structure.[29] This decision was consistent with the Canadian military practice of reacting to American developments in command organization instead of being proactive.

The CSC decided at the very least to recategorize Canadian defence roles to reflect the Cold War environment. Echoing the American Key West Agreement, each Canadian service was assigned "primary responsibility" to deal with particular potential threats to the country. The RCAF was responsible for air defence, the RCN for coast and seaward defence, and the army for reducing enemy lodgements on Canadian territory. Responsibility for administration and logistics remained with the respective service chiefs.[30] Although the army and the navy maintained their own regional/geographic command organization, when Air Marshal Wilfred Curtis became the Chief of the Air Staff in 1946, he insisted that the various roles of an air force necessitated that the RCAF be organized on a functional basis. After some initial resistance from MND Brooke Claxton, Curtis was finally able to convince him of the feasibility of a functional organization. The RCAF was then able to lay the foundation for the creation of one Canadian functional joint command organization, Air Defence Command.[31]

The RCAF demobilized quickly after the Second World War, maintaining only a minimal force, the Auxiliary, to provide protection against aerial attack. The RCAF Auxiliary had an authorized establishment of 4,500 personnel all ranks, and its 15 squadrons were located in large cities across the country such as Vancouver, Calgary, Winnipeg, London, Hamilton, Toronto, and Montreal. Most of its air personnel were fighter pilots from the Second World War, who flew P-51 Mustangs and DH.100 Vampire jet fighters. Mobile radar units on the ground manned by both Auxiliary and Regular Force personnel directed these aircraft engaged in interceptor roles. These measures were all rather rudimentary; as English and Westrop have noted, with no Regular

Force units to protect the country from aerial attack, "there was no national command and control organization for air defence" at this time.[32]

The pitiful Canadian air defence situation was perhaps best revealed in a 1948 emergency defence plan. Although only a small Soviet strategic bomber capability was anticipated, the plan stated bluntly that "there are no defences immediately available" to counter this threat. Besides the Auxiliary, the only air defences on hand in Canada were those of the RCN: four squadrons of fighter aircraft (eight machines per squadron) and the anti-aircraft guns of the naval ships "which happen to be in port."[33]

By the late 1940s, the RCAF had begun to address its shortcomings to deal with the increasing Soviet bomber threat by dedicating greater resources and doctrinal attention to the functional role of air defence. On December 1, 1948, it established No. 1 Air Defence Group and charged it with the responsibility to protect the country from air attack. The responsibilities of Air Defence Group were:

a) Preparation of an Air Defence plan for Canada including Anti-Aircraft defences.
b) Direct liaison with RAF and USAF on Air Defence procedures and doctrines.
c) Direct liaison with USAF on Air Defence planning for North America under the provisions of the Emergency Defence Plan.
d) Formulation of operational procedures and doctrines for day and night interceptor operations and tactics.[34]

Air defence received greater priority for funding in Parliament. Air Defense Group grew significantly, and on June 1, 1951, it was elevated to command status.[35]

The new RCAF Air Defence Command expanded greatly during the 1950s. Its headquarters was in St-Hubert, Quebec, and ADC reached its authorized strength of 19 fighter-interceptor squadrons (nine Regular Force and ten Auxiliary) by 1955.[36] ADC exercised "overall operational control" (not defined) over the forces assigned to it.[37] Soon afterwards, in April 1949, the Canadian Army's Anti-Aircraft Command moved its headquarters to St-Hubert, and its commander became responsible to the Air Officer Commanding RCAF ADC "for operational control of its units."[38] So by the early 1950s, air defence command and control in Canada was centralized under the RCAF, and in 1951 Air Defence Command began sending liaison officers to USAF Air Defense Command headquarters in Colorado Springs.[39] It was now time to coordinate with the American air defence system.

Coordinating Continental Air Defences

> *Until one man is given responsibility for the defence of both Canada and the US it is unrealistic to talk of giving him authority to control the entire operation.*
> – Group Captain K.L.B. Hodson, RCAF Headquarters[40]

In late 1950, the staffs of the Canadian and American air defence commands prepared a draft "Canada–United States Emergency Air Defence Plan" (EADP) to fulfill the air defence tasks of the EDP.[41] The plan recognized that the air defence battle would be fought over large parts of Canada and the United States, irrespective of international borders, and called for greater air defence integration. Each nation would retain operational control (not defined) of its aircraft when flying over its territory. However, the EADP stipulated that if the tactical situation required the interceptors of one country to enter the airspace of the other for reinforcement and/or to interdict enemy bombers, the host country should exercise some kind of operational control authority over these aircraft.[42]

The draft EADP met resistance when it came to the Canadian and American chiefs of staff for approval. The JCS and the CSC agreed that further integration of their air defence effort was desirable but that the command and control provisions required more study. Reflecting the Janowitzian societal/political imperative, Canadian and American military leaders recognized that the cross-border interception issue was a different matter entirely that required an official government-to-government agreement (discussed below).[43] On February 1, 1951, the PJBD produced a document, eventually known as the "Pinetree System," that called for the extension (especially northward) of the Canadian and American aircraft control and radar warning organizations and their consolidation into one bilateral air defence system.[44] This would require a massive program for building radar installations in order to plug current gaps in coverage, as well as a more integrated system of warning and interception.

Each government would retain command of national forces, but operational control would be "passed" (from whom was not stated, but likely from the commanders of the USAF ADC and RCAF ADC) to the respective group commanders "regardless of international boundaries" to carry out the air defence battle. "Operational control" was defined as "all functions of command solely operational except that deployments and dispositions of forces will be effected only after being mutually agreed upon by commanders concerned," and there was a provision that it could be delegated. The plan also stated that "overall operational control" (this time not defined) of the whole air defence system would be exercised by the commander of USAF ADC in Colorado Springs.[45]

The definition of operational control was rather simplistic in that it did not entail what "functions of command solely operational" actually were. However, by specifying that deployment and disposition of forces were to be carried out only after consultation by the respective commanders, it did confirm that national command authority for deploying forces remained with the services. The operational control delegation provision ensured that local commanders would have the authorization they needed to deal with the aerial threat in their area immediately, thus ensuring mission command and functional military efficiency of the air defence battle.[46] The last provision for "overall operational control" could be interpreted in two ways. It could be seen as a precursor to NORAD, whereby operational control of all air defence forces was under an overall CinC; but the more likely interpretation is that the emphasis was on *the operation* (i.e., the functioning) of the new integrated air warning and control system. In other words, the USAF ADC commander would have the radar reporting facilities at his disposal.

Emphasis was on the latter interpretation when the PJBD forwarded the Pinetree System proposal for approval as Recommendation 51/1.[47] The command and control provisions were too rudimentary and required further refinement, so in the interim the arrangements in the EDP Command Appendix would suffice. The AOC ADC would have "operational control" (not defined) over Canadian territory, and the commander of USAF ADC would have the same authority in American airspace.[48] The RCAF recognized that a single *integrated* system under one commander with operational control of the entire air defence battle was desirable, but it also understood that the time to implement it had not yet arrived. For the time being, there would be two coordinated systems, one Canadian and one American, with each air defence commander exercising operational control in his own airspace as per the EDP Command Appendix provisions.[49] In the meantime, the issues of mutual reinforcement and cross-border interception and engagement had to be addressed separately in specific government-to-government agreements. Both had important command and control implications.

Helping Out: Mutual Reinforcement and Cross-Border Interception

> *International boundaries cannot be respected when fighting an air battle.*
> – General Charles Foulkes, former Chairman, Chiefs of Staff Committee[50]

In the early 1950s, Canadian and American airmen increasingly saw the effort to protect their countries from aerial attack as a shared problem that required an integrated Canada–US air defence system.[51] Of particular concern was that interception, engagement, and mutual reinforcement might have to take place during a peacetime air defence emergency as well as during war, and would require fighter-interceptors from both countries to quickly cross into each other's airspace and for deployment at various RCAF and USAF bases. Such measures would be necessary to ensure that the air defence concept of "defence in depth" could be achieved: "an enemy should be attacked as far out as possible initially and the pressure on him increased as he neared his objectives by the employment of increasing numbers and varieties of weapons."[52] In North America, this meant pushing the air battle as far north as possible so that it could be fought farthest from the continent's vital centres and within the shortest amount of time of the continental radar network detecting an incoming bogey. It would also provide the warning time required for USAF Strategic Air Command to launch a nuclear counterstrike with its strategic bombers.[53] A/C Clare Annis, the RCAF's most prolific thinker and writer on air power at the time, likened "defence in depth" to a rugby game:

> The fighters are the line of the rugby game. Their duties are two-fold. It is hold the line and prevent the enemy bombers from getting through. We know that for as long as the line holds, the enemy bombers won't get through very far or very often. The second duty of the fighters is to wear down and finally crumple the enemy line. If that is achieved our bomber backfield [i.e., SAC] can roam at will.[54]

Air Defence Command interceptors formed the first line of defence against Soviet bombers, but limited RCAF aircraft resources necessitated USAF assistance. Mutual reinforcement and cross-border interception and engagement were therefore essential for "defence in depth." Both had important command and control implications.[55]

In late 1950, the RCAF approached the USAF about mutual reinforcement. The Canadian airmen stressed that an agreement between Canada and the United States should give the commanders of each nation's air defence command blanket permission to reinforce one another in an emergency.[56] The JCS agreed, and authorized the US Section of the PJBD to begin negotiations with the Canadian Section.[57] Air Marshal Curtis noted that a key issue was airspace sovereignty – in particular, "under whose command and control would such forces come?" "If," Curtis wrote, "such forces are an integral part of our Air Defence System they would have to come under our control [although] they would remain under the immediate command of the US subordinate commander."[58] This was consistent with the provisions in the EDP Command Appendix, but there were more factors involved.

Another essential concern was that the air defence system employed in the early 1950s required intricate cooperation between the pilot in the aircraft and the air controllers on the ground. Airborne radar equipment installed on interceptors was still quite rudimentary, having changed little since the Second World War. The detection and interception system was quite similar to the one that RAF Air Chief Marshal Sir Hugh Dowding's Fighter Command had employed during the Battle of Britain. As one American official historian explained,

> the radar nets were tied in with 14 control centers, which would direct the air battle in case of attack. The methods and equipment used at these centers were already approaching obsolescence. Radar data was transmitted to the control centers by conventional human telling [i.e., telephone]. Manual methods of operations were used for computing the tracks of hostile aircraft, assigning weapons, and vectoring fighter aircraft.[59]

When ground radars picked up an intruding aircraft, interceptors scrambled and the local Air Defence Control Centre (ADCC) in each sector (called Air Divisions in the United States) vectored them on to the enemy bomber.[60] Jockel terms this "tactical co-operation," though the pilot–air controller relationship arguably straddled the operational and tactical levels of conflict.[61] Indeed, Canadian and American officers consistently used the terms "operational control" and "tactical control" interchangeably.

There were other practical matters that needed to be resolved. The location of certain elements of the air defence system in Canada and the United States meant that the ADCCs of one nation would be required to vector interceptors from the other country. For instance, although both Canada and the United States had their own interceptor bases and ADCCs located near the border in the Great Lakes–St. Lawrence area, in the prairies the only ADCCs near the border were in the United States. As the RCAF's Air Member for Operations and Training (AMOT) Air Vice-Marshal Frank Miller noted, "it would be better if we used the control facilities rather than the international boundary as a line of demarcation between the control responsibility of the national commanders. For example, Canadian fighters operating over Canada but being controlled by a US [ADCC] would obviously be under the control of the American commander."[62] A formal arrangement for operational command and control was required. It had to be consistent with the EDP Command Appendix provisions and ensure effective coordination between interceptors and ground control centres.

In the spring of 1951, the RCAF and USAF devised a draft accord on command and control to address these concerns. It outlined the following arrangements: "When squadrons of one country are operating in the area of responsibility of the other, they will be under the control of the nearest Air Defence Control Centre of that area for all functions normally performed by an Air Defence Control Centre. For all other purposes, they will remain under the control of their parent organization."[63] The key aspect was the word "nearest." In areas such as the northwest and the Great Lakes–St. Lawrence Valley it meant that RCAF ADCCs would have control over USAF and RCAF interceptors. However, in areas such as the Prairies and Newfoundland, where there were no Canadian ADCCs, American Air Divisions would have control over USAF interceptors in both American and Canadian airspace. The "for all other purposes" clause meant that each nation would retain command over its own forces. This draft arrangement closely mirrored Curtis's feelings on the cross-border tactical control issue and was also consistent with the EDP Command Appendix provisions.[64]

In September 1951, Air Marshal Curtis submitted a draft proposal to his navy and army colleagues on the Chiefs of Staff Committee advocating that Canadian and American air defence commanders have the authority to redeploy each other's forces to provide mutual reinforcement. The command and control provisions in the EDP Command Appendix would apply: "any force located in Canada will operate under a commander designated by Canada, and that the forces of either country serving in the territory of the other will be under the immediate command of a Commander designated by the country furnishing the force."[65] The CSC agreed that air defence mutual reinforcement "was militarily desirable" but felt the RCAF should refer the matter to the PJBD to secure a formal governmental agreement.[66]

The PJBD moved quickly, producing Recommendation 51/6 on November 12. It stipulated that when the air defence commanders of each air force both agreed that the tactical situation warranted mutual reinforcement, they would have the authority in the event of war to redeploy interceptors to either country (see Appendix 1).[67] Although the recommendation made no mention of command and control authority, MND Claxton noted that the EDP Command Appendix provisions would apply (his memorandum was almost word-for-word from Curtis's September draft proposal).[68] The AOC Air Defence Command would be in charge in areas of the country covered by RCAF ADC squadrons: the west coast, central Canada, and eastern Canada.[69] Since there were no RCAF interceptors based in the Prairies or Newfoundland, Canada would have to designate a USAF commander.

The Canadian government approved Recommendation 51/6 on November 12, 1951, followed by the JCS and President Truman in January and March 1952 respectively. It allowed USAF interceptors to enter Canadian airspace and be redeployed to RCAF bases in a wartime emergency; it also permitted the stationing of RCAF fighters at US bases if the progress of the air battle so dictated.[70] As Jockel aptly put it, "the local air defence commanders would be able to draw upon the interceptors in their region, regardless of nationality, and use them to destroy hostile aircraft."[71] Recommendation 51/6 was therefore one important step in the Canada–US effort to coordinate and integrate their air defence systems. It still did not allow for mutual reinforcement during a peacetime air defence emergency, however; subsequent Canadian–American agreements concerning cross-border interception and engagement would go a long way towards remedying this deficiency.

In August 1950, not long after the outbreak of the Korean War, President Truman authorized the USAF to intercept and "destroy aircraft in flight within the sovereign boundaries of the United States which commit hostile acts, which are manifestly hostile in intent, or which bear the military insignia of the USSR, unless properly cleared or obviously in distress."[72] The USAF leadership had been clamouring for this "right to shoot" for years; it was now time to also exercise this authority in Canadian airspace to achieve air defence in depth. The RCAF only had enough interceptor resources to protect specific regions of Canada, leaving the air approaches through Newfoundland and the Prairies undefended. This gap in the air defences needed to be plugged.[73]

The resulting PJBD Recommendation 51/4 of May 9, 1951, gave reciprocal permission to the USAF and the RCAF to cross each other's borders and intercept unidentified aircraft (see Appendix 2).[74] It addressed Canadian concerns about sovereignty by including specific restrictions. USAF fighters could only intercept unidentified aircraft in Canadian airspace that were actually heading towards the United States and could not engage and destroy them until they crossed the border into American airspace.[75] These restrictions did not sit well with American airmen. Although the Canadian government quickly approved the recommendation on May 30, 1951, President Truman did not approve it until October due to USAF uneasiness with the restrictions, especially the lack of a "shoot order" in Canadian airspace.[76] After much hesitation, the American airmen only accepted the recommendation as an interim measure and immediately began efforts to negotiate a new arrangement.[77]

In December, the Canadian government gave the RCAF permission to engage an aircraft (force it to land or open fire on it) in peacetime, but this did not apply to any arrangements with the United States for cross-border interceptions.[78] To address this deficiency, in April 1952, USAF Headquarters devised a proposal to authorize RCAF and USAF interceptors to intercept potentially hostile unidentified aircraft in Canadian or American airspace in peacetime, regardless of the border.[79] The air defence commander "responsible for identification" would have the authority to order the interceptor to engage the intruder, depending on the following specific circumstances: "No attempt to be made to order intercepted aircraft to land nor to open fire unless the aircraft commit(s) a hostile act, is (are) manifestly hostile intent, or is (are) declared hostile by the Air Defense Commander responsible for identification."[80] This clause adhered very closely to RCAF Rules of Engagement (reproduced as Appendix 3).[81]

The actual nationality of the air defence commander "responsible for identification" was unclear. The RCAF interpreted it to mean "that *either* USAF or RCAF aircraft, directed *either* by a USAF or RCAF air defence controller, could order interception of aircraft over US or Canadian territory, and if it was deemed that a hostile act was being committed or intended, could order the aircraft being intercepted to land or be shot down."[82] Reflecting the Janowitzian societal/political imperative, Canadian airmen also felt that the proposed USAF provisions did not go far enough to address Canadian sovereignty concerns. In a September 8 counterproposal, they stipulated that American interceptors would be permitted to investigate unidentified aircraft in Canadian airspace only when a RCAF interceptor was not available. USAF aircraft would have to adhere to RCAF Rules of Engagement, and only the Air Officer Commanding RCAF ADC (or an officer who had been delegated this authority, i.e., a subordinate in a Canadian ADCC) could order the engagement. The provisions were reciprocal: the same requirements and restrictions would apply to RCAF interceptors investigating unidentified aircraft in American airspace. This would ensure national responsibility for interception in national airspace, effectively protecting sovereignty.[83] Tactical control and the ever-important authority to issue the "shoot order" in Canadian airspace would thus be in Canadian hands, while command would remain a national prerogative.

Although USAF and RCAF officers from each nation's air defence command headquarters continued discussing cross-border interception arrangements throughout late 1952 and into early 1953, a potential command and control problem delayed agreement on a new PJBD Recommendation. Much like the "designated by" clause in the EDP Command Appendix, the RCAF counterproposal had a similar provision allowing the national air defence commander to delegate tactical control authority to order an engagement to another officer. Once again, the nationality of this officer was not specifically mentioned, and this made Department of External Affairs officials uneasy. The DEA member of the PJBD, R.A. MacKay, noted that in areas of Canada where there were no RCAF aircraft it was likely there were also insufficient communications facilities; the Air Officer Commanding RCAF ADC would thus have to delegate engagement authority to a USAF officer. DEA approached DND about the matter. The Minister of National Defence (MND) Brooke Claxton noted that *because* these arrangements were consistent with the command and control principles in the EDP Command Appendix, the Canadian government should approve them.[84] Discussion on the new cross-border interception arrangements continued for several more months; it was not until October 1953 that the PJBD tabled them as Recommendation 53/1.

The Recommendation largely resembled the September 8, 1952, RCAF counterproposal, but with some important changes. There were minor differences in phrasing and added emphasis on certain wording. For instance, the ROE clause was clarified further to read (with the new parts in italics): "The Rules of Interception and Engagement of the country over which the interception or engagement takes place are to apply, *even though the intercepting aircraft is being controlled from the other country*."[85] Although Recommendation 53/1 included the provision allowing the national air defence commander to delegate tactical control authority to order an engagement to another officer, a phrase was added to allay Canadian concerns; it stated that this authority "should, to the greatest extent possible, be retained by the Air Defence Commander." Still, the recommendation also stated that when the operational situation warranted, the commander might delegate his authority "to a qualified officer not less in status than the senior officer in an Air Defence Control Center."[86]

As Jockel has noted, this last phrase "had been carefully inserted by the two air forces."[87] Although Canada was, in the words of the senior USAF PJBD member, "for political reasons" unwilling to specify that this "qualified officer" could be American *or* Canadian (as MacKay and Claxton alluded to above) this was implicit in the recommendation – something that the USAF and RCAF air defence commanders clearly understood.[88] Reflecting the functional imperative, to ensure that a potentially hostile aircraft was intercepted as quickly as possible, the air commander could delegate tactical control interception and engagement authority to the nearest ADCC, regardless of nationality. In parts of eastern and central Canada near the border and in the Vancouver area on the west coast, Canadian commanders would be able to order interceptions; but in the Prairies where there were no RCAF ADCCs, this authority would be delegated to a USAF officer at an American Air Division headquarters.

The Canadian government approved PJBD Recommendation 53/1 on November 3, 1953, and US president Dwight D. Eisenhower gave his assent the following month (it is reproduced as Appendix 4).[89] In the end, 53/1 complemented 51/6's mutual reinforcement provisions. As Jockel has noted, once an American interceptor deployed to Canadian airspace actually destroyed a hostile aircraft, "it would not be peacetime for long," and the RCAF and USAF air defence commanders "could then immediately draw upon emergency reinforcement authority granted them by Recommendation 51/6."[90] Both PJBD Recommendations thus ensured that the air defence battle could be fought effectively regardless of the Canada–US border while at the same time adhering to the EDP Command Appendix's provisions. Officers from one nation could exercise tactical control and operational control over the forces of the other nation, but command would always remain in national hands, thereby ensuring sovereignty and balance between the functional and societal/political imperatives.

Conclusion

As the Soviet atomic air threat grew between 1949 and 1953, Canada and the United States undertook significant developments in emergency command and control arrangements, air defence command organization, and cross-border reinforcement and interception agreements. As a functional exercise of sovereignty these measures did not erode the sovereignty of the host nation; also, they ensured that the air defence battle could be fought effectively and that air defence in depth could be achieved. Command remained in national hands, further reinforcing sovereignty for both nations.

The 1949–53 period saw a fundamental shift in the evolving Canada–US command and control relationship, as the cooperation–unity of command paradigm that had been in existence since 1941 was replaced with one centred on operational control. Further command and control centralization had been achieved through greater coordination of the two nations' air defence systems and mutual reinforcement and cross-border interception agreements. Although Canadian and American air defences were not yet integrated, by the end of 1953 the seeds for greater Canada–US command and control centralization had been sown.

7
The US Northeast Command

> *The special arrangements proposed for northeast Canada are admittedly equivocal but our military strength is so inconsistent with our political desires that no solution can be found at present to satisfy both Service and State.*
> – Air Commodore H.B. Godwin, RCAF Staff Officer, 1951

AFTER THE SECOND WORLD WAR, American forces remained in Newfoundland to develop and protect the bases they had leased from the British in 1941. Canada retained responsibility for the defence of the British colony, and this arrangement became permanent when Newfoundland joined Confederation as Canada's tenth province in 1949. This confirmed the need for Canadian and American forces to coordinate Newfoundland defence, and the command and control of these forces became a sensitive topic once again.

With the growth of RCAF Station Goose Bay in Labrador as a crucial base for Strategic Air Command bombers in the early 1950s and the increasing possibility that the Soviets would attack North America from the northeast, the air defence of Newfoundland became an American priority. The United States created Northeast Command to deal with this threat, and the USAF deployed fighter-interceptor forces to its bases in the new Canadian province. Canada, however, was already hard pressed to fulfill its air power commitments to NATO *and* defend Canada's northern approaches from aerial attack, and RCAF Air Defence Command was unable (some would say unwilling) to deploy scarce interceptor resources to Newfoundland. The practical task of defending the new province from air attack therefore fell to American forces.

US Northeast Command replaced the wartime Newfoundland Base Command, and like its predecessor had its headquarters at Fort Pepperrell near St. John's. It consisted of the American air force bases in Newfoundland (Harmon Field, Stephenville), Labrador (facilities leased at RCAF Station Goose Bay), and Greenland (Thule, leased from Denmark). It was responsible for continental air defence and for support to Strategic Air Command units deployed or staging in the northeast. As a unified command that reported directly to the Joint Chiefs of Staff, US Northeast Command fell outside the USAF's air defence command structure and thus beyond the jurisdiction of USAF Air Defense Command. This necessitated separate and special arrangements for air defence command and control outside regular USAF–RCAF channels, which raised political considerations regarding territorial rights and sovereignty. In particular, the continuous presence of US forces in Newfoundland persisted as a thorny topic for the Canadian government – and for opposition political parties. The two countries eventually reached mutually acceptable arrangements based on RCAF operational control of USAF forces operating in Newfoundland airspace. This solution was essentially window dressing to soothe Canadian political sensitivities; even so, it ensured Canadian sovereignty, military efficiency, and competent command and control.

What to Do with Newfoundland?

As the end of the Second World War drew near, the Canadian government faced the question of whether it would continue to defend Newfoundland and what stance it should take regarding the future of the US bases in the British colony. In January 1945 the government decided to maintain the status quo: Canada would retain its responsibility to defend Newfoundland, while the Americans would be limited to defending their leased bases.[1] The Americans concurred and recognized Canadian responsibility for the defence of Newfoundland in the Canada–US Basic Security Plan of 1946 (see Chapter 5). When Newfoundland became a Canadian province in the spring of 1949, this arrangement became permanent.[2] Newfoundland now being official Canadian territory, responsibility for its defence fell to the Canadian armed forces. The American bases remained, however, and this perpetuated the co-location of Canadian and American forces in Newfoundland. Canadian planners recognized that command and control would remain an ongoing issue, considering the buildup of American forces in Newfoundland in the late 1940s.[3] The American decision to stand up Northeast Command in 1949 brought the matter to the fore.

President Harry Truman signed off on the Unified Command Plan in December 1946, but Northeast Command was not stood up. The small number of postwar US forces in Newfoundland factored into this decision, but the most important reason was that the JCS felt that standing up this new unified command would involve "political difficulties involving the Canadian Government."[4] The location of Northeast Command on "foreign territory," noted Admiral Louis Denfeld, the US Navy Chief of Naval Operations, "would provide excellent propaganda for the communists and would generate misunderstanding and friction with Canada and the United Kingdom."[5] International developments, however, soon forced American planners to revisit the issue of activating Northeast Command. Concomitant to the changing constitutional position of Newfoundland in the late 1940s, there were now new questions about the bilateral command and control situation in the northeast.

In light of increasing international tensions with the Soviet Union and the growing possibility of an aerial attack on North America, in April 1949 the JCS agreed it would soon be necessary to stand up Northeast Command in Newfoundland. Given political sensitivities in Canada, the US Secretary of Defense instructed the US PJBD Section to inform their Canadian colleagues of American intentions. According to the Leased Bases Agreement of 1941, the United States had no obligation to notify or to seek approval from Canada about a new command organization on its bases in Newfoundland. However, American officials felt that "it would be most impolitic not to do so."[6] As a courtesy to their key ally, the Americans proactively ensured that Canadian officials were up-to-date on their plans.

In early May 1949, the chairman of the US PJBD Section, Maj.-Gen. Guy Henry, reassured his Canadian counterpart that Northeast Command would not infringe on Canadian sovereignty. He explained that the commander of Northeast Command

> would be responsible, insofar as US responsibilities and interests are concerned, for maintaining the security of the Northeast Command and for participating in the defense of Canada and the United States against attack through the Arctic regions. He would, of course, cooperate closely with appropriate Canadian authorities in the discharge of his peacetime responsibilities, and in an emergency would operate in accordance with combined plans.[7]

Northeast Command would enhance continental defence, Henry added, and was "a logical step" in bilateral efforts to implement the Canada–United States Emergency Defence Plan. Henry argued that since a Canadian officer was responsible for the overall defence of Newfoundland in the EDP, coordination with US forces in the area would be simplified if this Canadian officer had a single American commander to deal with, not only in an emergency but for mundane day-to-day matters as well.

An official Canadian response to this did not arrive until July, much to the annoyance of the Americans. The Canadian federal election in June 1949 accounted for the delay – in particular, the opposition Conservatives' pressure on the Liberal government regarding special rights the United States exercised in Newfoundland.[8] The Canadians sought reassurances that Northeast Command was not a territorial or regional command but an "administrative and tactical" peacetime command compatible with the provisions in the EDP. They were also wary that the name "Northeast Command" would imply US responsibility for the defence of the entire northeastern part of the continent, including Newfoundland, and in particular "that Canadian forces have come under

United States command in the Arctic."⁹ The Canadian Chiefs of Staff Committee therefore suggested the revised name "US Forces, Northeast."

To allay Canadian concerns, the Joint Chiefs outlined the terms of reference for the commander of Northeast Command. The new unified command was "primarily intended to provide a more direct operational control [not defined] by the Joint Chiefs of Staff" through the Unified Command Plan over all US forces stationed in Newfoundland and to aid in the development of bilateral defence plans and facilities.¹⁰ Nevertheless, the US PJBD Section, the CSC, and Department of External Affairs officials all felt that Northeast Command's terms of reference were too broad and obscure. The Canadians feared they would include wartime defence responsibilities for US Northeast Command, a matter that would have to be dealt with through intergovernmental discussions.¹¹ The Chiefs of Staff Committee sought further assurances that Northeast Command would be responsible for peacetime administration and operational control of US forces.

The US military leadership was frustrated with the time it was taking to bring this issue to a close. They reiterated that Northeast Command's purpose was to facilitate planning and the employment of assigned tactical forces, and they reassured their northern neighbour that the missions assigned would be consistent with Canadian ones. The JCS made it clear that Northeast Command would plan in concert with Canadian forces for the defence of the northeast, including Newfoundland and Labrador and the sea and air approaches therein, and that it would be in accordance with official Canada–US defence plans. As a compromise, the JCS offered the title of "United States Northeast Command" to emphasize that the new command was solely an American organization responsible for US forces.¹²

Although the Chiefs of Staff Committee was not entirely comfortable with the proposed name, it leaned towards accepting the JCS proposal. The Chief of the Air Staff, Air Marshal Wilfred Curtis, reminded his CSC colleagues that "this matter had been referred to the Canadian authorities as a courtesy and that the establishment of the command could have been carried out by the US authorities without any reference to Canada."¹³ Secretary of State for External Affairs Lester B. Pearson was even more frank in his recognition of the lengths the Americans had already taken to keep Canada informed. "I gather that it is the Canadian military view that it would be very difficult for us to suggest what a US organization should be called," Pearson noted, "especially when Washington has, at our suggestion, completely remodelled the functions proposed for the Command in negotiations extending over a period of a year."¹⁴ Still, the Chiefs of Staff Committee and the Canadian Cabinet did not give their final approval until May, though the CSC did express the Canadian government's appreciation for "the manner in which the United States authorities have taken into account Canadian views on the early proposals for the command."[15] With the way now clear, the JCS formally stood up US Northeast Command October 1, 1950, at Fort Pepperrell Air Force Base (AFB), St. John's, with USAF Maj.-Gen. Lyman T. Whitten as its CinC.[16]

During this entire process, the JCS, through the US Section of the PJBD, made certain that Canada was kept informed of American intentions. The US military leaders were not only patient with the Canadians but also very accommodating to Canadian wishes and sensitivities, even going so far as to change the name of the new unified command to reassure their northern neighbours. The JCS did not have to do this according to the Leased Bases Agreement and the United States' inherent right as a sovereign nation to organize its own forces as it saw fit. Nevertheless, the US military leadership decided to err on the side of caution – or in this case consultation – to keep its ally informed. Despite these American efforts, Canadian concerns persisted.

The RCAF and US Northeast Command

US Northeast Command was unlike most other American unified commands. Once established, unified commands usually consisted of an overall commander-in-chief to oversee operations and two or more component commands from different services to carry out detailed operations. US Northeast Command, however, had only one component command: Northeast Air Command. Maj.-Gen. Whitten was also its commander, giving him two "hats" as the unified command CinC and the commander of the USAF component command. Although some US Army anti-aircraft artillery forces were assigned to Northeast Air Command and a few naval liaison officers were on its staff, US Northeast Command was primarily an air force command. When established, however, it did not actually have any interceptor aircraft to carry out this mission; the arrival of two squadrons was not scheduled until July 1952.[17]

The Soviet Union's detonation of an atomic bomb in September 1949, fully four years ahead of intelligence estimates, and the outbreak of the Korean War in 1950, brought a new sense of urgency to air defence. Threat assessment deemed it increasingly likely that the Soviet Union would bomb targets in Newfoundland – notably the US leased bases – in any attack on North America. A crucial role for continental air defences was to protect the war-making capacity of Canada and the United States. Logically, this included the industrial, economic, and political centres of the two countries. The offensive power of the USAF's Strategic Air Command was another important part of American war-making capacity. Protecting the US bases in Newfoundland and the RCAF base at Goose Bay, Labrador, where authorities would locate SAC forces (following the signing of the Canada–US lease agreement in December 1952), became a crucial air defence task.[18] The USAF began accelerating plans to station interceptor squadrons at its bases, given that Canada's priorities to deploy interceptors to protect the vital areas in the Great Lakes–St. Lawrence area precluded the RCAF from furnishing fighter protection to the new province at that time.[19]

These developments set off alarm bells in the East Block of Parliament. With the USAF interceptors scheduled to begin active air patrols in Newfoundland airspace in the near future, Canadian officials began to fear that US Northeast Command would no longer be an "administrative rather than operational" command and that the air defence of Canadian territory in the northeast would fall to the United States by default.[20] When stories began to appear in the Canadian and American press implying that Canada had conceded air defence protection of the Maritimes to the USAF, political sensitivities became even more acute.[21]

In February 1951, the Canadian Joint Services Committee East Coast submitted a draft Canada–US defence plan for Newfoundland that alarmed the CSC. It proposed giving US Northeast Command "control and direction of the overall [sic] Area Air Defence system" of fighters and air control and warning resources.[22] The joint services committee had exceeded its terms of reference, the CSC noted, and the plan itself was inconsistent with the NATO defence plans that the Canada–US Regional Planning Group had already developed for the region. Maj.-Gen. Whitten's insensitivity to Canadian concerns, which were reminiscent of Maj.-Gen. Brant's attitude during the Second World War, did not help matters. The CSC reported that "from the tone" of talks between the CinC US Northeast Command and senior Canadian military commanders in the Maritimes, "it was apparent that Major General Whitten assumed that he was responsible for the defence [of] Northeastern Canada even though this was not in accordance with a previous agreement [i.e., the EDP] by which Canada was responsible for the defence of her own territory."[23] Canadian authorities became more concerned and began seeking arrangements with the United States to provide a Canadian contribution to US Northeast Command to protect Canada's interests.

Planners grappled with how to strike a compromise between protecting Canadian sovereignty and ensuring effective command and control in Newfoundland. In line with the EDP, the bilateral air defence effort would be focused on the "critical areas" in the province, notably US bases and installations. Although American forces were intended to protect these facilities, the concept of area air defence made it impractical to distinguish between defending installations and geographical areas. Air defences for American installations also protected the entire region and were therefore "an important, albeit a special, part of the Air Defence of Eastern Canada."[24]

These defences consisted of interceptor aircraft and anti-aircraft artillery but also included an air warning and control system. The network of radar installations in the northeast, scheduled for expansion as part of the Pinetree project, merged with the system along the St. Lawrence, forming an integrated early warning system for US facilities in both Newfoundland and eastern Canada. RCAF planners suggested that American and Canadian air defence forces in Newfoundland be grouped together in an "Air Defence division" of the USAF's Northeast Air Command. To maintain Canadian sovereignty, the responsibility for the air defence of Newfoundland had to remain with the RCAF, "regardless of what freedom of action is allowed the Northeast Air Command."[25] This required some form of agreement on command and control between Northeast Air Command and the RCAF's Air Defence Command.

The Canadian airmen felt the Pinetree System command and control arrangement (PJBD Recommendation 51/1) would work in the Northeast. "Overall control" should be vested in the commander of the USAF Air Defense Command "in close co-ordination and co-operation with" the Air Officer Commanding RCAF Air Defence Command. Individual group commanders should exercise actual "operational control" of the integrated air defence forces in the western and eastern areas.[26] This RCAF staff proposal reflected the desire of Canadian officials, notably those from DEA, to maintain Canadian responsibility for the defence of the new province by ensuring that the chain of command in Newfoundland went from US Northeast Command headquarters at Fort Pepperrell *through* RCAF Air Defence Command headquarters in St-Hubert before going to the JCS in Washington.[27]

Demonstrating a lack of appreciation for the Janowitzian societal/political imperative, Maj.-Gen. Whitten felt that the RCAF proposal placed disproportionate emphasis on political sovereignty issues at the expense of effective air defence. He worried that the JCS would misinterpret Canadian insistence on maintaining the sovereign responsibility to defend Newfoundland as a concrete commitment to provide air defences for the province. The RCAF did not have the resources to fulfill this responsibility, and Whitten feared that the JCS would be disinclined to deploy USAF forces to US Northeast Command in light of apparent Canadian promises to defend the region. In his view, the wording of an agreement on air defence responsibility in the

Northeast needed to "reassure both sides that the defense requirement will be met as well as to reassure the politician that he still has his job."[28] In his view, Canadian influence over air defence responsibilities in the region should be limited to the planning stages so that US Northeast Command had the freedom of action to execute complex operations. As a compromise, Whitten offered to have RCAF officers serve "in key positions" on the US Northeast Command's staff.

The Canadians had an even more ambitious idea in mind. The RCAF completed a revised version of its study on April 24, 1951, specifying that only all *American* air defence forces in the Newfoundland area should be grouped together into an air defence division under Northeast Air Command. This arrangement conspicuously omitted reference to RCAF units, which would only be attached to (and would not become a part of) the division "for purposes of operational control [not defined]." Most importantly, a RCAF officer would fill a new deputy CinC position in US Northeast Command. Unlike the Deputy CinC NORAD arrangements six years later, this RCAF officer would not be in charge in the absence of the CinC. Instead, he would coordinate planning of US and Canadian forces in the area and have direct access to the CSC to ensure that US Northeast Command activities in Newfoundland airspace were consistent with Canadian government policy.[29]

Air Vice-Marshal F.R. Miller, the Vice-Chief of the Air Staff, felt that the new RCAF study left too little authority for the Air Officer Commanding RCAF ADC. "How can he be responsible" for the air defence of Newfoundland, Miller asked, "if he has not the control and cannot say what US forces will be used?"[30] Group Captain K.L.B. Hodson, the RCAF's Director of Air Plans and a member of the JPC, was sympathetic to this critique. An arrangement for command and control in Newfoundland had to maintain, at the very least, an *appearance* of Canadian responsibility while at the same time ensuring effective protection of the province – and the important US bases located within its borders – from air attack.[31] The proposed RCAF deputy position would secure this appearance, as would provisions ensuring AOC ADC oversight on deployment and redeployment, operational practices, and USAF aircraft procedures to defend the Newfoundland area. "To make the best of an awkward situation," Hodson concluded, "real operational control" would be vested in the commander of Northeast Air Command, with "the semblance of overall responsibility invested" in the AOC ADC.

This was a convoluted and confusing arrangement at best, and it did not entail a substantive deputy CinC function. The Canadian Army's Director of Military Operations and Planning noted that "the position appears to be more that of a Canadian liaison officer with certain specifically assigned responsibilities."[32] Unsurprisingly, the Americans rejected the proposal. The JCS, uncomfortable with having any kind of bilateral Canada–US command at that time, decided that the RCAF deputy proposal was unwarranted. Even if a NATO command was eventually established in the area, US Northeast Command would be no more than a subordinate component command of it – not the principal command – and thus would not require a Canadian deputy.[33]

USAF Chief of Staff General Hoyt Vandenberg liked the idea of Canadian participation in US Northeast Command, arguing that it would soften Canadian apprehension over the US presence in Newfoundland and ensure greater operational coordination. When the JCS formally proposed this idea to its northern neighbour, however, the Canadians rejected it. Jockel observed that Canadian government officials, particularly in the East Block, felt uncomfortable with "the notion of Canadian officers serving in subordinate roles in an American command which was located on Canadian soil."[34] Canadian planners went back to the drawing board. This time the Joint Planning Committee took the lead and spent the better part of the winter of 1952 studying an alternative arrangement with a more joint flavour.[35]

Canadian authorities had grown weary of the idea of a RCAF officer serving in a subordinate role in a US command. In particular, DEA officials, adamant that Canada had at no point given up (or delegated) responsibility for the defence of Newfoundland to the United States, were uneasy about the idea of RCAF forces coming under the operational control of the commander of Northeast Air Command.[36] CSC chairman General Charles Foulkes stressed the need for Canadian command in Canadian territory:

> I think it is important to realize that the Canadian Government cannot accept that any American can have command over Canadian territory except over those particular portions which have been leased to the United States. Therefore, we must start from the assumption that the responsibility for the defence of Newfoundland and the Northeast approaches to Canada is a Canadian responsibility and cannot be delegated to anybody else. This is altogether different from any arrangements which may be made regarding a Supreme Commander [i.e., a NATO command] in time of war.[37]

Deeming it illogical to have a Canadian officer on the staff of a JCS unified command, Foulkes recommended Canada "develop some form of a cloak of Canadian control in the area." This cloak was RCAF operational control over USAF forces in Newfoundland.

RCAF Operational Control

> *Inasmuch as AOC ADC has not been given any forces nor have any forces been planned for him to carry out the responsibility of the air defence of Newfoundland and Labrador, his responsibility in this area is purely academic.*
> – Group Captain S.W. Coleman, Director of Air Plans, Air Force Headquarters[38]

At the March 4–5, 1952, meeting of the Permanent Joint Board on Defence, the USAF member announced his service's intention to deploy one squadron of 25 all-weather jet interceptor aircraft to Newfoundland to commence active air defence patrols.[39] The need to revisit the air defence command and control relationship in Newfoundland quickly became a higher priority in Canadian circles. After rejecting its previous proposals, the JPC once again had to try to find a compromise between Canadian sovereignty and effective air defence of the province.[40] Ideally, the RCAF should have been able to defend Newfoundland, but the magnitude of the national air defence mission combined with limited interceptor resources meant that Air Defence Command had to give precedence to the designated "critical areas" in the Great Lakes–St. Lawrence region. This left no RCAF fighters for Newfoundland.

Overseas commitments also played a factor, specifically the decision to send 12 fighter squadrons (1 Canadian Air Division) to Europe to fulfill NATO air defence needs.[41] Some senior Canadian officials were uncomfortable that Canada was sending RCAF fighters to defend Europe while the United States deployed USAF interceptors to Newfoundland to defend Canadian airspace.[42] Even MND Claxton was sympathetic to this line of thinking. "Personally," he wrote, "I would rather see an R.C.A.F. squadron there [i.e., in Newfoundland], even if this meant reducing our commitment to NATO by one, or weakening the air defence of some other part of Canada." However, the MND realized that "there is more involved."[43] Stripping interceptors from other areas of Canada would lessen the already thin air defences protecting the most crucial war-making capacity regions, and taking squadrons away from 1 Canadian Air Division would be in breach of Canada's NATO commitments at a time when they were needed most. This reflected the relative priority of NATO vis-à-vis North American air defence at the time. Canada had invested a huge amount of time, diplomatic and military effort, and money – including the construction and manning of new air bases on the European continent – to stand up the RCAF's air division.[44] Besides, Foulkes explained, the aircraft earmarked for Europe were the wrong kind: they were single-engine F-86E Sabre day fighters, not the two-engine all-weather night interceptors like the CF-100 Canuck that were required for the harsh conditions in the northeast.⁴⁵

Not that Newfoundland was worth defending – at least in the eyes of the Canadian military leadership. The American bases, especially those from which SAC forces were scheduled to deploy, were the only viable targets in Newfoundland needing fighter protection in the CSC's opinion.⁴⁶ The JPC agreed, adding that "from the standpoint of purely Canadian installations and activities this area has not warranted, vis-à-vis other important areas of Canada, a portion of Canada's limited air defence force."⁴⁷ Regardless, the nature and extent of American activities at the leased bases put Newfoundland in the enemy's crosshairs, and this necessitated some kind of air defence. It was clear that until the RCAF had sufficient numbers of interceptors – backed by the collective will of the Canadian military leadership – to permit a deployment of fighter squadrons to Newfoundland (not likely any time soon – the RCAF did not anticipate being able to deploy interceptors to Newfoundland until 1955 at the earliest), the USAF forces of US Northeast Command would be defending the region from air attack. If Canada was really concerned about sovereignty it would have acquired and deployed sufficient suitable aircraft in Canada to reduce the need to rely upon USAF forces.

The JPC adjusted its viewpoint on the US bases in an attempt to solve the sovereignty problem. Instead of regarding the bases as part of Strategic Air Command's mission, JPC would consider them part of the overall integrated continental air defence Pinetree System. Rather than concentrating on the *regional* aspect of the bases – their location in the Northeast/Newfoundland – the focus would be on the *functional* role of continental air defence. In this case, US Northeast Command would report to the Commanding General USAF Air Defense Command, a functional command organization, instead of to the JCS according to the largely regional Unified Command Plan. US Northeast Command's air defence forces would thus come under the operational control of the AOC ADC in St-Hubert.⁴⁸ This meant applying the command and control principles of the EDP to the RCAF–US Northeast Command situation in Newfoundland. Since the American bases and US Northeast Command were specifically excluded from the EDP, a revision of the Command Appendix would be necessary.⁴⁹

In the meantime, the new (as of March 20) US Northeast Command CinC, Lt.-Gen. Charles T. Myers, had gotten wind of the RCAF operational control proposal. In what M.H. Wershof, an official with DEA's Defence Liaison Division, described as a "complete reversal of the attitude adopted by his predecessor," Maj.-Gen. Whitten, Myers was quite willing to accept RCAF ADC operational control if it would help allay Canadian apprehension. Instead of being responsible to USAF ADC as the Canadians suggested, however, Myers recommended that US Northeast Command continue reporting to the JCS and that a special arrangement between himself and the AOC ADC for RCAF operational control over USAF forces be concluded.[50] This was good enough for the USAF leadership and the AOC ADC Air Vice-Marshal A.L. James. Canadian and American military planners began work on a revision to the EDP Command Appendix to reflect the new arrangement.[51]

In the meantime, at the September 25, 1952, PJBD meeting the USAF member told the Canadian Section that an American interceptor squadron would be stationed at Goose Bay as of 1 October. Though there was no requirement for the United States to seek Canadian permission for this deployment, the USAF member had informed his Canadian counterparts as a courtesy. This announcement caused concern in Ottawa, as Goose Bay was not included in the original 1940 United States–British Destroyers-for-Bases deal. Negotiations on a Canada–US agreement to permit the United States to lease part of RCAF Station Goose Bay were still not complete, and the USAF's announcement only muddied the waters. DEA officials and the MND in particular felt that Canada had at no time deferred responsibility for defending Goose Bay to the United States, and were upset with the optics that the stationing of a USAF squadron there without the consent of the Canadian government would entail. Canadian officials understood the importance of protecting SAC forces at Goose Bay, but they also realized that NATO commitments precluded the deployment of RCAF interceptors there for quite some time. The USAF announcement therefore provided further impetus to come to a satisfactory command and control arrangement with US Northeast Command.[52] The Canadian government shortly thereafter gave approval for the temporary deployment of the USAF interceptors to Goose Bay, but only "on the distinct understanding that as and when the Canadian Government desires it can take over [the] air defence role."[53] At the same time, since NATO relied upon the American strategic nuclear deterrent, Canadian agreement with US measures to defend SAC forces constituted indirect Canadian support of the overall NATO posture.

Canadian concerns meant that the arrival of the USAF aircraft was delayed by a few weeks. Finally, on November 2, 1952, 8 F-94B Starfire jet fighter-interceptors from the 59th Fighter-Interceptor Squadron (FIS) arrived at Goose Bay. They commenced operations on December 20 when the squadron began a standing 24-hour alert. The 61st FIS, flying 12 Starfires, arrived at Harmon Field to conduct air defence operations over Newfoundland in July 1953.[54] Since discussions on command and control were ongoing, as a temporary measure the AOC ADC was authorized to delegate authority to intercept and engage any hostile aircraft to Lt.-Gen. Myers of Northeast Air Command, but only according to RCAF Rules of Engagement.[55]

DEA officials still had concerns about how tangible RCAF operational control in Newfoundland would be. As Wershof noted, "if a Canadian commander has 'overall control' it should be real and not illusory. We want to know how it will work."[56] A formal JCS proposal went a long way to address these concerns. In addition to suggesting that the EDP Command Appendix provisions be applied to Newfoundland to give Canada operational control over US air defence forces, the document contained other concessions. Since the primary task of USAF forces in Newfoundland would be to defend the US bases, American authorities would retain authority for the deployment and redeployment of units within US Northeast Command. Movement of these forces within Canadian airspace, however, "would be coordinated with the Canadian military authorities insofar as possible," and the AOC ADC would have final authority regarding Rules of Engagement and ordering an interceptor to open fire on hostile aircraft.[57] The latter provision was an important part of the arrangement, as it presupposed the deployment of USAF forces to Newfoundland in peacetime (it was more likely that interceptions would take place before a formal declaration of war). The command and control principles of the EDP only applied in time of war, but the JCS's proposed interception arrangements guaranteeing both RCAF Rules of Engagement and the final "shoot order" authority extended the AOC ADC's operational control authority over American air defence forces in peacetime as well as war.

Canadian authorities were generally pleased with the JCS proposal, recognizing the lengths the Americans were willing to go to address Canadian concerns.[58] Canadian planners nonetheless still desired a specific section in the EDP Command Appendix outlining special operational control arrangements for USAF forces operating in Newfoundland airspace.[59] The resulting revised Command Appendix of November 21, 1952, specified that when operating over Canadian territory, American air defence forces would be considered to be employed in tasks implicitly for the EDP. In this instance, the AOC ADC would be granted "operational control" of US Northeast Command aircraft. Operational control was defined as "the power of directing, co-ordinating and controlling the operational activities of deployed units which may, or may not, be under command or operational command. It specifically excludes redeployment."[60] The keywords were "directing, co-ordinating and controlling." This definition of operational control (as with previous and future definitions of the term) centralized control over operational forces in the hands of one commander but left it up to the subordinate commanders to carry out tasks as part of their operational command authority over their forces. In this case, the CinC US Northeast Command would retain operational command as the double-hatted commander of the component Northeast Air Command.

The revised EDP Command Appendix also formalized the JCS's proposed peacetime interceptor deployment provisions. Although the CinC US Northeast Command had the authority to redeploy air defence forces within his command, "insofar as possible" this was to be done in coordination with the AOC ADC. If the American commander desired to redeploy his aircraft to a base in Canada outside the leased bases he had to secure RCAF approval. In any event, CinC US Northeast Command was required to inform the Air Officer Commanding ADC of the deployment of any American forces "into, out of, or within Canadian territory" in time of peace or war.[61] As Air Vice-Marshal Miller put it, since the Chief of the Air Staff delegated the RCAF's primary responsibility for air defence to the AOC ADC, the arrangement ensured that the CSC "will have, at all times, the complete picture regarding CINCNE's air defence forces in Canada."[62]

DEA officials were still not fully comfortable with the operational control arrangement for Newfoundland. As the DEA Member of the PJBD, Assistant Under-Secretary of State for External Affairs R.A. MacKay noted, the lack of efficient communications in Newfoundland would in his opinion prevent the RCAF commander at ADC headquarters in St-Hubert "from exercising *effective* operational control."[63] There was some truth to this statement, and the revised EDP Command Appendix addressed it: the AOC RCAF Air Defence Command and the CinC US Northeast Command would work out the detailed implementation of Canadian operational control themselves.[64] This arrangement would give the AOC Air Defence Command much leeway in how he would exercise his operational control authority over US Northeast Command forces; it would also potentially give the commander of Northeast Air Command considerable freedom of action in Newfoundland.

When the MCC formally submitted the revised EDP Command Appendix to the Canadian and American chiefs of staff on November 27, it also attached a memorandum spelling out the special situation in the northeast and why the separate command and control clauses were required. Since it was likely that any Soviet attack on Canadian and American vital centres in eastern North America would come through the air approaches to Newfoundland, US Northeast Air Command forces were "an important and special adjunct of the Canada–US air defense systems." The use of the term "adjunct" was intentional: it emphasized that the air defence command and control arrangement in Newfoundland was outside the coordinated Canada–US continental air defence system. The US Northeast Command chain of command, although going through RCAF ADC headquarters in St-Hubert, would still end at the JCS in Washington, not the USAF ADC commander in Colorado Springs, as the Canadians originally proposed in 1950.[65]

Given how slowly previous negotiations had progressed, approval of the revised EDP Command Appendix was relatively quick. The JCS gave their assent on December 10, 1952, followed one week later by CSC. The American and Canadian military leadership also instructed the USAF and RCAF chiefs to prepare directives detailing the new command and control arrangements in Newfoundland respectively for the CinC US Northeast Command and the AOC ADC.[66] The revised EDP Command Appendix formally came into effect on February 10, 1953, when it was approved by the Canadian Government.[67] This cleared the way for the RCAF and USAF leaders to issue their directives.

Putting RCAF Operational Control into Practice

> *Canadian leaders ... did, however, assert Canada's ultimate sovereignty and control over US forces in the northeast – even if this was more* de jure *than* de facto *at the time.*
> – David Bercuson[68]

According to RCAF Chief of the Air Staff (as of February 1, 1953) Air Marshal Roy Slemon's March 5 directive, the AOC RCAF ADC was to coordinate with the CinC US Northeast Command on providing detailed plans for the implementation of the new operational control arrangements as soon as possible and forward them to Air Force Headquarters in Ottawa for approval. Although the new EDP Appendix gave the AOC ADC operational control over USAF aircraft operating over Canadian territory, the CinC US Northeast Command would still "exercise command and operational command over all US Air Defence Forces operating in Canada." The American commander would ensure that his interceptors used RCAF Rules of Engagement "and all other operational procedures set by Canadian Air Defence Command" when operating in Canadian airspace.[69]

Slemon also provided precise definitions for the various command and control terms in the EDP Command Appendix. "Command," he explained, "implies full authority over all forces in all respects, including training, administration and the function of moving units. Command in this sense, is not exercised by one service over any other service in Air Defence." This definition is consistent with the modern definitions of national command and full command, a service prerogative (see the Introduction). It also adhered to the Canadian maxim, featured in all combined operations since the Second World War, that administration and discipline remained in national hands, which is also an aspect of national command. Next, Slemon defined "operational command" as "the functions of directing, supervising and controlling the training policy and operations of assigned units, including their redeployment in the area of Operational Command. It does not include the direct administration or logistical support of assigned units which although the responsibility of their respective National authorities should conform to the policy directives issued by the overall Operational Commander."[70] Although somewhat different from the modern definition of operational command, Slemon's version contained aspects of command that were not present in the EDP Command Appendix definition of "operational control."[71] Specifically, the authority over training policy and the redeployment of units was made apparent in Slemon's definition of operational command, reinforcing that administration was a national command prerogative.

In summary, the AOC ADC's operational control authority consisted solely of giving the CinC US Northeast Command directives on the effect that he wanted to achieve. So long as his aircraft conformed to RCAF Rules of Engagement and he consulted with the AOC ADC regarding redeployment, CinC US Northeast Command had substantial freedom of action for detailed operations in Canadian airspace. This arrangement balanced the societal/political and functional imperatives by maintaining Canada's responsibility to defend Newfoundland and ensuring effective command and control of air defence operations.

The USAF Chief of Staff's directive to the CinC US Northeast Command was simpler. It spelled out the same command and control arrangements as Slemon's document, and it designated Lt.-Gen. Myers as "the immediate commander of US forces assigned" to US Northeast Command. Meyers was to make the necessary arrangements to implement the operational control authority with the AOC ADC. The JCS approved the directive on April 2, 1953, forwarding it to Myers one week later.[72]

The AOC ADC and the CinC US Northeast Command signed the formal agreement on Newfoundland air defence command and control arrangements on April 21, 1953, at RCAF ADC headquarters in St-Hubert. It divided eastern Canada into four separate air defence sectors, 1, 2, and 3 RCAF Air Defence Control Centre Sectors and US Northeast Command 64th Air Division Sector (see Map 2).[73] In the latter sector, the CinC US Northeast Command exercised command and the AOC ADC operational control over all USAF air defence forces. The AOC ADC exercised command over any RCAF air defence units in the American sector. USAF aircraft in the 64th Air Division Sector were to adhere to RCAF Rules of Engagement, including interceptions, and "operational procedures for air defence forces are to be standardized, as far as possible, with those in effect in the Canadian air defence

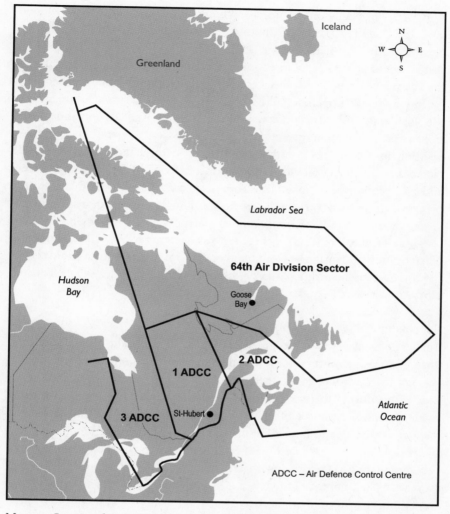

MAP 2   Command Boundaries, RCAF Air Defence Command and US Northeast Command, April 1953. *Map drawn by Mike Bechthold.*

system."[74] Provisions for deployment and redeployment (discussed above) were also included. All arrangements were in effect during peacetime as well as in wartime emergency.[75]

The agreement specifically stated that the AOC ADC was to exercise his operational control authority *through* the CinC US Northeast Command.[76] Since there were no RCAF interceptor forces in Newfoundland and no Canadian air force officer had been assigned to command the 64th Air Division Sector, the

AOC ADC had only one choice as a component commander in Newfoundland: the CinC US Northeast Command double-hatted as the commander of Northeast Air Command. As one US State Department official put it, the AOC ADC "would be ostensibly in command, although it was understood that, in the absence of Canadian defense forces, it would be appropriate and necessary for him to delegate operational responsibility to the American Commander in the Northeast."[77] On paper there was official RCAF AOC ADC operational control to ensure Canadian sovereignty, but in reality actual operational control in addition to operational command remained in the hands of the CinC US Northeast Command, provided that he adhered to RCAF Rules of Engagement and interception procedures. The American commander would remain the principal air defence commander in the Newfoundland area until such time as Canada decided to take over the entire air defence effort by deploying interceptor aircraft there. It never happened. With sovereignty for Canada and military efficiency for US Northeast Command operations secured, it was not surprising this agreement was renewed periodically until 1957, when all air forces in North America came under NORAD operational control.[78]

Conclusion

The United States was very cognizant of Canadian sensibilities about the presence of American forces in Newfoundland. American officials recognized it was not the Canadian government but the British who had negotiated the US base rights during the war, and that special arrangements would have to be made for command and control in the new Canadian province. The Americans were very proactive in informing the Canadians of their intentions regarding Northeast Command, something they did not have to do. They were also quite accommodating of Canadian wishes, making concessions on both the name of Northeast Command and the bilateral command and control arrangements – something else they did not have to do.

Securing a satisfactory arrangement for command and control in the northeast was so important to Canada that MND Claxton withheld submission of the Goose Bay lease agreement until the matter was resolved. As David Bercuson has noted, it was a clear exercise and assertion of Canadian sovereignty: "The signing of the Goose Bay lease was not a Canadian capitulation. Given the necessarily unequal nature of the Canadian–American defence partnership, the unwillingness of Canada to spend considerably more on defence, and Canada's acquiescence in SAC's role as NATO's deterrent punch, the negotiations produced a surprising degree of Canadian success."[79]

Reflecting a Janowitzian military appreciation of political concerns, this chapter has shown that the societal/political imperative of safeguarding Canadian sovereignty can be balanced with the functional imperative of ensuring competent command and control, efficiency in military operations, and (perhaps lost in the equation) air defence coverage of the area. If Canada was truly concerned about ensuring both *de facto* and *de jure* sovereignty, it should have provided fighter defences to protect Newfoundland instead of relying on the USAF. However, priorities elsewhere meant that US Northeast Command had to fulfill this requirement. Although critics may argue that having USAF aircraft protecting Newfoundland instead of the RCAF had a negative effect on Canadian pride as well as sovereignty, it must be remembered that one key feature of an alliance is that it helps an ally defend itself when and where needed, even in its own territory – something European members of NATO would agree with. The Canadian government was well aware of the steps taken in Europe that involved having "foreign" air forces operate in NATO countries (in fact, the RCAF was one of those foreign air forces). In this regard, the arrangements Ottawa was seeking in North America to protect Canadian sovereignty were more restrictive than the NATO norm in Europe.[80]

By securing a favourable operational control arrangement with the Americans, Canada was able to blunt the apparent damage done to Canadian pride and sovereignty, maintain responsibility for the air defence of Newfoundland, and have a voice in how US forces defended the province from enemy aerial attack. The operational control arrangement for US Northeast Command was a beneficial situation for both Canada and the United States that laid the foundation for the two countries' next bilateral air defence endeavour: NORAD.

8
Integrating North American Air Defences under Operational Control

> *Will the existing arrangements for command and control be adequate, and if not, what steps should Canada take to ensure that the air defence system operates with maximum effectiveness and at the same time Canadian interests are protected?*
> – "The Air Defence of North America,"
> Department of External Affairs Defence Liaison (1) Division

BY THE END OF 1953, the air defence of North America consisted of two coordinated systems – one Canadian and one American. Each country's national air defence commander retained authority over operations in his airspace as per the bilateral Emergency Defence Plan. The EDP also included a special arrangement for RCAF operational control of US Northeast Command interceptors operating in Newfoundland airspace. With the Soviet Union's test explosion of a thermonuclear bomb in August 1953, however, the existential threat of Soviet air attack led Canadian and American airmen to further integrate their continental air defences and centralize command and control under one overall authority.

The possibility that such integration would result in a senior USAF officer having "command of Canadian air defence forces" concerned Canada's Department of External Affairs. "Whether or not Canada is prepared to accept this surrender of its sovereignty in the interest of the defence of North America," officials concluded, "is the most difficult and the most important issue."[1] This sovereignty problem was solved by limiting the authority of an overall American air defence commander to operational control.

From 1953 to 1957, the bilateral air defence command and control relationship evolved from two coordinated but still separate national air defence systems into a single integrated binational command organization, the North American Air Defense Command. NORAD centralized authority for operational control of all continental air defences under an American Commander-in-Chief and a Canadian Deputy Commander-in-Chief (DCinC). The definition of operational control continued to evolve during this time period into the version that appears in the NORAD Agreement to this day. Reflecting the Janowitzian societal/political imperative, several RCAF officers and chairman of the Chiefs of Staff

Committee, General Charles Foulkes, appreciated Ottawa's political sensitivity to command and thus insisted on operational control for integrated Canada–US air defence. Their firmness was a vital factor in American acceptance of this arrangement for NORAD and in the maintenance of Canadian sovereignty.

Reassessing the Coordinated Systems

The Soviet Union's explosion of a thermonuclear device in August 1953 and development of long-range strategic M-4 Bison jet bombers significantly impacted American (and as a consequence, Canadian) strategic policy. Previously, the United States believed that North American critical points could survive atomic bomb attacks, which would only destroy structures in a one-mile radius; but this was not the case with thermonuclear bombs, which could obliterate anything within a radius of 10 miles or more. American strategic policy evolved from believing that the United States could win a war with atomic weapons to deterring the Soviets from attacking by threatening massive nuclear retaliation.[2] This was articulated in the Eisenhower administration's "New Look" Policy of October 1953, which was subsequently adopted in the US strategic concept NSC 162/2. Seeing spending on large conventional forces as weakening the US economy, Eisenhower redirected funding away from conventional forces and towards nuclear deterrence by expanding SAC.[3]

The "New Look" policy included greater funding for continental air defence. Whereas previously the United States had adopted an offensive strategy and had only modest interest in strategic defence, Eisenhower now felt that improved continental air defences were needed to protect the United States against Soviet nuclear capabilities. The growing vulnerability of US nuclear forces to a surprise Soviet attack also concerned the USAF leadership. Their fears that improvements in air defence would be at the expense of SAC were assuaged by the fact that funding was taken from the US Army's and US Navy's budgets to pay for them.[4] Although not revealed publicly, NSC 162/2 also included an option for pre-emptive nuclear strikes against the Soviet Union.

Destroying the Soviet Union's war-making capabilities was an important SAC objective, but its most important role was "blunting" the Soviet strategic bomber force by destroying aircraft on the ground before they could be launched. Indeed, as late as 1962 the American Single Integrated Operational Plan (SIOP) for strategic nuclear weapons contained pre-emptive options.[5] Moreover, MC 48, NATO's nuclear deterrence strategic concept of November 1954, had a first-use option in the event of a conventional Soviet attack in Europe.[6] If the United States got in the first nuclear blow in a pre-emptive strike, the role of North American air defences would be to deal with whatever nuclear bombers the Soviets had left to strike back with. If the Soviets launched the first attack, an improved air defence system would provide an effective tactical "tripwire" to give sufficient warning to SAC to launch a counterattack, warn the civilian population, and give interceptors the opportunity to destroy enemy bombers before they could strike North American vital points.[7] In November 1953, the Canadian government agreed to enhance its air defence collaboration with the United States.[8]

These strategic developments strengthened Canada's – and in particular the RCAF's – functional approach to protecting sovereignty. The new Mid-Canada radar line and the recently approved Distant Early Warning Line (to be completed in 1957) would increase the ability to detect Soviet bombers in the Arctic approaches to North America.[9] Moreover, having coordinated with the USAF on air defence from an early date and secured political and financial support from Ottawa, the RCAF now had an opportunity to make an even larger contribution to North America's strategic defence. To do so, however, would require integrating and centralizing continental air defence.

In 1953, the newly established Canada–US Military Study Group (MSG) concluded that the current scheme of coordinated national air defence systems was incapable of effectively dealing with the growing enemy threat to North America. American and Canadian airmen quickly grasped the need for a sophisticated and integrated continental air defence system to fight the air defence battle in the nuclear age.[10] In the spring of 1954, the commanders of USAF ADC and RCAF ADC established a Canada–US Air Defence Study Group (ADSG) to study the matter and report to the MSG. Since the RCAF had begun posting officers to USAF Air Defense Command headquarters in Colorado Springs in 1951, Canadian and American airmen saw the value of closer collaboration with like-minded air force officers; this paved the way for the establishment of the ADSG, which consisted of personnel from staff of the Canadian and American air defence commands and was tasked with finding the most effective means of providing for the air defence of North America.[11]

In the minds of American airmen, the biggest obstacle to North American air defence integration was the Canadian political leadership. Ottawa's support of the existing coordinated system was too "complacent, or conservative," and driven by the desire "to maintain a public posture of national sovereignty"; this resulted in "a lack of appreciation of the nature of the air atomic threat."[12] This stood in direct opposition to the American airmen's ultimate objective, which was to centralize the two nations' air defence efforts under an integrated Canada–US continental air defence command.[13] What the USAF felt privately, Rep. Sterling Cole (R–NY) made public during a speech in New York City in April 1954. He called for the establishment of a unified Canada–US North American Continental Defence Organization consisting of army, naval, and

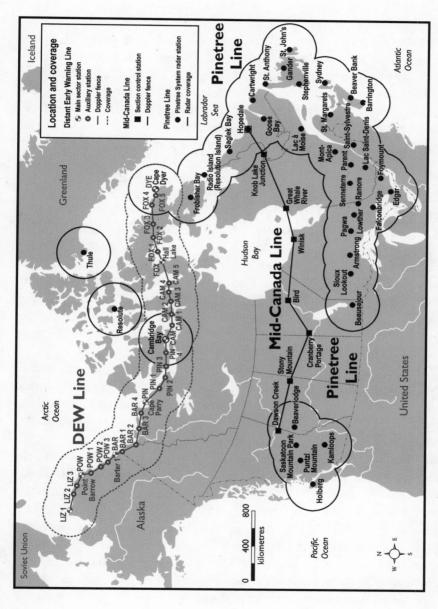

MAP 3 The Pinetree, Mid-Canada, and Distant Early Warning (DEW) radar lines, 1958. *Map drawn by Mike Bechthold.*

air force units from both countries.¹⁴ Writing privately to US Secretary of Defense Charles Wilson, Cole opined that "without such unity of command, I frankly do not believe our two countries can exploit all possibilities for creating an effective continental defense."¹⁵

Wilson referred the matter to the JCS, who admitted they preferred a Canada–US command but concluded it was simply not feasible at the time. Years of bilateral defence planning experience had shown that Canadian military planners were handcuffed by the political restraints faced by their government. Ottawa would be opposed to bilateral command and would likely only consider (if anything at all) a NATO command. This was out of the question in the minds of the Joint Chiefs, who preferred a purely bilateral continental defence relationship to a multilateral NATO one. Pressing the issue with Canadian authorities, the JCS warned, risked jeopardizing the existing coordinated air defence arrangements with Canada.¹⁶ The Secretary of Defense concurred, putting the matter to rest – at least in US circles.

Congressman Cole's public remarks irritated many Canadian officials. MND Claxton sent a strongly worded letter to Secretary Wilson and JCS Chairman Admiral Arthur Radford, who on many occasions thereafter had to reassure Canadian officials that the United States was satisfied with the current system of air defence coordination.¹⁷ Regardless, the Canadian Chiefs of Staff, now anxious about the air defence integration issue, instructed the Joint Planning Committee to explore various options for forming a combined command for North American defence.¹⁸

The resulting JPC study acknowledged the weakness of the coordinated air defence system and recognized the need for peacetime centralized command and control. Warning that the United States might approach Canada about appointing a CinC for air defence of the Canada–US NATO region, the JPC recommended a proactive approach to protect Canada's sovereignty and ensure Canadian participation in continental air defence. The JPC outlined the following four possible options: (1) a Supreme Allied Commander Canada–US responsible to the NATO Standing Group; (2) a CinC for Canada–US continental defence forces responsible to the NATO Canada–US Regional Planning Group (CUSRPG); (3) a CinC Air Defence Canada–US responsible to CUSRPG; and (4) a CinC Air Defence Canada–US responsible "co-equally" to the Canadian and US chiefs of staff. For each option the CinC would be an American officer. The focus of the study was on command organization; detailed specific command and control arrangements would be determined at a later time.¹⁹

The JCS deemed the first two alternatives unfeasible due to their multiservice and multilateral characteristics. Option 1 called for CUSRPG to be replaced with a NATO continental defence command under a supreme commander like other NATO commands. He would have authority over Canada–US maritime, land, and air forces and would report directly to the NATO Standing Group. Option 2 was similar but had a NATO CinC reporting directly to CUSRPG. Both options suffered from the fact that the aerial threat to North America required a functional air defence solution that did not warrant a multi-service joint air, sea, and land Canada–US command. The United States preferred a bilateral continental defence relationship with Canada only, not a multilateral one with other NATO allies whose security measures were suspect and who might, American airmen feared, try to gain influence over Strategic Air Command.[20]

The Canadian planners favoured Option 3 because it focused solely on air defence. Centralization of continental air defences under one CinC would ensure flexibility, coordination, and "optimum utilization of the resources available." However, American opposition to a NATO command for North America precluded a CinC responsible to CUSRPG as a viable option; the Americans would instead prefer the purely bilateral Option 4. The appointment of a CinC Air Defence Canada–US responsible to both nations' chiefs of staff would centralize air defence under one authority to ensure greater operational efficiency and avoid potential NATO interference. With no NATO connection, however, Canada would be deprived of the multilateral "safety in numbers" that it enjoyed in the European NATO commands. Canada would instead have to deal with the Americans on a purely bilateral – and in Canadian eyes unequal – basis. Still, the JPC admitted that Option 4 was acceptable "from a military point of view" and should therefore have Canadian support if the Americans insisted on it. The air defence effort in North America was becoming further integrated, and it might be impossible for Canada to resist this trend. The JPC recommended quickness in addressing the integration issue, warning that to postpone action until an emergency "might provide less favourable results for Canada."[21] Unfortunately, the Canadian military leadership did not heed this advice.

CSC chairman General Foulkes advised a cautious approach, given that "informal talks" with his JCS counterpart revealed no American desire for immediate action. Foulkes's colleagues on the CSC agreed; further study by the RCAF and the ADSG would suffice for the time being.[22] The JPC report was proactive on air defence integration, but the Canadian military leadership reverted to its traditional reactive ways. It would not take long for the United States to give Canada something to ponder.

In 1954, the JCS decided that the Soviet thermonuclear threat required national air defence centralization and stood up the Continental Air Defense

Command (CONAD) in September. The JCS made the air force CONAD's executive agency and assigned USAF ADC commander General Benjamin Chidlaw as its CinC. CONAD was superimposed on the existing USAF Air Defense Command structure, and Chidlaw wore another "hat" as commander of the component USAF Air Defense Command (i.e., much like Lt.-Gen. Charles T. Myers was CinC US Northeast Command and commander of the component Northeast Air Command in Newfoundland – see Chapter 7).[23] As a JCS joint command, CONAD included the US Army Anti-Aircraft Command and Naval Forces Continental Air Defense Command as component commands to advise CinC CONAD on army and naval matters.[24] Other JCS commands (including US Northeast Command) were to support CinC CONAD "in accordance with plans approved by the JCS and mutual agreements by the commanders concerned to insure that plans for and operations of the elements of the early warning system would be responsive to the needs of CINCONAD." In an emergency these forces would come under CONAD.[25]

CONAD's authority to exercise "operational control over all forces assigned or otherwise made available by the JCS or other authority" was weak and convoluted. CinC CONAD's terms of reference defined operational control as "authority to direct the tactical air battle, control fighters, specify conditions of alert, station early warning elements, and deploy the command combat units."[26] Too much was left unsaid, especially regarding non-USAF resources, and this proved to be a consistent irritant for CONAD throughout its first two years.[27] CONAD operational control did not include discipline, logistics, and determining the original composition of component forces, which remained the individual US services' national command authority. This was consistent with American and Canadian command and control culture and practice. The CONAD arrangements would set an important precedent for NORAD three years later.

Another important change in American air defence policy was the centralization of American air defences under CONAD. By appointing a USAF officer as CinC CONAD and assigning him operational control over all joint air defence forces, the JCS finally formally recognized the primacy of the USAF in the air defence of the United States. Although CinC CONAD's convoluted operational control authority was not ideal and CONAD was not responsible for defending Alaska and the US bases in Newfoundland, much had been accomplished towards air defence integration. It was now time to integrate Canadian air defences into a continental system.

Proposals for Air Defence Integration

After intense study of the aerial threat, the ADSG concluded in the autumn of 1954 that the requirement to make quick decisions in the modern air defence battle necessitated a combined Canada–US air defence command under one overall commander. This command would be responsible to both governments and ought to be stood up in peacetime to be ready in an air defence emergency.[28] The MSG concurred and brought these conclusions to the Canadian and American military leadership.[29]

The special April 6 meeting of the Chiefs of Staff Committee was pivotal in convincing Canadian authorities of the need for integrated continental air defence. It was a rather large gathering in Ottawa that day. Besides the chiefs themselves, those in attendance also included C.M. Drury, the Deputy MND; R.B. Bryce, the Secretary to the Cabinet; R.A. MacKay, the Associate Under-Secretary of State for External Affairs; Air Vice-Marshal C.R. Dunlap, Chairman of the Canadian Section MSG; and a few other officers from ADC Headquarters in St-Hubert and Air Force Headquarters in Ottawa. A/C Clare Annis, an Air Defence Command staff officer and a member of the ADSG, conducted the briefing.[30] Annis was an authority on air power and one of only a few RCAF airmen who had written and spoken publicly on the subject of North American air defence.[31] He had an intimate understanding of the air defence integration issue and was the ideal person to give the briefing. Stressing in great detail the urgent need for North American air defence integration, Annis explained that Canada currently faced two main problems: a Soviet military threat; and a potential American defence-against-help threat to Canadian sovereignty "resulting from the US reaction to the Russian military threat."[32] Canada thus needed to act quickly by accepting the establishment of and participation in a continental air defence command. Air Marshal Slemon was convinced. Firmly believing in the need for an overall air defence commander, he set out to make it a reality.

Before Slemon could forward any formal proposals to government authorities, however, Canadian political sensitivities had to be overcome. Although cool to the idea of integrating Canadian and American air defences under one overall commander, Canadian government officials, notably those from the Department of External Affairs, were not completely opposed to it. Military planners therefore needed to find the right command and control *authority* for a commander that would balance safeguarding Canadian sovereignty with ensuring efficient continental air defence. The solution was avoiding the term "command" and using the principle of operational control.

Although Joseph Jockel originally noted that the operational control idea was an American one,[33] recently declassified records show that it was actually the brainchild of a Canadian, Air Marshal Slemon. The CAS had previously distinguished the differences between operational control and command in his 1953 US Northeast Command directive to the AOC ADC (see Chapter 7).

Slemon deduced that assigning similar operational control authority to an overall continental air defence commander would be "politically acceptable to both countries." He explained this reasoning in a December 2 brief for CSC chairman General Foulkes:

> We are giving consideration now on the military level to the working out of a system of operational control which will avoid the use of the term 'command.' 'Command' infers control of logistics, which is not necessary, and creates a great many political difficulties, particularly in peacetime. However[,] we are completely convinced that operational control of the whole system should be vested in one commander and we are attempting to work out proposals which will allow the operational control of the system without the necessity of adopting a unified system of command.[34]

Slemon's brief – and subsequent discussions about integrating North American air defences under operational control – is a good example of the military being conscious of political sensitivities to command and working to find a solution to remedy it that also ensured military effectiveness and efficiency, thereby balancing the functional and societal/political imperatives.

Foulkes agreed with Slemon's reasoning. In a December 5 top secret meeting of Canadian and American military and External Affairs officials, Foulkes explained that the sensitive nature of continental air defence integration in Canada meant that any proposals for an overall North American commander "would be couched in terms of operational control in order to avoid the politically awkward term of command."[35] American officials were open to this approach.

JCS chairman Admiral Radford in particular agreed that Slemon's concept of operational control held much promise. Concluding that Canadian officials would find the establishment of a Canada–US air defence command unacceptable at that time, the JCS decided that the formal proposal to Canada "should be limited to the peacetime integration of operational control of Canadian and US warning systems and air forces assigned to continental air defences."[36] The wording was crucial. The key aspect was differentiating between "a command" and the exercise of command and control authority. In the new proposal, the focus would shift away from the establishment of a combined command since it – and the very mention of the word "command" – was so politically sensitive in Canada. The spotlight would instead be on the actual command and control *authority* to be exercised, operational control, which would not include the word "command." Operational control authority could be exercised by a new bilateral entity established to oversee all Canadian and American air defences, or it could be assigned to an existing command organization such as CONAD. Regardless of the arrangement, operational control would ensure functional military efficiency because it was centralized, provided effective command and control, and guaranteed the rational utilization of the continent's air defence resources.[37] Exactly what kind of entity would exercise operational control and what specific authority it would entail remained to be determined.

The Joint Strategic Plans Committee (JSPC), the main planning organ of the JCS, made the first attempt at clarifying the issue. Operational control would include "the composition of subordinate forces, the assignment of tasks, the designation of objectives, and the authoritative direction necessary to accomplish the mission of providing air defense for Canada and the continental United States. Operational control in this respect would not include such matters as administration, discipline, internal organization, unit training and logistics."[38] Though generally similar to the British 1944 definition of operational control, this JSPC version differed in two significant ways. First, it omitted the words "those functions of command" that appeared in the 1944 version, which was understandable since Canadian and American planners were careful to avoid any mention of the word "command" in Canada–US integrated air defence proposals.[39] The second difference was that the JSPC version did not include the phrase that operational control "shall always be exercised where possible by making use of normal organisation Units assigned, through the responsible Commanders."[40] The reason for this omission was that the JSPC had yet to address the relationship between the overall air defence commander and his subordinate component commanders, and especially the delegation of operational control authority. Nonetheless, by specifically stating that operational control "would not include such matters as administration, discipline, internal organization, unit training and logistics," the JSPC definition adhered to Canadian and American practice that authority over these was a service prerogative under national command. The JSPC definition of operational control laid out important principles and precedents that the JCS and Secretary of Defense agreed would form the basis for commencing formal negotiations with the Canadians to integrate continental air defences.[41]

In February 1956, the JCS and the CSC agreed to establish the Ad Hoc Study Group (AHSG) to study the issue and report to the Canada–US Military Study Group. Because negotiations would likely raise "very delicate political matters," the AHSG was to limit its focus only "to the problems of operational control," not the actual establishment of a command.[42] The Ad Hoc Study Group consisted of officers from the USAF, US Army, USN, RCAF, and Canadian Army, and it began examining operational control integration in the summer of 1956.[43] The absence of DEA or State Department members was not an oversight. East Block was aware of the military discussions but decided "it would be best not to have political officers associated directly with this process."[44] The RCN declined to assign a member to the AHSG because the Chief of the Naval Staff felt that his service's participation was unnecessary. The Canadian Army was also reluctant to have a representative, but the CGS, Lt.-Gen. H.D. Graham, decided to appoint a token staff officer since the US Army was sending one and in case the subject of possible army involvement in continental defence arose.[45] The reluctance to participate on the part of both the RCN and the Canadian Army was symbolic of the pre-eminence of air power and professional air forces in continental defence during the early Cold War.

In the meantime, CONAD was undergoing changes that would significantly impact the AHSG operational control proposal. General Earle Partridge, the new CONAD CinC, was frustrated with the limitations on his operational control authority. During the summer of 1956 the JCS thus revamped CONAD's command structure and terms of reference to provide more effective command and control. They removed Partridge's additional "hat" by making Air Defense Command a separate component command under CONAD. Partridge remained CinC CONAD, and Lt.-Gen. Joseph H. Atkinson took over USAF ADC as Partridge's component commander.[46] The JCS also broadened CONAD's mission by assigning responsibility for the air defence of Alaska and the US bases in Newfoundland to CinC CONAD. As a result, effective September 1, 1956, the JCS disestablished US Northeast Command and reassigned its responsibilities for air defence in that region to CONAD.[47]

Revised terms of reference provided CinC CONAD with more effective command and control authority and greater leeway in coordinating operations. CONAD's authority was expanded to include "the responsibility to determine procedures for conducting the air battle, for exercising operational control of all assigned forces, and for directing engagement and disengagement of weapons." To resolve the Army Anti-Aircraft Command decentralized control problem once and for all, operational control of all army anti-aircraft forces was centralized under CinC CONAD; this included authority to assign individual batteries to particular vital points.[48] The terms of reference also included a new definition of operational control: "those functions of command involving composition of subordinate forces, assignment of tasks, designation of objectives, and direction necessary to accomplish the air defence mission."[49] This was very similar to the JSPC's definition a few months earlier.

Clarifying CONAD operational control authority over American national air defences had a significant effect on the development of the bilateral Canada–US air defence command and control relationship. The disestablishment of US Northeast Command and the assigning of its air defence responsibilities to

CONAD left open the issue of RCAF operational control in Newfoundland. Although for the time being the AOC ADC retained operational control over USAF forces in the area, this was a temporary arrangement that would eventually have to be addressed in light of ongoing AHSG discussions. CONAD's terms of reference set another precedent for the exercise of operational control over air defence forces and reinforced the idea that one overall commander should manage the air defence battle. Having one authority for American air defence with effective operational control authority would therefore make integration with the Canadian air defence system much easier to accomplish. Consequently, the JCS instructed the AHSG to utilize the new CONAD terms of reference "as general guidance" for bilateral negotiations.[50]

In October 1956, the AHSG presented its final study to the MSG, which made minor revisions largely at the request of the Canadian chairman, Air Vice-Marshal C.R. "Larry" Dunlap.[51] As Vice-Chief of the Air Staff, Dunlap was the RCAF member of the PJBD, and it was his responsibility to brief the PJBD and the CSC on integrated air defence operational control.[52] Dunlap, who later in his career would be CAS and then Deputy CinC NORAD, was considered one of Slemon's "boys," meaning that he and the Chief of the Air Staff were on the same page when it came to air defence matters.[53] Dunlap had the best interests of the RCAF (and its air defence mission) in mind, and this proved to be crucial with regard to the RCAF's ability to protect Canadian sovereignty.

The MSG's final air defence integration proposal, dated December 19, 1956, recommended that an overall continental air defence commander be established and that he exercise operational control over the air defences of both countries based on the NATO and JCS practice of unified command. Reference to NATO, however, was only in regard to the *type* of command; there was to be no formal connection between NATO and the new air defence commander. "Drawing a parallel," as Jockel noted, "is as far as the study group went."[54] The MSG highlighted CONAD as a model for the new overall air defence commander. The CinC CONAD's relationship with his subordinate commanders was "based on principles established during World War II for joint and combined commands, and is now used in NATO. It provides for the forces of each service to be assigned to and commanded by a commander of the same service although they are under the operational control of joint commanders."[55] CinC CONAD would don another "hat" and become the new overall air defence commander, called Commander-in-Chief Air Defence Canada–United States (CINCADCANUS). The new CinC would be responsible to the Canadian and US chiefs of staff, who were in turn responsible to their respective political authorities. The proposal also called for the establishment of a Deputy Commander-in-Chief who would be in charge in the absence of the CinC. The

MSG recommended that these officers "should not normally be from the same nation," which meant there would always be an American CinC and a Canadian DCinC. A new headquarters would be established consisting of personnel from both Canada and the United States, making it a completely bilateral effort.[56]

CINCADCANUS would delegate operational control (defined below) to subordinate component commanders of geographical areas yet to be determined.[57] Each service providing forces to the component commands would retain national command. This provision was particularly important for Canada, as RCAF Air Defence Command would be one of the component commands. By this time, however, ADC was no longer a joint command, for the Canadian Army had disassociated itself from area air defence under ADC in 1955 to focus on in-close point defence.[58] This made things easier for the CINCADCANUS arrangement, as RCAF ADC would be the only Canadian component command to be incorporated into the new command structure. The Air Officer Commanding Air Defence Command would be responsible to CINCADCANUS in Colorado Springs for operational control but would still be responsible to the Chief of the Air Staff in Ottawa for operational command, thus reinforcing national command of RCAF forces through the chain of command.[59]

The MSG proposal also reinforced the Canada–US cross-border reinforcement arrangements in PJBD Recommendation 51/6. During peacetime any permanent changes of station would have to be outlined in CINCADCANUS deployment plans, which would have to be submitted to higher authority for approval. In an air defence emergency, however, component commanders, through operational control that CINCADCANUS delegated to them, would be permitted "temporary reinforcements from one area to another, including crossing the international boundary to meet operational requirements."[60]

Although planners had originally proposed the definition of operational control from CONAD's terms of reference for CINCADCANUS, they decided instead to utilize the operational control definition from the 1952 US Northeast Command arrangement (see Chapter 7). The reason for this decision is unclear, though there are a few possible explanations.[61] One has to do with a Janowitzian awareness of political issues. The CinC CONAD terms of reference's inclusion of the phrase "those functions of command" when describing operational control authority may have been too politically sensitive for Canadian authorities wishing to avoid the word "command." Another explanation is that the Canadian members of the AHSG were also familiar with the 1952 US Northeast Command arrangement's definition of operational control, which was the one Air Marshal Slemon had suggested for air defence integration in his December 2, 1955, brief for General Foulkes. Operational control in the

MSG proposal was defined as "the power of directing, coordinating and controlling the operational activities of deployed units which may, or may not, be under the command of the authority exercising operational control."[62] A separate clause stated that "command of forces of one nationality, which includes such matters of logistic support, administration, discipline, internal organization and unit training, should be exercised by national [i.e., component] commanders responsible to their national authorities for these aspects of their commands."[63] This was consistent with the principle of national command as a service prerogative.

The MSG proposal did not specifically recommend that a unified command be established – only that it was the "best method" of dealing with the Canada–US continental air defence mission. Emphasis was on the *authority* of operational control and on the *person* who would exercise it, not on the establishment of a unified command, which was still too politically sensitive.[64] The new overall air defence commander would have the benefits that the CinC of a unified command possessed to ensure military efficiency but would not be the head of a binational Canada–US command. This was, to say the least, an awkward arrangement. It would not last.

Realizing Integrated Operational Control

> *The JCS correctly perceived that Canada's political leadership would consider the issue of a joint command to be a hot potato.*
> – Kenneth Schaffel, USAF official historian[65]

The MSG submitted the December 19, 1956, proposal as its Eighth Report, formally recommending approval for an overall air defence commander exercising operational control over integrated Canada–US continental air defences. The JCS gave its assent on February 6, 1957, and US Secretary of Defense Charles Wilson endorsed the proposed air defence arrangements on March 16, which according to American procedure constituted official government approval.[66] The Canadian government's decision proved to be much slower, however.

The CSC approved the air defence proposal in February 1957 and secured the assent of MND Ralph Campney in early March. Air Marshal Slemon and General Foulkes prepared a formal memorandum for the minister to introduce to the Cabinet Defence Committee for final government approval. On March 4, General Foulkes informally indicated to the Chief of Staff of the USAF that he did not anticipate having any difficulties in obtaining official Canadian government endorsement.[67] However, delays soon set in. The Cabinet Defence Committee was supposed to discuss the proposal on March 15, but this meeting was cancelled and rescheduled for a later date. In April, Prime Minister Louis St. Laurent dissolved Parliament and called a federal election for June 10. He did not want the air defence proposal to become an election issue, so he tabled a decision on the matter until after the vote. Although St. Laurent was confident his Liberal Party would be returned to another majority government, to his – and most of Canada's – surprise he lost the election to John Diefenbaker's Progressive Conservatives.[68] With the Americans growing impatient with Canadian delays, on June 12, 1957, Foulkes implored the outgoing Liberal administration to approve the air defence proposal before leaving office. St. Laurent declined. With the election loss, he felt that his government no longer had the mandate to make a decision and that it should instead fall to the incoming Diefenbaker government.[69]

Rebuffed, Foulkes sought approval for the air defence proposal from the new Tory government, which assumed power on June 21.[70] Prime Minister Diefenbaker's visit to London for a Commonwealth conference shortly after taking office, however, further delayed a government decision until July. In the meantime, Foulkes drafted a fresh memorandum for the new MND, George Pearkes, to submit to Diefenbaker upon his return to Canada.[71] Pearkes discussed the matter with Diefenbaker on July 24, and that afternoon the prime minister approved the integration of the operational control of Canada–US air defences.[72] Foulkes informed his American colleagues shortly thereafter, and on July 31, the Cabinet confirmed the appointment of Air Marshal Roy Slemon as the Deputy CinC of what was now being called the Canada–United States Air Defence Command.[73]

The next day, Canada and the United States announced that they were "setting up a system of integrated operational control" of the two nations' air defences under a new "integrated command" in Colorado Springs responsible to each nation's chiefs of staff. Although the joint press release did not include the definition of operational control from the MSG proposal, it emphasized that "other aspects of command and administration will remain a national responsibility" (see Appendix 5 for the full text). The message was clear: Canadian air forces would at no time come under US command. To re-emphasize this expression of sovereignty, the Canadian government announced in a separate press release that Air Marshal Slemon would assume duties as the Deputy CinC of this new integrated air defence command.[74]

On September 12, 1957, the North American Air Defense Command (the name change is explained below) was officially stood up at Colorado Springs. The CinC CONAD, General Earle Partridge, donned another "hat" as the new CinC NORAD with the authority to exercise operational control over Canadian and American air defence forces. Air Marshal Slemon became NORAD

Deputy CinC and would exercise operational control in Partridge's absence.⁷⁵ Operational control over integrated Canada–US air defence forces had become a reality.

Establishing NORAD's binational organizational structure proved an easy task. The new command was placed on top of the existing CONAD structure, and RCAF Air Defence Command was simply plugged in as an additional component command. NORAD and CONAD remained separate entities: a binational command with General Partridge wearing his CinC NORAD "hat" exercising operational control over Canadian and American air defences for continental air defence; and a national JCS command with General Partridge wearing his CinC CONAD "hat" exercising operational control over American air defences defending the United States.⁷⁶

NORAD operated on an interim basis from its establishment in September 1957 until May 12, 1958, when an official governmental exchange of notes, known as the NORAD Agreement, confirmed it.⁷⁷ CinC NORAD's operational control authority (which the DCinC would exercise in his absence) as stated in both the NORAD Agreement and the Terms of Reference that the JCS and CSC gave to Partridge and Slemon in June 1958 was consistent with the definition of operational control in the MSG proposal of December 19, 1956.⁷⁸ For the sake of clarity, the NORAD Agreement made a transparent distinction between the authority of operational control and what remained national command responsibility:

> "Operational control" is the power to direct, coordinate, and control the operational activities of forces assigned, attached or otherwise made available. No permanent changes of station would be made without approval of the higher national authority concerned. Temporary reinforcement from one area to another, including the crossing of the international boundary, to meet operational requirements will be within the authority of commanders having operational control [this was consistent with PJBD Recommendation 51/6]. The basic command organization for air defence forces of the two countries, including administration, discipline, internal organization and unit training, shall be exercised by national commanders responsible to their national authorities.⁷⁹

The revised Canada–US Emergency Defence Plan of July 1, 1958, also confirmed NORAD's operational control authority over combined RCAF–USAF forces.⁸⁰

The creation of NORAD in 1957 was controversial. Others have documented this well,⁸¹ so only a brief summary is in order. The main issue was procedure: Prime Minister Diefenbaker approved the creation of NORAD himself without first consulting his government colleagues in a Cabinet Defence Committee (which he had not yet formed after his election win), or in the Department of External Affairs, which he suspected was largely loyal to the former Liberal regime (he called them "Pearsonalities") and thus not to be trusted.[82] Officials at DEA were perturbed that there was no formal government-to-government agreement or exchange of notes between Canada and the United States upon the establishment of NORAD in September 1957 – in particular, one that included a clause on consultation between Canada and the United States.[83] The formal signing of the NORAD Agreement in May 1958 remedied these matters.[84]

It is very important to note that DEA's concern over NORAD's establishment in September 1957 did not have to do with command and control. A number of officials from External Affairs had been involved at various stages in the development of the final MSG proposal and did not oppose its operational control provisions. DEA *was* concerned that there was no precedent for peacetime operational control of Canadian forces by another country in Canadian territory or airspace. Department officials recognized that there was a precedent with NATO for foreign operational control of Canadian units, but pointed out that this was an expeditionary arrangement that did not include continental defence forces operating in Canada.[85]

There were actually two precedents for operational control over another nation's air forces in North America, although no one was cognizant of them in 1957. First, the formation of the Canadian Northwest Atlantic Command in 1943 had given the Air Officer Commanding-in-Chief RCAF Eastern Air Command operational control over American maritime air forces during the last two years of the Battle of the Atlantic. Second, the US Northeast Command arrangement of 1953 had given the AOC ADC operational control over USAF interceptors operating in Canadian airspace. Neither of these arrangements foresaw the establishment of a binational command like NORAD; instead, they only consisted of the *assignment* of the command and control *authority* of operational control over one nation's forces to the operational commander of a pre-existing command organization from the other nation.[86]

Canada's National Defence Act was unclear regarding international command organizations. Although paragraph 18 granted the Minister of National Defence the authority to establish military commands, this only included individual Canadian air force, navy, army and joint tri-service national commands.[87] NORAD set a new precedent for an international (in this case binational) command responsible for the defence of Canadian territory and airspace. What had also helped make NORAD a reality was a change in Canadian apprehensions regarding the word "command."

Due to political sensitivities within St. Laurent's Liberal administration, planners had avoided reference to the word "command" in their discussion of air defence integration. Their focus was on the *person* exercising operational control authority, the CinC, not an *entity* such as a binational command. By spring 1957, however, correspondence on air defence integration increasingly began making reference to the establishment of a formal Canada–US command.[88] The word "command" was less politically controversial with the Tories, and by the time they came to power "command" appeared much more freely in the correspondence. By the end of July, Prime Minister Diefenbaker's Cabinet was referring to a "Canada–United States Air Defence Command," and this was the wording that appeared in the August 1, 1957, joint press release with the United States.[89]

Why the word "command" in general and the establishment of a formal binational command in particular became more acceptable in political circles is unclear. Jockel emphasizes the awkward arrangement in the original MSG integrated air defence proposal and argues that practicality was the main reason: Canadian and American officials had "bowed to the inevitable" and "in a bow to reality" had accepted a binational command.[90] Another possible explanation lies in the appointment of George Pearkes as the new MND. Pearkes had won a prestigious Victoria Cross for valour during the First World War and was the Canadian Army's General Officer Commanding-in-Chief Pacific Command at the rank of major-general during the Second World War. After retiring from the army in 1945, he successfully ran for federal office as a Progressive Conservative, and since 1955 he had been a staunch advocate for Canada–US air defence integration as Diefenbaker's defence critic.[91] As a former operational commander who had dealt with the Americans for Canada–US operations in the Aleutians during the war, he was more inclined towards a formal Canada–US command under an overall CinC – and he described it in these terms to his boss, Diefenbaker.[92] Whatever the reason, by late July 1957 the focus of the integrated air defence proposal had switched from the person to the entity, consisting of a binational command with a CinC exercising operational control.

The name NORAD was the brainchild of General Earle Partridge, NORAD's first CinC. Partridge had never been comfortable with the titles ADCANUS and CINCADCANUS, finding the last four letters particularly awkward. He suggested instead that the new entity be named the North American Air Defense Command or NORAD for short. "From the public relations point of view," Partridge explained, "ADCANUS is cumbersome [and] hard to say or to fit in documents without appearing amusing. NORAD in contrast is punchy[,] forceful and conveys the idea with simplicity."[93] The JCS and CSC agreed, and NORAD became the name of the new command when it was stood up on September 12, 1957.

The joint press release of August 1, 1957, described NORAD as an "integrated command."[94] This designation was in a sense true in that the new command integrated Canadian and American air defence efforts under one CinC. The first draft of this press release, however, referred to NORAD as a "unified command."[95] The Americans, not the Canadians, insisted that the designation be changed to "integrated command" to make it consistent with the wording in the December 1956 MSG air defence integration proposal.[96] Regardless of what designation appeared in the final joint press release, NORAD was a unified command modelled on CONAD that exercised operational control over both nations' air defence forces. It was thus a hybrid of the British operational control system and the American unified command system of command and control (see Chapter 2).

NORAD was definitely not a NATO command. NORAD was (and still is) purely binational, not multilateral like NATO commands. The August 1, 1957, press release specifically noted that NORAD reported to each nation's chiefs of staff, who in turn reported to their governments, not NATO.[97] Unfortunately, to curb criticism of the NORAD approval procedure, Prime Minister Diefenbaker incorrectly tried to link NORAD to NATO when defending his government's actions. The JCS protested vehemently against this characterization; although NORAD had some similarities with NATO commands in terms of the type of command it was (i.e., a unified command), that was where the comparison ended. However, several officials, including General Foulkes, External Affairs Minister Sidney Smith, and Prime Minister Diefenbaker, quickly forgot (or ignored) and then blurred this distinction, and all stressed incorrectly that there was a direct and formal link between NORAD and NATO.[98] As a consequence, the JCS was forced to set the record straight with their Canadian allies, emphasizing once and for all that NORAD was a binational command, not a multilateral NATO one. This was a major embarrassment for Diefenbaker and his government.[99]

The crux of the problem was that in the correspondence regarding the establishment of NORAD there was too much emphasis on NATO commands as a model type for NORAD as a unified command, and not enough on the American JCS commands, notably CONAD.[100] The bilateral nature of NORAD and its placement on top of the existing CONAD structure made Foulkes and Diefenbaker nervous. They did not want the public to see the new command exercising operational control over Canadian air forces as being overly American, so they emphasized a multilateral connection to NATO. This was a classic case of Canadian apprehension over the bilateral defence relationship with the United States, in which Canada was always the junior partner, in comparison to the more "comfortable" multilateral approach (such as NATO), where Canada had an equal say and safety in numbers. To connect NORAD to NATO emphasized multilateralism over bilateralism, which appealed to many Canadian officials.[101] In "categorical rebuttals" to Foulkes's and Diefenbaker's argument that NORAD was a NATO command, however, the JCS and the NATO Secretary General correctly explained that NORAD was a binational command responsible to each nation's chiefs of staff, not a multilateral NATO command responsible to the NATO Standing Group.[102] To this day, NORAD remains a purely binational Canada–US command.

NORAD protected Canadian – and American – sovereignty because the operational control authority that its CinC (and deputy in his absence) exercised did not include aspects of command. Authority over the original composition of assigned forces, as well as logistics, administration, and discipline, remained a national command service prerogative. Although CinC NORAD could develop plans and make recommendations to the Canadian and American chiefs of staff for the development, composition, and deployment of forces, they had to adhere to existing USAF and RCAF procedures, and it was up to the JCS and CSC to accept and implement them since they had command over the national forces in question. Component commanders coordinated any approved plans and proposals and were responsible to their services for command, administration, training, and support for their forces, as well as "detailed planning, programming and specific siting for air defence units."[103] In 1957, the Canadian component commander was the Air Officer Commanding RCAF Air Defence Command, Air Vice-Marshal L.E. Wray.[104] Today (as of writing), it is the Commander of 1 Canadian Air Division in Winnipeg, Maj.-Gen. Christian Drouin.

The DCinC NORAD position also effectively protected Canadian sovereignty. As Prime Minister Diefenbaker noted in his memoirs, "the appointment of a Canadian as Deputy Commander-in-Chief NORAD would give Canada a desirable measure of responsibility in any decisions that might have to be taken to defend North America against Soviet attack."[105] Even Jules Léger, the Under-Secretary of State for External Affairs (and future Governor General of Canada), recognized the significance of the position: "the fact that his [the NORAD CinC's] deputy is a Canadian and that Canadian officers are integrated into the combined headquarters offer further guarantees that Canadian interests will be given proper attention."[106]

Air Marshal Slemon's recollections of his experiences as DCinC NORAD illustrate the importance of the deputy CinC position. General Partridge had told him shortly after NORAD was stood up, "Roy, I'm supposed to be the

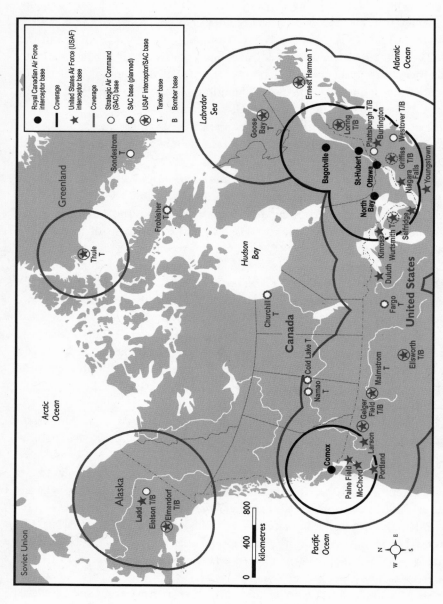

MAP 4   RCAF and USAF Interceptor Deployment and Coverage, 1958. *Map drawn by Mike Bechthold.*

Commander in Chief of NORAD and you're supposed to be the Deputy Commander in Chief. When I go out on a trip, inspecting units or go away to have a little fun, you have the responsibility and the authority."[107] This clear understanding of the relationship between the NORAD CinC and his deputy, especially when the former was absent from headquarters, was essential for NORAD's success. It is one that continued with successive NORAD commanders and their deputies. Slemon's recollections of his relationship with Partridge's successor General Laurence S. Kuter are revealing: "he, unfortunately, was sick about a third of the time, in hospital and so on. So I was in the hot seat. But by General Partridge having taken this policy there was no problem, I just carried on."[108]

From the beginning, General Partridge went out of his way to ensure Canadian involvement in NORAD as equal partners. Slemon's recollections are again revealing: "although we were a little partner making a relatively small contribution to the operational capability of the joint effort, our views were considered in exactly the same light as our partners, the Americans." When Partridge organized NORAD headquarters he made sure to include Canadian officers among some of the most important positions. For example, he appointed Air Vice-Marshal Keith Hodson as the first NORAD Deputy Chief of Staff for Operations, a position Slemon considered "the guts of our joint effort." A Canadian officer has subsequently always held this position at NORAD headquarters.[109] Today (as of writing), Maj.-Gen. Christopher Coates holds this position as J3 (Operations) NORAD.

RCAF Chief of the Air Staff Air Marshal Hugh Campbell's comments regarding the NORAD set-up are also telling. As he remarked to General Foulkes in March 1958, "the allocation to Canada of some of the key senior positions in the [NORAD] Headquarters is particularly advantageous since the incumbents can personally monitor and guard Canadian interests while, at the same time, they will gain experience of future value to the RCAF."[110] This practice of having places of importance in NORAD Headquarters allocated to Canadians continues today in Colorado Springs. It has given Canada what Joel Sokolsky has termed "a seat at the console," allowing Canadian officers to safeguard Canadian sovereignty while at the same ensuring that Canada has a functional "piece of the action" for continental air defence by fulfilling an important operational role in the defence of the continent.[111] This was an important position for Canadian personnel in NORAD during the Cold War and continues to be today. At no time was this advantage more evident than on the morning of September 11, 2001: it was a Canadian who was in charge at NORAD Headquarters in Cheyenne Mountain when the first airliner stuck the World Trade Center in New York City.[112]

Conclusion

> *The hard fact [was] that NORAD was advancing Canadian interests and making Canadian sovereignty more, not less secure.*
> – Prime Minister John Diefenbaker[113]

The establishment of NORAD in 1957 was the culmination of over ten years of Canadian and American airmen's efforts to integrate North American air defences under effective command and control. Throughout the long process the Americans were cognizant and respectful of Canadian sensitivities and Canadian officers were also aware of political concerns. This was especially true regarding the efforts planners took to avoid the term "command," the subsequent focus on operational control, and the agreement to include a Canadian Deputy CinC for NORAD. As a consequence, there was a clear balance between the functional imperative of military effectiveness and the societal/political imperative of the military obligation of responsibility to the community as required by military professionals according to today's Canadian doctrine and Janowitzian CMR theory.[114]

As a binational command responsible to the JCS and CSC, NORAD was consistent with Canada–US command and control arrangements dating back to ABC-22 in that the chain of command for operational commanders was always to the Canadian and American chiefs of staff.[115] With Canadian and American air defences integrated and centralized under operational control, NORAD ensured (and today still ensures) military efficiency and effective command and control. At the same time, command remained a national service prerogative under the RCAF and Canadian personnel played (and continue to play) a key "seat at the console" role as members of the NORAD team. Canadian sovereignty was safeguarded in 1957 and continues to be safeguarded today.

Conclusion

The command of armed forces was the "acid test of sovereignty" in Canada's continental air defence relationship with the United States from 1940 to 1957. Canada safeguarded its sovereignty by ensuring that its forces never came under American command. Concurrently, Canadian planners negotiated bilateral arrangements that protected the continent from enemy attack in an efficient and effective manner. These arrangements – derived from Canadian, American, and British command and control culture and practice – were unique to the North American air defence mission. Each country's efforts and willingness to compromise within their respective service cultures affected how the bilateral relationship grew and evolved from the Second World War to the formation of NORAD in 1957.

From the time Canada and the United States began to collaborate for continental defence in 1940, Canadian military officials steadfastly resisted attempts to bring their forces under American command. Through their efforts, responsibility for administration, discipline, and logistics remained in Canadian hands. The service chiefs of staff exercised national command authority on behalf of the government – a practice that continues to this day. This prerogative reflected a fundamental principle of Canadian command over its armed forces, which it refused to delegate or assign to a foreign commander. Thus, Canada's maintenance of national command authority was an effective and crucial means of protecting sovereignty in its continental defence relationship with the United States.

Canada was also able to retain operational command over its forces for continental air defence by focusing on actual operations: the task at hand and how to utilize the means allocated to accomplish it. A service chief delegated operational command to his service's operational-level commanders. He could also grant operational command (or operational control) to a commander from a different service for joint operations or to a foreign commander in command of bilateral or multilateral forces. The RCAF Chief of the Air Staff's delegation of operational control of Air Defence Command forces to the CinC NORAD is an excellent example. Operational and service culture issues factored heavily in the decision to refuse or permit one's forces to come under the operational command or operational control of a foreign commander. The RCAF's successful resistance to American pressure to place its maritime patrol forces under US unity of command during the Second World War due to concerns about American maritime air power organization and doctrine demonstrated this vividly.

Canadian commanders retained responsibility for the disposition and deployment of Canadian air defence forces and their use in specific operations. Canada only contemplated granting operational command of Canadian forces to the United States in the "Black Plan" of 1940, which postulated a major enemy assault on North America. Even in this "worst-case scenario," Canadian planners secured important concessions from the Americans. These included consultation and the right of the Chiefs of Staff Committee and operational commanders to appeal to higher authority if they disagreed with an American directive, thus ensuring that Canada would have a say in the deployment and movement of its forces. Fortunately, this scenario never came to pass. Canada and the United States instead based their wartime continental defence operations on a second bilateral plan, ABC-22, which was predicated on a more favourable strategic situation.

ABC-22 included a provision for unity of command that some American officials insisted was warranted. Nonetheless, Canadian military officials steadfastly insisted that the enemy threat to the continent was never serious enough to demand such unity, and successfully argued that mutual cooperation would suffice. The Canadian chiefs of staff continually refused to consider implementing unity of command unless the American chiefs of staff specifically requested them to do so. The Canadian military leadership's insistence on dealing directly with their American counterparts reinforced the CSC's national command authority and the primacy of the Canadian service chiefs, acting on behalf of the Canadian government, to decide on command and control arrangements with the United States.

When mutual cooperation proved ineffective for the maritime war against German U-boats off the east coast, Canada and the United States implemented the British operational control system for the Canadian Northwest Atlantic Command in April 1943. In doing so, each nation preserved command over its forces. Operational control did not contain authority inherent in national command or operational command, and its characteristics became important considerations when postwar Canadian planners sought a way to coordinate their air defence efforts with the Americans without having to place Canadian forces under US command.

Even though the 1946 Canada–US Basic Security Plan postulated a more dire enemy threat from Soviet bombers armed with atomic weapons than previously contemplated, the command and control relationship between Canadian and

American forces remained rooted in mutual cooperation. Although the Air Interceptor and Air Warning Appendix to the BSP anticipated Canadian–American continental air defence efforts coordinated by operational control, the specific nature of this relationship was never fully explained. Mere mention of operational control, however, showed that Canadian and American airmen were interested in this command and control principle.

A permanent paradigm shift in the Canada–US command and control relationship occurred in the late 1940s and early 1950s in response to the growing Soviet atomic bomb threat. Canada and the United States moved away from mutual cooperation and towards operational control. This change became more evident from the arrangements in the Canada–US Emergency Defence Plan, the Pinetree System, the cross-border interception PJBD Recommendations, and the US Northeast Command agreement, all of which culminated in the establishment of NORAD in 1957. Although the definition of operational control continued to evolve, a consistent feature was a lack of authority over forces inherent in national command and operational command, both of which remained in Canadian hands. The RCAF Chief of the Air Staff, Air Marshal Roy Slemon, explicitly recommended operational control for integrated Canada–US air defence because it limited the authority of an overall air defence commander, did not mention the word "command," and did not contain operational command and national command authority. NORAD operational control thus assuaged Canadian political concerns that an American continental air defence CinC would amount to a surrender of Canadian sovereignty. It safeguarded Canadian sovereignty and ensured RCAF input into the disposition, deployment, and detailed operations of its forces.

Maintaining Canadian sovereignty in continental air defence with the United States was indeed a crucial goal of Canadian officials. Planners and commanders from 1940 to 1957 faced the major dilemma of negotiating command and control arrangements that balanced sovereignty concerns with framing efficient military operations to protect North America from enemy attack. Accomplishing this goal ensured a proper balance between the Janowitzian societal/political and functional imperatives, which testified to the professionalism of these military personnel. It also exemplified the important "human element" in command and control.

Professional interaction and cordial working relationships among military personnel were essential for effective command and control from 1940 to 1957. Canadian experiences with Maj.-Gen. Brant of the US Army in Newfoundland during the Second World War showed that one individual's truculent attitude could negatively affect overall cooperation and operational efficiency. This experience sharply contrasted the situation at NORAD Headquarters after its establishment in September 1957, where CinC NORAD General Earle Partridge's close professional relationship with his RCAF deputy, Air Marshal Roy Slemon, proved crucial in the organization's first months. The rapport between the two airmen ensured the efficient stand-up of NORAD and also fostered implicit trust that was essential when Slemon exercised operational control of continental air defences in Partridge's (and later General Kuter's) absence. Partridge's faith in Slemon and his proactive and positive attitude towards Canada's role in NORAD as an equal partner with the United States also resulted in excellent opportunities for other Canadian personnel to get "a piece of the action" in the Canada–US continental air defence effort. In consultation with Slemon, Partridge earmarked key positions in NORAD's organizational structure and staff for Canadians. This practice continues today. RCAF personnel have had and continue to have a "seat at the console," from which they safeguard Canada's sovereignty and make a meaningful contribution to the Canada–US continental air defence effort.

The advantages of strong professional relationships went beyond the operational level. The professional transnational "fraternity of the uniform" allowed RCAF and USAF planners to develop common views on air defence in the 1950s that benefited their air power service institution. For Canada, these airmen included Larry Dunlap, Frank Miller, Keith Hodson, Roy Slemon, and Clare Annis. All these individuals held high-ranking positions in the RCAF and/or NORAD, ensuring that air force leadership had an intimate understanding of continental air defence planning imperatives during the early Cold War. These airmen were able to protect and advance Canadian interests while developing a common viewpoint with their American counterparts regarding the best means to achieve air defence integration and centralized command and control.

A positive attitude towards the continental air defence relationship with the United States also allowed Canada to protect its sovereignty. Instead of a belligerent approach, the Canadian military after 1940 took a constructive, functional approach by coordinating with its southern neighbour. Accordingly, Canada was able to protect its sovereignty from unauthorized independent American intervention in Canadian territory to "help" Canada defend itself. The protection of Canadian citizens from enemy threats was in Canada's national interest, and the decision to secure a "piece of the action" with the United States in continental air defence accomplished this goal. Arranging effective command and control of continental air defences with the United States thereby allowed Canada to avoid an intolerable "defence against help" situation.

The United States' approach to continental air defence during the early Cold War was a factor that played into discussions on the future of air defence and Canada's role and relative importance. Whereas American strategic culture placed less priority on strategic defence, Canadian strategic culture placed greater emphasis on continental air defence for the related goals of protecting both North America and the SAC deterrent from attack. Since strategic defence was a high priority for Canada in terms of political and material commitment, Ottawa devoted greater political and military attention to continental air defence arrangements than Washington. This early focus on strategic defence helps explain how and why Canada was able to use its functional approach to protect Canadian sovereignty – especially when the United States began giving greater attention to continental air defence integration after the Soviet Union's explosion of a hydrogen bomb.

The United States recognized Canadian sensitivities regarding command and control arrangements. Although the American planners were adamant that unity of operational command was necessary during the Second World War, they did not force their viewpoint on the Canadians, instead acquiescing to the Canadian position on mutual cooperation. When the situation in the North Atlantic called for more centralized command and control of maritime operations by late 1942, the Americans acknowledged that Canada was providing the bulk of the forces and agreed that USN and USAAF aircraft should come under the operational control of RCAF Eastern Air Command. When standing up Northeast Command in Newfoundland in the early 1950s, the US Joint Chiefs of Staff also kept the Canadians informed and addressed their concerns, which included changing the command's name to US Northeast Command. The JCS extended this courtesy in good faith and in the interest of cordial and respectful relations with Canada, even though the American bases in Newfoundland were considered US territory. The Americans recognized Canadian sensitivities to the word "command" and apprehensions that RCAF air defence forces might come under American command during the mid-1950s. A pragmatic American approach proved essential in establishing operational control authority for CinC NORAD and creating the Canadian Deputy CinC NORAD position.

In order to ensure effective organization and command and control, air forces should be organized functionally according to their primary mission.[1] During the Second World War, Eastern Air Command evolved into a functional maritime air power organization to counter growing German U-boat attacks on shipping. The establishment of the Canadian Northwest Atlantic Command in 1943 enhanced this functional role by elevating the Air Officer Commanding Eastern Air Command to CinC status and centralizing command control over all maritime patrol resources (regardless of nationality) under his operational control. Canada and the United States' establishment of functional air defence command organizations after the Second World War further centralized national air defence, resulting in greater operational efficiency. The United States' establishment of CONAD, a functional unified command, in 1954 continued this trend and was a model for NORAD, which was placed on top of the existing CONAD structure when established in 1957. Because the RCAF had centralized Canadian air defences under the functional Air Defence Command during the early 1950s, Canada was able to easily "plug in" ADC to NORAD as a component command under the operational control of the CinC NORAD.

The defence of North America – be it a purely functional mission to protect the continent from aerial attack or a multi-service effort to defend against air, land, and sea threats – demands a bilateral approach. The Americans in particular were opposed to any multilateral approach that allowed the involvement of non–North American allies in continental air defence. Canada and the United States thus saw the task of defending the continent as a shared responsibility, and they formed a "special relationship" from which Canada continues to accrue certain advantages. These include protection for Canadian citizens and territory, access to superior American intelligence resources, having the ear of one of the world's most powerful countries, and opportunities for Canadian military personnel to have a functional "piece of the action" through "a seat at the console" in defence endeavours with the United States. James Fergusson in particular notes that

> NORAD was the key to Canada's five core interrelated strategic defence interests as they concerned North America and the US relationship – the expropriation of US defence resources for the defence of Canada, access to US defence thinking and planning, influence opportunities to ensure Canadian defence interests were taken into account in US defence plans, protection of Canadian sovereignty from US unilateral defence undertakings, and reassuring the United States of Canada's commitment to continental defence.[2]

Canada's functional approach to having a "seat at the console" and a NORAD Deputy CinC position to safeguard sovereignty also proved essential in the two best known cases where the two national governments disagreed: the Cuban Missile Crisis and the ceasefire alert during the Yom Kippur War. In these crisis situations Canada was able to decide on its own if, when, and how its forces would participate.[3]

NORAD as a binational command organization has also given Canada rare command and control authority over American forces – something that is quite unique in US military culture (and, it could be argued, underappreciated by Americans).[4] While it is true that American forces did not come under

Canadian command, it is unclear whether it is widely known in the United States that on several occasions USAF forces have been under the operational control of the RCAF NORAD Deputy CinC/Commander – including on 9/11. Indeed, it could be argued that the Canada–US NORAD relationship has impacted other military arrangements between the two countries and led to additional opportunities for Canadian commanders to exercise command and control over American forces. For example, in the maritime operations that supported the Afghanistan campaign, some USN ships were placed in a task force under RCN command, while during the 2011 Operation Unified Protector campaign in Libya the NATO theatre commander was a RCAF officer – and former NORAD Deputy Commander – Lt.-Gen. Charlie Bouchard.[5]

In retrospect, the Canada–US continental air defence command and control relationship from 1940 to 1957 went a long way towards addressing Canadian sovereignty concerns. This book has demonstrated a distinct link between command and sovereignty: maintaining command of Canadian military forces is an effective way to protect Canadian sovereignty. Such considerations are especially vital today given the increased importance the Canadian government has assigned to the role of the Canadian Armed Forces in the protection of Canadian sovereignty.[6]

In the time period examined, the bilateral Canada–US continental air defence relationship evolved from a compromise cooperation–unity of command system during the Second World War to a hybrid operational control–unified command system with the establishment of NORAD in 1957. Operational control ensured that Canada retained command of its forces and protected Canadian sovereignty. It also proved to be an efficient means of command and control for bilateral air defence operations, thus ensuring the societal/political–functional imperative balance. Operational control became the cornerstone of NORAD and remains the foundation of Canada–US continental air defence command and control culture and practice to this day.

Appendices

Appendix 1
PJBD Recommendation 51/6, 12 November 1951

That when the Air Defence Commanders of the United States and Canada agree that mutual reinforcement of their Air Defence Forces is necessary in the light of the tactical situation:

a. The Canadian Air Defence Commander should have the power, in the event of war, to authorize the redeployment of U.S.A.F. Air Defence Forces to Canada and the redeployment of R.C.A.F. Air Defence Forces to the United States;

b. The U.S. Air Defence Commander should have the power, in the event of war, to authorize the redeployment of U.S.A.F. Air Defence Forces to Canada.

Source: PJBD Recommendations, DHH 79/35.

Appendix 2
PJBD Recommendation 51/4, 9 May 1951

That aircraft controlled by the Air Defence System of the United States or of Canada engaged in intercepting unidentified aircraft crossing the border between Canada and the United States be permitted to fly over the territory of both countries as may be required to carry out effective interception. These flights would be conducted under the following provisions:

a. Investigations by U.S. military aircraft over Canadian territory would only occur in the case of an aircraft headed for the Canada–United States border from the Canadian side whose flight plan had not been transmitted to the U.S. authorities; or which was off course, and then only in the event that the actions of the aircraft gave rise to a reasonable interpretation of intention to cross the international boundary; the activities of Canadian military aircraft over U.S. territory would be similarly restricted.

b. Close investigation with all due precaution, or interrogation, would be performed solely on unidentified multi-engine aircraft for the purpose of obtaining electronic or visual identification. No attempt would be made to order an intercepted aircraft to land, nor to open fire except when the intercepted aircraft is over the national territory of the air force performing the interception.

c. Investigating aircraft would not approach closer, in accordance with Civil Aeronautics Authority and Department of Transport standards, than is necessary to establish identification.

d. Translation of the general principles and limitations of the agreement into operational instructions of the two air forces would be performed by a Canadian–United States team, and

e. These arrangements will remain in force until modified by agreement or terminated by either Government.

This Recommendation was superseded by PJBD Recommendation 53/1.

Source: PJBD Recommendations, DHH 79/35.

Appendix 3
RCAF Rules of Engagement

RCAF Directive, "Authority to Intercept and Engage Hostile Aircraft"

Introduction

1. It is necessary to take steps to prevent hostile aircraft flying over Canada in the period prior to a formal declaration of war or the declaration of a national emergency. For this reason it is desirable for the Air Defence Commander to have powers to intercept unidentified aircraft and, under certain conditions, to engage them with fighters or anti-aircraft artillery.
2. The purpose of this directive is to lay down the powers of the Air Defence Commander and the extent and conditions under which he may delegate these powers.
3. Although it is necessary to ensure that any unfriendly aircraft entering certain areas of Canada is identified, it is also necessary to infringe as little as possible upon the operational freedom of civil aviation. Fighter interceptions should, therefore, be carried out only after all other methods have failed to establish identification.

Powers of the Air Defence Commander

4. The following are to be the powers of the Air Defence Commander prior to the declaration of war or of a national emergency:
    a) He may authorize the interception of aircraft in order to establish identification if other methods of obtaining identification have failed.
    b) If the aircraft is identified as hostile, by its appearance or because it perpetrates *a hostile act*, or if its pattern of behaviour is sufficiently suspicious to justify a belief that it has hostile intentions, the Air Defence Commander may authorize its engagement by fighters or anti-aircraft artillery.
    c) If the aircraft cannot be identified by any means whatever the Air Defence Commander may, if circumstances warrant, authorize action to force the aircraft to land or engage it with fighter or anti-aircraft artillery.

Responsibility of the Air Defence Commander

5. The Air Defence Commander is to be responsible for:
    a) The promulgation of detailed instructions governing the interception of unidentified aircraft and the engagement of hostile aircraft by fighter or anti-aircraft artillery. These regulations must clearly define the

responsibilities and powers of ADCC [Air Defence Control Centre] commanders and controllers, GCI [Ground Control Intercept] controllers, the commanders of air defence units, anti-aircraft artillery commanders and fighter pilots. Detailed instructions are also to be promulgated for naval personnel engaged in the Air Defence of Canadian land targets or of naval vessels in Canadian ports.

b) The delineation of Canadian Air Defence Identification Zones (CADIZ), Inner Artillery Zones (IAZ), Gun Defended Areas (GDA), AA Defended Airfields and unrestricted areas.

c) Operational coordination with other agencies to implement this directive.

Delegation of Authority by the Air Defence Commander

6. Interception of Unidentified Aircraft. The Air Defence Commander may delegate authority for ordering the interception of aircraft for identification purposes.

7. Engagement of Hostile Aircraft. The Air Defence Commander may delegate authority to engage aircraft which are definitely identified as hostile. The factors mentioned in para 9 (below) must influence the degree to which this authority is delegated in actual fact.

8. Engagement of Unidentified Aircraft. An occasion may arise in which all methods fail to establish the identity of an aircraft. Circumstances at the time of the incident must determine whether action is to be taken to force the aircraft to land or, if this is unsuccessful, to engage it. Similarly, these circumstances must determine the extent to which the Air Defence Commander will delegate the authority for ordering action to be taken against such an aircraft.

Circumstances Determining the Course of Action

9. A decision to engage an aircraft which appears to be hostile, or one which cannot be identified by any means, and the amount of authority to order such action which the Air Defence Commander may delegate must depend on many factors. Among these are the following:

a) The international situation.

b) The capability of the defence system for detecting and recognizing hostile aircraft.

c) The degree of importance of the target area threatened.

d) The standard of recognition training of the units.

e) The amount of control exercised over, and the amount of information available on, the movements of friendly aircraft.

    f) The availability of IFF [Identification Friend or Foe] equipment to military and civil aircraft and its effectiveness.

10. The extent to which authority may be delegated in certain specific instances is given in Appendices 'C' and 'D' to this directive.

Precautions to Be Taken by the Air Defence Commander

11. The Air Defence Commander is to take *every precaution possible* to prevent untoward incidents such as, for example, attacks on friendly aircraft. He is, in particular, to ensure that:
    a) All pilots, controllers, anti-aircraft commanders, commanding officers of ships in Canadian harbours and other personnel concerned are fully cognizant of the provisions of this directive and of the amplifying instructions to which he is to issue, and of the limitations of the authority to delegate to them.
    b) All RCAF personnel concerned are fully proficient in aircraft recognition and are in particular, able to recognize instantly civil and military aircraft of Canadian, British, United States, Western European and Russian origin. Although not responsible for the recognition training in the other services the Air Defence Commander should be informed of the standard of recognition attained in the other services.

Appendix "A"   Definitions

Unidentified Aircraft

1. Any Aircraft which penetrates, or approaches with the obvious intention of penetrating, a Canadian Air Defence Identification Zone, and which cannot be identified by positive means, is to be regarded as "unidentified." For the purpose of these instructions this term will include guided missiles and gliders.

Hostile Act

2. An unidentified or unrecognized aircraft shall be considered as committing a hostile act if, *without previous notification being received*, it carries out any of the following actions:
    a) An attack on friendly personnel, ground targets, ships or aircraft with bombs, rockets or other weapons.
    b) The opening of bomb-bay doors when the aircraft is approaching a vital area.
    c) The dropping of parachutists greater in number than the crew of a similar type of friendly aircraft.
    d) Mine-laying operations.

3. This list is not to be regarded as comprehensive and may be modified as circumstances warrant.

Suspicious Act

4. Behaviour of an unidentified aircraft which might be regarded as a *suspicious act* includes:
    a) The dropping of flares at night over friendly territory or over shipping by an unidentified aircraft without prior notification being received.
    b) Diving or low flying over friendly personnel, vital areas or ships.
    c) Taking evasive action or other action designed to avoid detection ...

Appendix "C" Rules of Engagement of Fighter Aircraft

Procedure for Interception and Recognition

1. The following regulations are to apply to the interception and recognition of unidentified aircraft within a Canadian Air Defence Identification Zone:

    a) The intercepting aircraft is to fly no closer to the intercepted aircraft than is necessary to recognize it positively. Close investigation or interrogation is to be performed solely on unidentified aircraft for the purpose of obtaining electronic or visual identification. If close surveillance is required the approach is to be cautious. Dangerous or reckless flying for the purpose of obtaining recognition is not to be tolerated.

    b) When more than one fighter aircraft is employed on the interception only one pilot is to be responsible for effecting recognition. The remaining aircraft are to remain in a position from which surveillance of the intercepted aircraft can be maintained.

    c) The interceptor pilot is to keep the GCI controller informed of marginal conditions of visibility. No attempt is to be made at visual identification when the visibility at the height of interception is less than one-half mile except when under GCI control and when specifically ordered by the Air Defence Commander or a delegated officer.

    d) The interceptor pilot is to attempt, by observation, to determine the type, model, name and number, ownership and registration or serial number of the intercepted aircraft. As much information as possible is to be reported by R/T [Radio Telephone] to the GCI controller in the following form:

    > Call sign of GCI Controller
    > From
    > Call sign of aircraft.

    Visual report (number of aircraft, type, model, ownership, name or number, serial or registration number, position, course and altitude, and distance).

    e) If the aircraft cannot be identified positively as friendly the interceptor aircraft is to maintain surveillance and continue reporting to the GCI controller until the controller gives further instructions.

    f) If the GCI controller received information from an interceptor pilot that the intercepted aircraft cannot be recognized, this information is to be passed to the Air Defence Commander immediately for a decision concerning further action unless the Air Defence Commander has delegated authority for ordering further action to be carried out.

g) An interceptor pilot, while maintaining surveillance of an unidentified aircraft, is to take positive offensive action against the aircraft without further instructions *ONLY* if the intercepted aircraft commits a hostile act as defined in Appendix 'A.'

2. If an interception of an aircraft which cannot be identified as friendly takes place outside an Air Defence Identification Zone, the pilot of the interceptor is to notify a GCI or an Air Traffic Control Centre through a DOT [Department of Transport] ground station. As much information as possible should be provided including the position, course and altitude of the intercepted aircraft. This information is to be passed by the ground station to the nearest ADCC and a check on the flight plan data is to be initiated in order that the aircraft can be identified if possible. The interceptor pilot is to maintain surveillance of the unidentified aircraft where practicable pending instructions from the ADCC through the available communications facilities.

Source: RCAF Directive, "Authority to Intercept and Engage Hostile Aircraft," 22 November 1951, Cabinet Defence Committee Fonds, DHH 2002/03, Series I, File 59.

Appendix 4   PJBD Recommendation 53/1, 1 October 1953

(This Recommendation supersedes Recommendation 51/4)

Aircraft controlled by the Air Defence System of the United States, or of Canada, engaged in intercepting unidentified aircraft during peacetime, shall be permitted to fly over the territory of either country as may be required to carry out effective interception. These flights will be carried out under the following provisions:

a. Investigations of unidentified aircraft by United States military aircraft over Canadian territory will only occur when it is not possible for a Canadian military aircraft to carry out the investigation; the activities of Canadian military aircraft over United States territory will be similarly restricted. For the purpose of this agreement, an unidentified aircraft is an aircraft which flies within the air defence identification zone in apparent violation of the rules for operation within such zone. When the pattern of behaviour of an aircraft is sufficiently suspicious to justify a belief that it has hostile intentions, it may also be considered to be an unidentified aircraft.
b. In accordance with published civil and military regulations, investigating aircraft will not approach closer than is necessary to establish identification. Investigation or interrogation will be performed solely on unidentified aircraft for the purpose of obtaining electronic or visual identification.
c. The Rules of Interception and Engagement of the country over which the interception or engagement takes place are to apply, even though the intercepting aircraft is being controlled from the other country.
d. The engagement of an aircraft is to be carried out only on orders issued by the Air Defence Commander of the country over which the engagement is to take place, or by an officer who has been delegated the requisite powers. The authority to issue orders to engage an unidentified aircraft should, to the greatest extent possible, be retained by the Air Defence Commander. However, when circumstances so necessitate, he may delegate such authority to a qualified officer not less in status than the senior officer in an Air Defence Control Center.
e. Translation of the general principles of this arrangement into coordinated operational instructions will be carried out by the Air Defence Commanders concerned.
f. This arrangement will remain in force until modified by mutual agreement, or until terminated by either Government.

Source: PJBD Recommendations, DHH 79/35.

Appendix 5 Canada–U.S. Press Release on the Establishment of NORAD, 1 August 1957

The Secretary of Defense of the United States, Honorable Charles E. Wilson, and the Minister of National Defence of Canada, the Honourable George R. Pearkes, announced today that a further step has been taken in the integration of the air defence forces of Canada and the United States. The two governments have agreed to the setting up of a system of integrated operational control of the air defence forces in the Continental United States, Alaska and Canada under an integrated command responsible to the Chiefs of Staff of both countries. An integrated headquarters will be set up in Colorado Springs and joint plans and procedures will be worked out in peacetime, ready for immediate use in case of emergency. Other aspects of command and administration will remain the national responsibility. This system of integrated operational control and the setting up of a joint headquarters will become effective at an early date. This bilateral arrangement extends the mutual security objectives of the North Atlantic Treaty Organization to the air defences of the Canada–United States Region.

Source: Press Release by the Secretary of Defense of the United States and the Minister of National Defence of Canada, 1 August 1957, Raymont Collection, DHH 73/1223/84.

Notes

Introduction

1 Richard Gimblett, "The Canadian Way of War: Experience and Principles," in *Canadian Expeditionary Air Forces*, ed. Allan D. English, Bison Paper no. 5 (Winnipeg: Centre for Defence and Security Studies, University of Manitoba, 2002), 9–20.
2 Douglas L. Bland, *The Military Committee of the North Atlantic Alliance: A Study of Structure and Strategy* (New York: Praeger, 1991), 45.
3 Quoted in C.P. Stacey, *Canada and the Age of Conflict*, vol. 2 (Toronto: University of Toronto Press, 1984), 349.
4 See discussion of terminology below; see also Chapter 1 for formal military definitions of command and control and for a discussion of theories of command and control.
5 Canada, Department of National Defence, *Strong, Secure, Engaged: Canada's Defence Policy* (Ottawa: Minister of National Defence, 2017), 14. The policy statement also indicates that Canada will "work with the United States to ensure that NORAD is modernized to meet existing and future challenges."
6 Michael Byers, "Canadian Armed Forces under US Command," report commissioned by the Simons Centre for Peace and Disarmament Studies, Liu Centre for the Study of Global Issues, University of British Columbia, May 6, 2002. Parts of the report were also published in Byers, "Canadian Armed Forces under United States Command," *International Journal* 58, 1 (Winter 2002–3): 92–93, 89–114.
7 Byers, "Canadian Armed Forces under US Command," 91.
8 Allan English, *Command and Control of Canadian Aerospace Forces: Conceptual Foundations* (Ottawa: Minister of National Defence, 2008), 37. On Canadian military culture, see Allan English, *Understanding Military Culture: A Canadian Perspective* (Montreal and Kingston: McGill-Queen's University Press, 2004).
9 This custom was consistent with Allied command and control practice during the Second World War: "The basic principle of Allied cooperation," C.P. Stacey has noted, "was that 'administration and discipline' should remain under national control, being quite distinct from operational command." Stacey, *Arms, Men and Governments: The War Policies of Canada, 1939–1945* (Ottawa: Queen's Printer, 1970), 354. See also Allan English and Colonel John Westrop (Ret.), *Canadian Air Force Leadership and Command: The Human Dimension of Expeditionary Air Force Operations* (Trenton: Canadian Forces Aerospace Warfare Centre, 2007), 8.
10 Bland, *The Military Committee*, 51–52; Directorate of History and Heritage, Department of National Defence, Ottawa (DHH) 87/47, Colonel R.L. Raymont, *The Evolution of the Structure of the Department of National Defence 1945–1968*, Report of the Task Force on Review of Unification of the Canadian Armed Forces (November 30, 1979), 24, and Appendix A, 21; Gen. Charles Foulkes, "The Complications of Continental Defence," in *Neighbors Taken for Granted: Canada and the United States*, ed. Livingston T. Merchant (Toronto: Burns and MacEachern, 1966), 108; Major-General Daniel Gosselin, "Canada's Participation in the Wars of the Early 20th Century: Planting the Seeds of Military Autonomy and National Command," *Canadian Military Journal* 7, 2 (Summer 2006): 65;

Richard Evan Goette, "The Struggle for a Joint Command and Control System in the Northwest Atlantic Theatre of Operations: A Study of the RCAF and RCN Trade Defence Efforts during the Battle of the Atlantic," MA thesis, Queen's University, 2002, 49–50, 121–24; Stacey, *Arms, Men, and Governments*, 354; Richard H. Gimblett, "Command in the Canadian Navy: An Historical Survey," *Northern Mariner* 14, 4 (October 2004): 45–46. The Royal Canadian Air Force's modern definition of "Full Command" is also along similar lines as "national command": Full Command is "the military authority and responsibility of a commander to issue orders to subordinates. It covers every aspect of military operations and administration and exists only within national services. It applies to all levels from the Chief of the Defence Staff down to the unit commander. Since it is applicable to national service only, alliance or coalition commanders cannot have full command over forces of other nations." RCAF, *Canadian Forces Aerospace Command Doctrine*, B-GA-0401-000/FP-001, March 2012 (Trenton: Canadian Forces Aerospace Warfare Centre, 2012), 6–7.

11 Stacey, *Arms, Men, and Governments*, 67, 120–1; Philippe Lagassé, "The Crown's Powers of Command-in-Chief: Interpreting Section 15 of Canada's *Constitution Act, 1867*," *Review of Constitutional Studies* 18, 2 (2013): 210–16; Canada, Department of National Defence, *National Defence Act* (1950), ss. 4, 19, 20; Douglas Bland, *The Administration of Defence Policy in Canada, 1947–1985* (Kingston: Ronald P. Frye, 1987), 4; Peter Haydon, *The 1962 Cuban Missile Crisis: Canadian Involvement Reconsidered* (Toronto: Canadian Institute of Strategic Studies, 1993), 88–89, 93, 96; Gosselin, "Canada's Participation," 65–76.

12 DHH 87/47, Raymont, *The Evolution of the Structure*, Appendix A, 21; Canada, DND, *NDA* (1950), ss. 4, 19, and 20; Haydon, *The Cuban Missile Crisis Revisited*, 88–89, 96. Operational command focuses on conducting operational-level activities such as campaign planning and the movement of units to accomplish a mission. RCAF, *Canadian Forces Aerospace Command Doctrine*, B-GA-0401-000/FP-001, 7

13 Assistant Chief of the General Staff (ACGS), Memorandum, "Note on a Meeting of Permanent Joint Board on Defence Held at New York, 19–20 Dec 41," to Chief of the General Staff (CGS), December 22, 1941, DHH 314.009 (D116).

14 Ibid.

15 For examples, see Stacey, *Arms, Men, and Governments*, 116, 134, 389–90 (regarding Lt.-Gen. Ken Stuart); Lt. (N) Richard Oliver Mayne, "Keeping Up with the Jones's: Admiralship, Culture, and Careerism in the Royal Canadian Navy, 1911–1946," research paper for the Canadian Forces Leadership Institute, September 2002, 33–34; and Lt.-Cdr. Richard Oliver Mayne, "Vice-Admiral George C. Jones: The Political Career of a Naval Officer," in *The Admirals: Canada's Senior Naval Leadership in the Twentieth Century*, ed. Michael Whitby, Richard H. Gimblett, and Peter Haydon (Toronto: Dundurn Press, 2006), 125–55 (regarding VAdm G.C. Jones).

16 English and Westrop, *Canadian Air Force Leadership and Command*, 23, 28. The best examples include the RCAF's experiences in Europe during the Second World War and as part of the North Atlantic Treaty Organization (NATO) during the Cold War.

17 Joseph Jockel, "Canada in NORAD, 1957–2007: A History," presentation given as part of the Queen's Centre for International Relations [hereafter QCIR] National Security Seminar Series, January 24, 2007; Jockel, *Canada in NORAD 1957–2007: A History* (Montreal and Kingston: McGill-Queen's University Press in association with the Queen's Centre for International Relations and the Queen's Defence Management Program, 2007), 3–4. See also Haydon, *The 1962 Cuban Missile Crisis*, 77, 91–92; Stéphane Roussel, *The North American Democratic Peace: Absence of War and Security Institution-Building in Canada–US Relations, 1867–1958* (Montreal and Kingston: McGill-Queen's University Press,

2004), 211–12. On civil/military relations, see the classics by Samuel Huntington and Morris Janowitz: Samuel P. Huntington, *The Soldier and the State: The Theory and Politics of Civil–Military Relations* (Cambridge, MA: Belknap Press of Harvard University, 1957); and Morris Janowitz, *The Professional Soldier: A Social and Political Portrait* (New York: Free Press, 1960). For a civil/military relations and professionalization perspective on the RCAF, see Rachel Lea Heide, "The Creation of a Professional Canadian Air Force, 1916–1946," PhD diss., History, Carleton University, 2010.

18 Adam Chapnick, "Principle for Profit: The Functional Principle and the Development of Canadian Foreign Policy, 1943–1947," *Journal of Canadian Studies* 37, 2 (Summer 2002): 68.

19 Joel Sokolsky, "A Seat at the Table: Canada and Its Alliances," *Armed Forces and Society* 16, 1 (Fall 1999): 21–22.

20 W.A.B. Douglas, *The Creation of a National Air Force: The Official History of the Royal Canadian Air Force*, vol. 2 (Toronto: University of Toronto Press and Department of National Defence, 1986), 390.

21 Nils Ørvik, "Defence against Help – A Strategy for Small States?," *Survival* 15, 5 (September–October 1973): 228; Ørvik, "The Basic Issue in Canadian National Security: Defence against Help / Defence to Help Others," *Canadian Defence Quarterly* 11, 1 (Summer 1981): 8–15. Quote from former.

22 Maj. Louis Grimshaw, "On Guard: A Perspective on the Roles and Functions of the Army in Canada," MA thesis, War Studies, Royal Military College of Canada, 1989, 26–27, 47, 59–60; P. Whitney Lackenbauer, "Right and Honourable: Mackenzie King, Canadian–American Bilateral Relations, and Canadian Sovereignty in the Northwest, 1943–48," in *Mackenzie King: Citizenship and Community: Essays Marking the 125th Anniversary of the Birth of William Lyon Mackenzie King*, ed. John English, Kenneth McLaughlin, and P. Whitney Lackenbauer (Toronto: Robin Brass Studio, 2002), 161; Ørvik, "The Basic Issue in Canadian National Security"; Donald Barry and Duane Bratt, "Defense against Help: Explaining Canada–U.S. Security Relations," *American Review of Canadian Studies* 38, 1 (Spring 2008): 63–89; P. Whitney Lackenbauer, "From 'Defence against Help' to 'A Piece of the Action': The Canadian Sovereignty and Security Paradox Revisited," University of Calgary Centre for Military and Strategic Studies, Occasional Paper no. 1, May 2000, 1–3, 13; Philippe Lagassé, "Nils Ørvik's 'Defence against Help': The Descriptive Appeal of a Prescriptive Strategy," *International Journal* 65, 2 (Spring 2010): 463–74.

23 Maj.-Gen. Maurice Pope to Col. J.H. Jenkins, April 4, 1944, Library and Archives Canada (LAC), Record Group (RG) 25, vol. 5749, file 52C(s), pt. 1.

24 Grimshaw, "On Guard," 27, 62. See also Lackenbauer, "From 'Defence against Help' to 'A Piece of the Action,'" 4–5.

25 In the context of the early Cold War period, James Fergusson has described the Canadian defence dilemma as Canada having three basic choices: "It could ally with the Soviet Union, which, of course, made no political or moral sense at all. It could opt for neutrality, knowing full well that in the case of war both the United States and the Soviet Union would violate its neutrality unless Canada developed a large, sophisticated, and very expensive independent air defence capability to deter both. Finally, it could ally with the United States and cooperate in the air defence of North America against the Soviet Union." Fergusson, *Canada and Ballistic Missile Defence*, 11.

26 See the following for examples of the negative "satellite" and "protectorate" perspectives on Canadian sovereignty: Donald Creighton, *The Forked Road: Canada 1939–1957* (Toronto: McClelland and Stewart, 1976); John W. Warnock, *Partner to Behemoth: The Military Policy of a Satellite* (Toronto: New Press, 1970); Shelagh Grant, *Sovereignty or*

Security? Government Policy in the Canadian North, 1936–1950 (Vancouver: UBC Press, 1988); Anne Denholm-Crosby, Dilemmas in Defence Decision Making: Constructing Canada's Role in NORAD 1958–1996 (New York: St Martin's Press, 1998); James M. Minifie, "Peacemaker or Powdermonky: Canada's Role in a Revolutionary World (Winnipeg: Universal Printers, 1960); Jon B. McLin, Canada's Changing Defense Policy, 1957–1963: The Problems of a Middle Power in Alliance (Baltimore: Johns Hopkins University Press in cooperation with the Washington Center of Foreign Policy Research, School of Advanced International Studies, Johns Hopkins University, 1967).

27 These are the words of New York Mayor Fiorello La Guardia, who during most of the Second World War was the American chairman of the Canada–US Permanent Joint Board on Defence (PJBD). LaGuardia to President Franklin Roosevelt, May 28, 1942, Franklin D. Roosevelt Papers, Franklin D. Roosevelt Library, Hyde Park, New York, Official File 4090, copy in C.P. Stacey Papers, Accession no. B90-0020, box 30, file "FDR Library," University of Toronto Archives, Toronto.

28 On coercion and sovereignty, see Eric James Lerhe, At What Cost Sovereignty? Canada–US Military Interoperability in the War on Terror (Halifax: Centre for Foreign Policy Studies, Dalhousie University, 2013), 84; and Stephen D. Krasner, Sovereignty: Organized Hypocrisy (Princeton: Princeton University Press, 1999), 26–27, 36–37.

29 On balancing sovereignty and security concerns and the collaborative approach to Canada–US relations in the Arctic, see Ken Coates, P. Whitney Lackenbauer, William R. Morrison, and Greg Poelzer, Arctic Front: Defending Canada in the Far North (Toronto: Thomas Allan, 2008); Lehre, At What Cost Sovereignty, 33, 100.

30 Lackenbauer, "From 'Defence against Help' to 'a Piece of the Action,'" 5.

31 Air Commodore (A/C) W.I. Clements, RCAF Member, Joint Planning Committee, to Secretary, Joint Planning Committee, May 26, 1954, LAC, RG 24, vol. 21422, file CSC 1855:8, copy obtained through DND ATIP, A-2007-00234. Emphasis in original.

32 See, for example, Group Capt. M. Lipton, "The Wisdom of Our Air Defence Policy," R.C.A.F. Staff College Journal 1 (1956): 28–32. Canadian military officers were active in the political realm in carrying out their professional role as advisers to government; as Rachel Heide put it, military officers "should attempt to influence the political process by providing advice about what is in the best interest of the military in order to fully protect the nation." Heide, "The Creation of a Professional Canadian Air Force," 26.

33 Lackenbauer, "Right and Honourable," 153–54.

34 Martha Maurer, Coalition Command and Control: Key Considerations (Washington: National Defence University and Harvard University Program on Information Resources Policy, 1994), 10; Canadian Forces Operations, Chapter 8, 801, 8–1. Quote from former. On the development of combined command organizations, see Sean M. Maloney, Securing Command of the Sea: NATO Naval Planning, 1948–1954 (Annapolis: Naval Institute Press, 1995).

35 Maurer, Coalition Command and Control, 10. This is echoed by Canadian scholars Douglas Bland and Daniel Gosselin: Bland, The Administration of Defence Policy, 18; Gosselin, "Canada's Participation," 68.

36 Maurer, Coalition Command and Control, 107. It could be argued, however, that more recent operations in which NATO has been involved have led to a larger number of disagreements and restrictions that have limited efficient military operations. Recent conflicts in Kosovo, Afghanistan, Libya, and Iraq/Syria bear this out.

37 Ibid., 10. The cultural differences of other government departments (OGD) and non-governmental organizations (NGO) in today's Comprehensive Approach and Joint, Interagency, Multinational, and Public (JIMP) are efforts even more complex. See Peter Gizewski and Lt.-Col. Michael Rostek, "Towards a JIMP-Capable Land Force," Canadian

Army Journal 10, 1 (March 2007): 55–72; Heather Hrychuk, "Combating the Security Development Nexus? Lessons Learned from Afghanistan," *International Journal* 64 (Summer 2009): 825–42; Michael H. Thomson, Barbara D. Adams, Courtney D. Hall, Andrea L. Brown, and Craig Flear, *Collaboration within the JIMP (Joint, Interagency, Multinational, Public) Environment* (Toronto: Defence Research and Development Canada, August 2010); Michael H. Thomson, Barbara D. Adams, Courtney D. Hall, Andrea L. Brown, and Craig Flear, *Collaboration between the Canadian Forces and the Public in Operations* (Toronto: Defence Research and Development Canada, March 2011).

38 Paul Mitchell, *Network Centric Warfare and Coalition Operations: The New Military Operating System* (New York: Routledge Global Security Studies, 2009), 76–77; Maj. Colin Marks, "Taking Off in NORAD's Bi-National Flying Fortress: Is Canada the Co-Pilot, Navigator, or Passenger?," Masters of Defence Studies Directed Research Project, Canadian Forces College, 2016, 14; Fergusson, *Canada and Ballistic Missile Defence*, 14. Former RCAF officer Lt.-Gen. Angus Watt, who served as the NORAD Deputy Chief of Staff for Operations (and later became Commander of the RCAF), described NORAD as a binational command organization in 2005 in the following way: "It's the Canadians and Americans together, doing the job *together*, and *together* providing information to Canada and the US at the same time. So it's *totally* integrated, seamlessly and totally integrated and authority is delegated to Canadians which you don't see the Americans doing anywhere else." Quoted in Mitchell, *Network Centric Warfare and Coalition Operations*, 77. Emphasis in original. I also thank Adam Chapnick for his insights.

39 The example of the RCAF during the Second World War is illustrative. In contrast to the situation of the RCAF Overseas, the RCAF's Home War Establishment fielded a "national air force." This is one of the main reasons Volume 2 of the RCAF Official History, which addresses the RCAF at home during the interwar period and the Second World War, is titled *The Creation of a National Air Force*. Douglas, *Creation of a National Air Force*, ix; English and Westrop, *Canadian Air Force Leadership and Command*, 22–23, 28; Gosselin, "Canada's Participation in the Wars of the Early 20th Century," 68.

40 Ibid., 69; Stacey, *Canada and the Age of Conflict*, vol. 2, 289.

41 Eliot A. Cohen, *Supreme Command: Soldiers, Statesmen, and Leadership in Wartime* (New York: Anchor Books, 2003), 242. Here Cohen is capturing Huntington's ideas on corporateness. Huntington, *The Soldier and the State*, 10, 16–18.

42 Canada, Department of National Defence, *Duty with Honour: The Profession of Arms in Canada* (Kingston: Canadian Forces Leadership Institute, 2009), 17, 20–21, 55. On the distinctive culture of Canada's air force, navy, and army, see English, *Understanding Military Culture*, 89–97; and Alan Okros, "Leadership in the Canadian Military Context," CFLI monograph 2010-1, Canadian Forces Leadership Institute, Canadian Defence Academy, Kingston, November 2010, 27–32.

43 Huntington, *The Soldier and the State*, 13.

44 Joel J. Sokolsky, "Exporting the 'Gap': The American Influence," in *The Soldier and the State in the Post Cold War Era*, ed. Albert Legault and Joel Sokolsky (Kingston: Queen's Quarterly Press, 2002), 213; Lehre, *At What Cost Sovereignty?*, 37; Matthew Paul Trudgen, "The Search for Continental Security: The Development of the North American Air Defence System, 1949 to 1956," PhD diss., History, Queen's University, 2011, 14. Quote from Sokolsky.

45 This was an important stewarding-the-profession role that senior Canadian air force officers played as RCAF institutional leaders. On the role of institutional leaders stewarding the profession, see Canada, DND, *Duty with Honour*, 14, 19, 56.

46 English, *Understanding Military Culture*, 95, 121; Jack Granatstein, "The American Influence on the Canadian Military, 1939–1963," *Canadian Military History* 2, 1 (1993):

69–70; Richard Goette, "A Snapshot of Early Cold War RCAF Writing on Canadian Air Power and Doctrine," *Royal Canadian Air Force Journal* 1, 1 (Winter 2012): 53; Andrew Richter, *Avoiding Armageddon: Canadian Military Strategy and Nuclear Weapons, 1950–63* (Vancouver: UBC Press, 2002), 57. Jockel has written that "as airmen," RCAF and USAF officers shared an outlook that "created a similar identity and even an emotional bond." Joseph Jockel, *No Boundaries Upstairs: Canada, the United States and the Origins of North American Air Defence, 1945–1958* (Vancouver: UBC Press, 1987), 56. See also Bruce Barnes, "'Fighters First': The Transition of the Royal Canadian Air Force, 1945–1952," MA thesis, War Studies, Royal Military College of Canada, 2006.
47 Fergusson, *Canada and Ballistic Missile Defence*, 11.
48 See Randall Wakelam, *Cold War Fighters: Canadian Aircraft Procurement, 1945–54* (Vancouver: UBC Press, 2011); and Jockel, *No Boundaries Upstairs*.
49 Lehre, *At What Cost Sovereignty?*, 28; Agreement on the Air Standardization Coordination Committee (Washington, DC, February 1948), Appendix, para. 1; Lt.-Col. Christopher England, "Air and Pace Interoperability Council (ASIC) and the RCAF," Masters of Defence Studies Directed Research Project, Canadian Forces College Toronto, 2016, 1–2, 10–16. Significantly, the first USAF member of the ASCC was Maj.-Gen. Earle Partridge, who would later become the first commander-in-chief of NORAD. Partridge thus had an early strong association with and connection to the RCAF. I thank Lt.-Col. Chris England for bringing this fact to my attention.
50 James Eayrs, *In Defence of Canada*, vol. 3: *Peacemaking and Deterrence* (Toronto: University of Toronto Press, 1972), 122; Jockel, *No Boundaries Upstairs*, 56, 93; Leonard V. Johnson, *A General for Peace* (Toronto: James Lorimer, 1987), 33; Ray Stouffer, *Swords, Clunks, and Widowmakers: The Tumultuous Life of the RCAF's Original 1 Canadian Air Division* (Trenton: Canadian Forces Aerospace Warfare Centre, 2015), 166; Matthew Trudgen, "Good Partners or Just Brass Intrigue: The Transnational Relationship between the USAF and RCAF with Respect to the North American Air Defence System, 1947–1960," in *Sic Itur Ad Astra: Canadian Aerospace Power Studies*, vol. 1: *Historical Aspects of Canadian Air Power Leadership*, ed. William March (Ottawa: Minister of National Defence, 2009), 76; C.P. Stacey, *A Date with History: Memoirs of a Canadian Historian* (Ottawa: Deneau Publishers, 1983), 258. "Colonial" quote from latter.
51 Trudgen, "Good Partners or Just Brass Intrigue," 76–77; Richard Goette, "Air Defence Leadership during the RCAF's 'Golden Years,'" in *Sic Itur Ad Astra*, vol. 1, ed. March, 58–59.
52 It is thus hoped that this book's "human element" emphasis on effective working relationships between individuals serves as a useful example of how to minimize joint and inter-allied friction in order to maximize unity of effort in modern security and defence endeavours. On the importance of this, see the observations by Britain's Lt.-Gen. Sir Michael Jackson: Mike Jackson, "The Realities of Multi-national Command: An Informal Commentary," in *The Challenges of High Command: The British Experience*, ed. Gary Sheffield and Geoffrey Till (Camberley: Strategic and Combat Studies Institute, 1999), 139–45.
53 Stouffer, *Swords, Clunks, and Widowmakers*, 3.
54 Eayrs, *In Defence of Canada*, vol. 3, 58.
55 Goette, "Air Defence Leadership during the RCAF's 'Golden Years,'" 51–64; Stouffer, *Swords, Clunks, and Widowmakers*, 166; Granatstein, "The American Influence on the Canadian Military," 70.
56 Stouffer, *Swords, Clunks and Widowmakers*, 3. On the tendency of the Canadian services to focus on their alliance partners, see Joel Sokolsky, "A Seat at the Table: Canada and Its

Alliances," *Armed Forces and Society* 16, 1 (Fall 1989): 11–35; see also Hugues Canuel, "Canadian Civil-Military Relations in the Early 'Command Era,' 1945–1955: Forging a Normative Prescription through Rational Analysis," *Canadian Military History* 24, 2 (Spring 2015): 120–21; and English, *Understanding Military Culture*, 122.
57 Gosselin, "Canada's Participation," 74–75; Bland, *The Administration of Defence Policy in Canada*, 6–7. La Guardia's "pride and little brother" description of Canadian resoluteness above, page 9, also comes to mind. On "nationalistic pride" and nationalist motivations on the part of Canadian politicians and DEA officials in continental defence relations with the United States, see for example, Grant, *Sovereignty or Security*, xvi; and Trudgen, "The Search for Continental Security," 7–13.
58 Gen. Charles Foulkes, "The Complications of Continental Defence," in *Neighbors Taken for Granted: Canada and the United States*, ed. Livingston T. Merchant (Toronto: Burns and MacEachern, 1966), 107. I thank an anonymous reviewer for contributing to this line of thought.
59 Goette, "Early Cold War RCAF Writing," 52–53; Richter, *Avoiding Armageddon*, 16–18, 42; Stouffer, *Clunks, Swords, and Widowmakers*, 3–4; Jockel, *No Boundaries Upstairs*, 7–8.
60 Barnes, "Fighters First"; Trudgen, "The Search for Continental Security," 15.
61 Eayrs, *In Defence of Canada*, vol. 3, 122.
62 James Jackson to author, December 19, 2015. Jackson illustrated this point with an amusing anecdote:

> "Keith Hodson had picked up a copy of the USAF's doctrine manual on one of our trips to Alabama [i.e., location of the USAF Air University], and he dropped by my office one day keen on our producing the same thing and asking my opinion. I found it an impenetrable mess of baffledegab [sic]. I recall one section that read something like 'the purpose of manned strategic airborne and/or ballistic systems is the transportation of appropriate weaponry under tactically optimum conditions to pre-selected or opportune targets.' I sent back a translation, 'Bombers are meant to bomb.' I heard nothing more of the doctrine manual."

63 Canada, Department of National Defence, *Canadian Forces Joint Publication CFJP 3.0 – Operations* B-GJ-005-300/FP-001, September 2011 (Ottawa: Canadian Forces Warfare Centre, 2011), Chapter 1, 0107–0110, 1–2.
64 G.E. (Joe) Sharpe, Brig.-Gen. (Ret.), and Allan D. English, *Principles for Change in the Post–Cold War Command and Control of the Canadian Forces* (Winnipeg: Canadian Forces Training Material Production Centre, 2002), 33.
65 *Canadian Forces Joint Publication 3.0 Operations* B-GJ-005-300/FP-001, Chapter 1, 0101, 1–2, Chapter 7, 0701, 7–13; English, *Conceptual Foundations*, x. British and American military officials finalized these definitions of "joint" and "combined" at the Arcadia Conference in January 1942, though personnel continued to utilize the terms interchangeably throughout the war. Louis Morton, *Strategy and Command: The First Two Years: The War in the Pacific, United States Army in World War II* (Washington, DC: Office of the Chief of Military History, Department of the Army, 1962), 164.

Chapter 1: Command and Control, Sovereignty, Civil-Military Relations, and the Profession of Arms

1 In the literature, command and control is often abbreviated to C2 or C^2.
2 Canada, DND, *Duty with Honour*, 4, 27; Ross Pigeau and Carol McCann, "Reconceptualizing Command and Control," lecture delivered for the Joint Command and

Staff Program, Canadian Forces College, 29 November 2013. With permission. On unlimited liability, see Richard A. Gabriel, *The Warrior's Way – A Treatise on Military Ethics* (Kingston: Canadian Defence Academy Press, 2007), 75–76.
3 Canada, DND, *Duty with Honour*, 52–53; Cohen, *Supreme Command*, 242; Huntington, *The Soldier and the State*, 11.
4 Command and control theorist Ross Pigeau has also described this as "bureaucratic" command and control. Ross Pigeau, e-mail to author, February 14, 2016.
5 Gosselin notes that authority "consists of the formal power that a person has because of the position in the organization," and orders that come from a commander "in an authoritative position are followed because persons in higher positions have legal authority over subordinates in lower positions." Maj.-Gen. (Ret.) Daniel Gosselin, "Introduction to Concepts of Strategic Command and Civil-Military Relations," lecture to National Security Program 7, Canadian Forces College, March 11, 2015. With permission.
6 *Canadian Forces Joint Publication 3.0 Operations* B-GJ-005-300/FP-001, Chapter 3, 0304, 3-1; Royal Canadian Air Force, *Canadian Forces Aerospace Command Doctrine*, B-GA-0401-000/FP-001, March 2012 (Trenton: Canadian Forces Aerospace Warfare Centre, 2012), 4.
7 C. Kenneth Allard, *Command, Control, and the Common Defense* (New Haven: Yale University Press, 1992), 16–17.
8 *Canadian Forces Aerospace Command Doctrine*, B-GA-0401-000/FP-001, 4.
9 Pigeau and McCann, "Re-conceptualizing Command and Control."
10 *Canadian Forces Aerospace Command Doctrine*, B-GA-0401-000/FP-001, 4. Note that command and control are both verbs and nouns – they are both actions and something identifiable (i.e., in doctrine).
11 Thomas P. Coakley, *Command and Control for War and Peace* (Washington, DC: National Defense University Press, 1992); Maj. Alex Day, "The Right Tool for the Job? Using the McCann and Pigeau Framework to Unify Canadian Forces Command and Control Doctrine," Masters of Defence Studies Directed Research Project, Canadian Forces College, May 2007, 8. I thank Ross Pigeau for exposing me to Coakley's theories.
12 Coakley, *Command and Control for War and Peace*, 8–9, 41–42.
13 Ibid., 27–29. Some have even gone so far as to call this "command and control warfare." See Allard, *Command, Control, and the Common Defense*, 5.
14 Coakley, *Command and Control for War and Peace*, 32–33; Gary Sheffield, "The Challenges of High Command in the Twentieth Century," in *The Challenges of High Command*, ed. Sheffield and Till, 5; Carl Builder, Steven Bankes, and Richard Nordin, *Command Concepts* (Santa Monica: RAND National Defense Research Institute, 1999), 6.
15 Coakley, *Command and Control for War and Peace*, 33–34; Maurer, *Coalition Command and Control*, 24; Sheffield, "The Challenges of High Command," 6. On Boyd's specific OODA loop theory, see Phillip S. Meilinger, "The Historiography of Airpower: Theory and Doctrine," *Journal of Military History* 64, 2 (April 2000): 495–96; and Lt.-Col. David K. Fadok, "John Boyd and John Warden: Airpower's Quest for Strategic Paralysis," in *The Paths of Heaven*, ed. Col. Phillip S. Meilinger (Maxwell Air Force Base, Alabama: Air University Press, 1997), 357–70.
16 Builder, Bankes, and Nordin, *Command Concepts*, iii, xi–xiv, 1–4, 11–14, 135–36; Ross Pigeau, "Human Perspectives on Strategic Command," lecture to National Security Program 6, Canadian Forces College, April 9, 2014. With permission. Builder, Bankes, and Nordin explain that it is necessary to "separate the art of command and control from the hardware and software systems that support C2."

17 Eliot A. Cohen and John Gooch, *Military Misfortunes: The Anatomy of Failure in War* (New York: Macmillan, 1990), 21. I thank Eric Ouellet and Ross Pigeau for their insights on Cohen and Gooch's organizational perspective on command and control.
18 Ibid., 7–8, 13–14, 21–28; Pigeau, "Human Perspectives on Strategic Command."
19 Pigeau and McCann, "Re-conceptualizing Command and Control"; Ross Pigeau and Carol McCann, "What Is a Commander?," in *Generalship and the Art of the Admiral: Perspectives on Canadian Senior Military Leadership*, ed. Bernd Horn and Stephen J. Harris (St Catharines: Vanwell, 2001), 79–104; Day, "The Right Tool for the Job," 3. Quote from latter. I thank Ross Pigeau for imparting his additional thoughts on the Pigeau–McCann theory to me over the past few years.
20 Government of Canada, "The Chain of Command," Chapter 4 from *Dishonoured Legacy: The Lessons of the Somalia Affair: Report of the Commission of Inquiry into the Deployment of the Canadian Forces to Somalia*, vol. 1 (Ottawa: Queen's Printer, June 30, 1997), 67.
21 Pigeau and McCann, "Re-conceptualizing Command and Control," 55; Pigeau, "Human Perspectives on Strategic Command."
22 Pigeau and McCann, "Re-Conceptualizing Command and Control," 56; Pigeau, "Human Perspectives on Strategic Command."
23 Pigeau and McCann, "Re-Conceptualizing Command and Control," 54–55; Pigeau, "Human Perspectives on Strategic Command."
24 Physical competency consists of a commander's physical abilities that are mandatory for any operational task, such as flying an aircraft. Intellectual competency consists of skills and abilities necessary for "planning missions, monitoring the situation, for reasoning, making inferences, visualizing the problem space, assessing risks and making judgements." Emotional competency consists of the skills of resilience, hardiness, and the ability to cope under stress (i.e., emotional "toughness"). Interpersonal competency consists of skills of interaction, trust, respect, and effective teamwork and requires "articulateness, empathy, perceptiveness and social understanding on the part of the individual in command." Pigeau and McCann, "What Is a Commander?," 84–85. Many of these competencies – especially intellectual – address the expertise attribute of the Canadian military profession of arms ethos. See the Introduction and also Canada, DND, *Duty with Honour*, 17–19.
25 Pigeau and McCann note that legal authority is significant for the military, going "well beyond [the legal authority assigned to] any other private or government or organization." This is so because the legal authority a military has allows it to "enforce obedience among its members" and "to place these members in harm's way if the operational needs of the mission demand it." Pigeau and McCann, "What Is a Commander?," 85.
26 Extrinsic responsibility "involves the obligation for public accountability" while intrinsic responsibility "is the degree of self-generated obligation that one feels towards the military mission." Pigeau and McCann, "What Is a Commander?," 86–87.
27 *Duty with Honour*, 21–22, 17–19, 42.
28 Krasner further identifies the "fundamental rules" of conventional sovereignty as "recognition of juridically independent territorial entities and non-intervention in the affairs of other states." Stephen D. Krasner, "Sharing Sovereignty: New Institutions for Collapsed and Failing States," in *Leashing the Dogs of War: Conflict Management in a Divided World*, ed. Chester A. Crocker, Fen Osler Hampson, and Pamela Aall (Washington, DC: United States Institute of Peace Press, 2007), 653. On the ongoing International Relations (IR) theory debate on the concept of sovereignty, see for instance, Mohammed Ayoob, "State Making, State Breaking, and State Failure," in ibid., 95–114; Stephen D. Krasner,

Sovereignty: Organized Hypocrisy (Princeton: Princeton University Press, 1999); Hent Kalmo and Quentin Skinner, *Sovereignty in Fragments: The Past, Present, and Future of a Contested Concept* (Cambridge: Cambridge University Press, 2010); Stephen D. Krasner, ed., *Problematic Sovereignty: Contested Rules and Political Possibilities* (New York: Columbia University Press, 2001); Kalevi Holsti, *The State, War, and the State of War* (Cambridge: Cambridge University Press, 1996), 82–122; Rosemary E. Shinko, "Sovereignty as a Problematic Conceptual Core," *The International Studies Encyclopedia*, vol. 11, ed. Robert A. Denemark (Malden: Blackwell, 2010), 6515–33; and Richard Little, "Sovereignty," in *Encyclopedia of International Relations and Global Politics*, ed. Martin Griffiths (New York: Routledge, 2005), 768–69, 775–76. I am indebted to Pierre Pahlavi and Chris Spearin for educating me about the IR discourse on sovereignty, and especially the work of Stephen Krasner.

29 This book therefore focuses on state sovereignty (the realist and liberal schools) instead of individual sovereignty (the constructivist or postmodern school). Nonetheless, it should be noted that individuals and groups in this study played a large part in the protection of Canadian sovereignty through the retention of command. Both individual military officers and collective professional military organizations/institutions (i.e., senior leaders within the RCAF), acting as agents and representatives of the state, exercised Canadian sovereignty on behalf of the state. The concept of the exercise of state sovereignty is elaborated later in this chapter.

30 Shinko, "Sovereignty as a Problematic Conceptual Core," 6516–19, 6521; Jack Donnelly, "Sovereignty," in *The Oxford Companion to International Relations*, ed. Joel Krieger (Oxford: Oxford University Press, 2014), 299, 302; "Sovereignty," in *The Dictionary of World Politics*, ed. Graham Evans and Jeffrey Newnham (Toronto: Simon and Schuster, 1990), 369–70; Little, "Sovereignty"; Shelagh D. Grant, *Polar Imperative: A History of Arctic Sovereignty in North America* (Toronto: Douglas and McIntyre, 2010), 10.

31 *Canadian Forces Aerospace Command Doctrine*, B-GA-0401-000/FP-001, 4. The obedience aspect of command is nicely captured in Douglas Bland's definition of "command authority," which he describes as "the legal right and requirement to issue orders and obey them." Bland, *The Administration of Defence Policy in Canada*, 4. This is echoed by Martha Maurer's emphasis on authority and the importance of an order actually being carried out: "The authority of the order exists only when the order is carried out by the forces." Maurer, *Coalition Command and Control*, 69. See also the Somalia Inquiry Report, which articulates command as "the legal authority to issue orders and to compel obedience." Government of Canada, "The Chain of Command," Chapter 4 of *Dishonoured Legacy*, vol. 1, 67.

32 Donnelly, "Sovereignty," 299.

33 *United States Constitution*, Article II, Section 2, http://context.montpelier.org/document/175?gclid=CIS7nqLq8s0CFQmRaQodmoMCPw.

34 Allard, *Command, Control, and the Common Defense*, 26–27, 58, 105; Ray S. Cline, *United States Army in World War II: The War Department, Washington Command Post: The Operations Division* (Washington: Office of the Chief of Military History, Department of the Army, 1951), 1–2. The President and the Secretary of Defense (or their duly deputized successors) are referred to in the United States as the "National Command Authority." After the Goldwater–Nichols Act of 1986, the JCS was formally removed from the US chain of command. Department of National Defense Directive 5100.30, "World-Wide Military Command and Control System (WWMCCS)," issued by Deputy Secretary of Defense David Packard, December 2, 1971, https://biotech.law.lsu.edu/blaw/dodd/corres/pdf/d510030wch1_120271/d510030p.pdf; Joint Publication 1, *Doctrine of the Armed Forces of the United States*, March 25, 2013 (Washington, DC: Joint Chief of Staff,

2013), II-9. The chain of command is also codified in US law. *10 U.S. Code 162*, "Combatant Commands: assigned forces; chain of command," Legal Information Institute, Cornell University Law School, https://www.law.cornell.edu/uscode/text/10/162.

35 Philippe Lagassé, "The Crown's Powers of Command-in-Chief: Interpreting Section 15 of Canada's *Constitution Act, 1867,*" *Review of Constitutional Studies* 18, 2 (2013): 210–11, 190; Lagassé, "Accountability for National Defence: Ministerial Responsibility, Military Command, and Parliamentary Oversight," *IRPP Study* 4, March 2010, 6–7. Quote from former.

36 Lagassé, "The Crown's Powers of Command-in-Chief," 210–11, 213, 220. In the time period studied in this book, there were two British monarchs who were the commander-in-chief of the Canadian military: King George V and his daughter Queen Elizabeth II.

37 Lagassé makes this distinction at several points in his article. Although the legislative power of the Canadian Parliament – and this could also be applied to the US Congress – can restrict or enable the capabilities of the Canadian military through regulations and especially by granting or withholding funding, parliamentary authority does not equate to legal authority of command over the military, which is solely vested in the executive. Lagassé, "The Crown's Powers of Command-in-Chief," 190–91.

38 Ibid., 191–93, 204–7, 213, 215; Lerhe, *At What Cost Sovereignty*, 3.

39 Gosselin, "Canada's Participation," 67. During the time period studied in this book, the chain of command consisted of the chiefs of the three Canadian services (the RCAF Chief of the Air Staff, the RCN Chief of the Naval Staff, and the Canadian Army Chief of the General Staff), who were members of the Chiefs of Staff Committee (which in 1951 included a chairman to coordinate, though not command or control, the three services) as indicated in the National Defence Act. Bland, *The Administration of Defence in Canada*, 4.

40 Pigeau and McCann, "What Is a Commander?," 86–87; Canada, DND, *Duty with Honour*, 14–16. See also the discussion of military professionalism later in this chapter. Gosselin additionally notes: "Because of the severe consequences of war, command authorities must leave no doubt as to which commander is responsible for which decisions and actions – or inaction – and to which superior officer a commander must account in the performance of his responsibilities." Gosselin, "Canada's Participation," 67.

41 Lerhe, *At What Cost Sovereignty*, 11, 83, 102. Lerhe's definitions of internal and external sovereignty are legal and are based on L. Oppenheim, *International Law: A Treatise*, 8th ed., vol. 1, ed. H. Lauterpacht (London: Longman, 1974), 286–89.

42 Donnelly, "Sovereignty," 301; Little, "Sovereignty," 775–76; Lerhe, *At What Cost Sovereignty*, 111n. Even then, there are limits placed on such yielding of sovereignty, as all international agreements, including alliances, have opt-out clauses, where states are free to leave the agreement. This too is a functional exercise of sovereignty. I thank an anonymous reviewer for highlighting this point.

43 This is well articulated by Daniel Gosselin: "to achieve a greater war effort and ensure unity of command, nations might delegate restricted sovereignty over military forces to a lead nation in a coalition." Gosselin, "Canada's Participation," 69.

44 Byers, *Canadian Armed Forces under U.S. Command*, 6–8. Eric Lerhe has devised an excellent methodology for determining sovereignty violations/losses and gains centred on the issue of Canadian–American military interoperability, and clearly demonstrates that there is more to gain than to lose. Lerhe, *At What Cost Sovereignty*.

45 Lackenbauer, "From 'Defence against Help' to 'a Piece of the Action,'" 5.

46 This is termed "international legal sovereignty" – also known as Westphalian or Vattelian sovereignty. Krasner, "Sharing Sovereignty," 654–55; Lerhe, *At What Cost Sovereignty*, 111; Little, "Sovereignty," 769; A. James, "Sovereignty," in *The Dictionary of World Politics:*

A *Reference Guide to Concepts, Ideas and Institutions*, ed. Graham Evans and Jeffrey Newnham (Toronto: Simon and Schuster, 1990), 369.
47 David Bercuson, "SAC vs. Sovereignty: The Origins of the Goose Bay Lease, 1946–1952," *Canadian Historical Review* 72, 2 (1989): 221–22.
48 That is, this book examines CMR theory as it directly relates to command and control, not overall CMR between the Canadian military and government during the 1940–57 time period. Two of the best writings that examine this subject during the early Cold War period are Hugues Canuel, "Canadian Civil–Military Relations in the Early 'Command Era,' 1945–1955: Forging a Normative Prescription through Rational Analysis," *Canadian Military History* 24, 2 (Spring 2015): 103–26; and Peter Kasurak, *A National Force: The Evolution of Canada's Army, 1950–2000* (Vancouver: UBC Press, 2013). For an excellent examination of Canadian civil/military relations and RCAF professionalization, see Rachel Lea Heide, "The Creation of a Professional Canadian Air Force, 1916–1946," PhD diss., History, Carleton University, 2010. Recent works by Ray Stouffer and Bert Frandsen also contain some interesting insights on the popularity of air power with the Canadian government: Stouffer, *Swords, Clunks, and Widowmakers*; Bertram Frandsen, "The Rise and Fall of Canada's Cold War Air Force, 1948–1968," PhD diss., History, Wilfrid Laurier University, 2015.
49 Canada, DND, *Duty with Honour*, 2.
50 I thank an anonymous reviewer for sharing these ideas. As Chapters 4 and 7 show, there were, however, some instances where American operational-level commanders did not understand and appreciate this broader context – but were overruled (and in one case reprimanded) by superiors who did.
51 Huntington, *The Soldier and the State*, 11, 80–85. See also Heide, "The Creation of a Professional Canadian Air Force," 11–13.
52 Cohen, *Supreme Command*, 4, 242–46.
53 Morris Janowitz, *Sociology and the Military Establishment* (Beverly Hills: Sage Publications, 1959), 131–33; Janowitz, *The Professional Soldier: A Social and Political Portrait* (Illinois: Free Press of Glencoe, 1960), 15, 233–36, 303–4, 342–43; Heide, "The Creation of a Professional Canadian Air Force," 13–14; Cohen, *Supreme Command*, 231. Cohen also emphasizes the political consequences of military actions, citing military theorist Carl von Clausewitz's famous dictum that "war is not merely an act of policy but a true political intercourse, carried on with other means." Cohen, *Supreme Command*, 7–8; Carl von Clausewitz, *On War*, ed. and trans. Michael Howard and Peter Paret (Princeton: Princeton University Press, 1982), Book 1, Chapter 1, 87.
54 For a discussion of military effectiveness and the ability to produce favourable military operations, see Stephen Biddle, "Military Effectiveness," *International Studies Encyclopedia*, vol. 7, ed. Robert A. Denemark (Malden: Blackwell, 2010), 5140–41; and A.R. Millett and W. Murray, eds., *Military Effectiveness*, 3 vols. (Boston: Allan and Unwin, 1988).
55 Gosselin, "Canada's Participation," 66. The subject of military efficiency has particularly been a concern of the air force; as Al Okros has put it, "the Air Force buys into the cult of efficiency." Alan Okros, "Leadership in the Canadian Military Context," CFLI Monograph 2010-1, Canadian Forces Leadership Institute, Canadian Defence Academy, Kingston, November 2010, 27.
56 Cohen, *Supreme Command*, 12–13, 209. I am indebted to an anonymous reviewer for highlighting the applicability of Cohen's CMR theory for me.
57 Canada, DND, *Duty with Honour*, 6; Canada, DND, *Leadership in the Canadian Forces: Conceptual Foundations* (Kingston: Canadian Defence Academy and Canadian Forces Leadership Institute, 2005), 130. Samuel Huntington also articulates the functional and

societal/political imperatives. See Huntington, *The Soldier and the State*, 2, 14. Although *Duty with Honour* was published relatively recently, it is relevant for examining historical Canadian military professionalism from 1940 to 1957 because it codified Canadian profession of arms doctrine for the first time. Indeed, *Duty with Honour* did not appear until the early twenty-first century because the Canadian military as an organization had not taken the time to codify it previously. Canada, DND, *Duty with Honour*, 2; conversation with Al Okros, August 26, 2016. Quote from former. As head of the Canadian Forces Leadership Institute, then Captain Okros was part of the team that wrote *Duty with Honour*. I thank him for his insights.

58 Peter Feaver, *Armed Servants: Agency, Oversight, and Civil–Military Relations* (Cambridge, MA: Harvard University Press, 2003), 12. This concept is also consistent with Eliot Cohen's above contention that it is the civilian politician leaders of government who exercise supreme command.

59 Feaver describes the relationship this way: "Working is doing things the way civilians want, and shirking is doing things the way those in the military want." Feaver, *Armed Servants*, 12–13, 54–68. Quote from page 60.

60 Ibid. Canadian political scientist Douglas Bland has described the early Cold War period as the "Command Era," when there was more uniformity of defence matters among the Canadian political leadership and military brass. Bland, *The Administration of Defence Policy in Canada*, 1–6. See also Canada, DND, *Duty with Honour*, 65–66.

Chapter 2: Command and Control Culture and Systems

Epigraph: LAC, MG B5, Brooke Claxton fonds, vol. 221, Unpublished Memoirs of Brooke Claxton, 821.

1 The command and control situation with Royal Canadian Navy forces tasked with the maritime role in the defence of shipping was a different matter. It is discussed below and also in Chapter 4.

2 On the significance of the Canadian Corps, see Paul Dickson, "The End of the Beginning: The Canadian Corps in 1917," in *Vimy Ridge: A Canadian Reassessment*, ed. Geoffrey Hayes, Andrew Iarocci, and Mike Bechthold (Waterloo: Laurier Centre for Military Strategic and Disarmament Studies and Wilfrid Laurier University Press, 2010), 31–49; Shane B. Schreiber, *Shock Army of the British Army: The Canadian Corps in the Last 100 Days of the Great War* (Westport: Praeger 1997); Desmond Morton, "'Junior but Sovereign Allies': The Transformation of the Canadian Expeditionary Force, 1914–1918," in *Canada's Defence: Perspectives on Policy in the Twentieth Century*, ed. B.D. Hunt and R.G. Haycock (Toronto: Copp Clark Pitman, 1993), 31–43; and Terry Copp, "The Military Effort, 1914–1918," in *Canada and the First World War: Essays in Honour of Robert Craig Brown*, ed. David Mackenzie (Toronto: University of Toronto Press, 2005), 35–61.

3 Specifically, in one course paper McNaughton argued for maintaining "a direct channel of responsibility to the Canadian Government," constitutional responsibility, and accountability to the Canadian government. Stacey, *Arms, Men, and Governments*, 210–11.

4 Gen. Charles Foulkes, "The Complications of Continental Defence," in *Neighbors Taken for Granted: Canada and the United States*, ed. Livingston T. Merchant (Toronto: Burns and MacEachern, 1966), 107. The animosity towards the British flowed in part from the experience of having large numbers of Canadian soldiers being killed under British command due to British generals' lack of imagination in conducting warfare on the Western Front, and also from the British execution of some Canadian soldiers for desertion. On such executions, see Teresa Iacobelli, *Death or Deliverance: Canadian Courts Martial in the Great War* (Vancouver: UBC Press, 2013).

5 Stacey, *Canada and the Age of Conflict*, vol. 2, 133–35.
6 Brereton Greenhous, Stephen J. Harris, William C. Johnston, and William G.P. Rawling, *The Crucible of War, 1939-1945: The Official History of the Royal Canadian Air Force, Volume III* (Toronto: University of Toronto Press in cooperation with the Department of National Defence and the Canadian Government Publishing Centre, Supply and Services Canada, 1994), 25.
7 *Visiting Forces (British Commonwealth) Act*, George V c. 21, April 12, 1933.
8 This was best captured by Stacey:

> It may be said that with the enactment of the Visiting Forces (British Commonwealth) Act by the Canadian Parliament in 1933, complete control and punishment in the Canadian forces finally passed to the Canadian government. This meant, in effect, that all disciplinary matters affecting members of the Canadian forces serving abroad were controlled by the Canadian government – either directly, throughout its own officers, or indirectly, by authority *delegated* to the appropriate officer of another Commonwealth country.

See Stacey, *Arms, Men, and Governments*, 211, 248. Quote from page 248. Emphasis in original. The *Visiting Forces Act* did not apply to the Royal Canadian Navy until after the Second World War. Ibid., 324.

9 *Visiting Forces (British Commonwealth) Act*, 1933; Memorandum on the Visiting Forces (British Commonwealth) Act, 1933, January 16, 1942, DHH 181.006 (D601); Stacey, *Arms, Men, and Governments*, 211–12; 254–55; C.P. Stacey, *Official History of the Canadian Army in the Second World War*, vol. 1: *Six Years of War: The Army in Canada, Britain and the Pacific* (Ottawa: Queen's Printer, 1956), 255; C.P. Stacey, *Official History of the Canadian Army in the Second World War*, vol. 3: *The Victory Campaign: The Operations in North-West Europe 1944-1945* (Ottawa: Queen's Printer, 1960), 32; Paul D. Dickson, "Colonials and Coalitions: Canada–British Command Relations between Normandy and the Scheldt," in *Leadership and Responsibility in the Second World War: Essays in Honour of Robert Vogel*, ed. Brian P. Farrell (Montreal and Kingston: McGill–Queen's University Press, 2004), 237–38; Greenhous et al., *Crucible of War*, 25; Michael Stuart Weisenfeld, "Stumbling Forward: The Canadianization of the Royal Canadian Air Force during the Second World War: How the RCAF Overseas Eventually Came into Its Own," Master of Defence Studies thesis, Royal Military College of Canada, 2013, 33; Stacey, *Canada and the Age of Conflict*, vol. 2, 289. General Foulkes notes that although the Visiting Forces Act "was not especially designed to deal with conditions of a major war[,] it was the only legal instrument available as a basis for the many command arrangements that were necessary in World War." Foulkes, "The Complications of Continental Defence," 107.

10 Stacey, *Arms, Men, and Governments*, 211–13, 217, 254–55; Greenhous et al., *Crucible of War*, 25; Stacey, *Canada and the Age of Conflict*, vol. 2, 291; Foulkes, "The Complications of Continental Defence," 107–8. The assignment of Canadian troops that were to come under British operational command was articulated in an "Order of Detail" that the Senior Combatant Officer, Canadian Army Overseas, supplied to the British commander. In each theatre, the senior Canadian commander also received specific instructions placing him under British operational command and providing him with a right to appeal to the Canadian government if there was any disagreement with the British general officer. On the right to appeal to higher authority, see the discussion of the coalition supreme command system (below).

11 One exception was in 1941, when the 1st Canadian Corps was assigned to defend a section of the Sussex coast and was placed "in combination" with the British South Eastern

Command. Lt.-Gen. Harry Crerar, Commander of the 1st Canadian Corps, thus came under the operational command of General Bernard Montgomery, CinC South Eastern Command. Stacey, *Canada and the Age of Conflict*, vol. 2, 289.

12 Stacey, *Arms, Men, and Governments*, 222–23; Dickson, "Colonials and Coalitions." See also Stacey, *The Victory Campaign*. Canadian general officers headed the First Canadian Army, which included "five major-generals commanding divisions plus three lieutenant-generals for the army and the two corps." Stacey, *A Date with History*, 258. A detailed study of how the "in combination" expeditionary command and control arrangement worked for the Canadian Army during the Second World War is still lacking and would be a fruitful area of future research.

13 Plans for uniquely Canadian units had been dashed by the end of the war in November 1918. On the Canadian contribution to the air war during the First World War, see S.F. Wise, *Canadian Airmen and the First World War: The Official History of the Royal Canadian Air Force*, vol. 1 (Toronto: University of Toronto Press and Department of National Defence, 1980).

14 Samuel Kostenuk and John Griffin, *RCAF Squadron Histories and Aircraft, 1924–1968* (Toronto: A.M. Hakkert, 1977), 75; Weisenfeld, "Stumbling Forward," 36–37; Richard Oliver Mayne, "A Test of Resolve: Article XV, The British Commonwealth Air Training Plan, and a Crusade for National Recognition," *RCAF Journal* 5, 2 (Spring 2016): 20; Stacey, *Arms, Men, and Governments*, 152. Stacey in particular was quite critical of this numbering of RCAF squadrons, remarking that it "had the effect of suggesting that the renumbered squadrons were in fact units of the Royal Air Force rather than of the Royal Canadian Air Force." Ibid., 259.

15 *Visiting Forces (British Commonwealth) Act*, 1933; Stacey, *Arms, Men, and Governments*, 221; Stacey, *Canada and the Age of Conflict*, vol. 2, 289. The only two exceptions were when, as mentioned above, the 1st Canadian Corps was placed "in combination" with British forces to defend the Sussex coast (again, highlighting the geographical practice) as part of the British Army's South Eastern Command, and for the emergency defence of Britain. In both cases, Canadian forces would be placed "in combination" with British forces and could come under British operational command. However, this latter emergency defence provision was never put into effect during the war because there was no major land attack on Britain that required an army response. As Chapters 2 and 3 show, a similar "emergency" provision was provided for (but also never enacted) for the bilateral Canada–US defence plan ABC-22.

16 Greenhous et al., *Crucible of War*, 25–27, 34–35; Weisenfeld, "Stumbling Forward," 33–35; Stacey, *Arms, Men, and Governments*, 254–55, 261, 267–68. English and Westrop capture this best: "operational command of RCAF units and formations was transferred to the RAF and exercised by the RAF Command to which the units were assigned ... [and] the RAF assigned and re-assigned RCAF squadrons to RAF commands and to RAF subordinate formations as operational requirements dictated." English and Westrop, *Canadian Air Force Leadership and Command*, 22–23. As with the history of the Canadian Army, an in-depth study of Canadian expeditionary air force command and control is lacking and would be a fruitful area of study for future historians.

17 English and Westrop, *Canadian Air Force Leadership and Command*, 23.

18 Stacey, *Arms, Men, and Governments*, 255, 260–61; Greenhous et al., *Crucible of War*, 26–27.

19 The exception was Tiger Force, an independent RCAF bomber command organization that would have been deployed to the Pacific if not for the dropping of the atomic bombs on Hiroshima and Nagasaki, which ended the war. Greenhous et al., *Crucible of War*, Chapter 4; Heide, "The Creation of a Professional Canadian Air Force," 234–73.

20 See Greenhous et al., *Crucible of War*, Chapters 8 to 10 and 17. Additionally, 15 of the 28 squadrons and half the ground crew in the 2nd Tactical Air Force's No. 83 Group that served in the Normandy campaign in 1944 were Canadians. Lt.-Col. Paul Johnston, "McNaughton and the Evolution of Canadian Tactical Air Power: A Cautionary Tale of the Limits to Junior-Alliance-Partner Innovations," *RCAF Journal* 5, 4 (Fall 2016): 18. For an excellent focused account of No. 83 Group in Northwest Europe, see Paul Johnston, "2nd TAF and the Normandy Campaign: Controversy and the Under-Developed Doctrine," MA thesis, War Studies, Royal Military College of Canada, 1999. Significantly, No. 83 Group supported Second British Army while No. 84 Group, a fully RAF organization, supported the First Canadian Army in Normandy. This was a lost opportunity to develop jointness and air/land integration between Canada's air force and army; Stacey in particular called the arrangement "objectionable." Stacey, *Arms, Men, and Governments*, 296. Ray Stouffer has recently opined that the No. 83 Group arrangement reflected the RCAF leadership's institutional interests, which trumped any possible jointness with the Canadian Army, and was repeated with the formation and location of the RCAF's 1 Canadian Air Division in NATO during the early 1950s. Stouffer, *Swords, Clunks, and Widowmakers*, 10–11.

21 Stacey, *A Date with History*, 258; English and Westrop, *Canadian Air Force Leadership and Command*, 23.

22 Ibid., 23. See also Stacey, *Arms, Men, and Governments*, 268, which quotes Air Marshal Wilfred Curtis's recollections that the posting of RCAF officers to senior command and staff positions was "resisted in the [RAF] Commands." Curtis, as later chapters will show, played a leading role in bringing the RCAF closer to the USAF in terms of the continental air defence mission and also culturally.

23 English and Westrop, *Canadian Air Force Leadership and Command*, 23. English and Westrop are quoting C.P. Stacey's memoirs: Stacey, *A Date with History*, 257. Stacey also points out, with some justification, that if Canada was truly serious about establishing its own "national air force" completely under Canadian control, it should have paid for it instead of letting the British take financial responsibility for Canadian airmen deployed overseas. Stacey, *Arms, Men, and Governments*, 262; Weisenfeld, "Stumbling Forward," 45–46.

24 Stacey, *A Date with History*, 258.

25 K.L.B. Hodson, "The RCAF Air Division in Europe: Address to the United Services Institute, London, ON, 15 December 1954," copy at the Keith Hodson Memorial Library, Information Resource Centre (IRC), CFC.

26 Stouffer, *Swords, Clunks, and Widowmakers*, 3, 11, 165. Historian Jack Granatstein notes that "senior RCAF officers had been short-changed in their search for high command in World War II by the RAF; feelings of bitterness toward the British may well have persisted." Granatstein, "The American Influence on the Canadian Military," 72n.

27 Foulkes, "The Complications of Continental Defence," 110; Galen Roger Perras, *Stepping Stones to Nowhere: The Aleutian Islands, Alaska, and American Military Strategy, 1867–1945* (Vancouver: UBC Press, 2003), 192–93; F/L F.J. Hatch, "The Aleutian Campaign" (two parts), *The Roundel* 15 (May–June 1963): 18–23.

28 Stacey, *Six Years of War*, 500–1; Galen Roger Perras, "Canada as a Military Partner: Alliance Politics and the Campaign to Recapture the Aleutian Island of Kiska," *Journal of Military History* 56 (July 1992): 440–43; Stacey, *Arms, Men, and Governments*, 390–391. Quote from latter. Stacey notes that no occasion arose where the Canadian commander had to exercise these powers.

29 British Joint Staff Mission, "Higher Direction and Command in War," November 19, 1941, DHH 72/145. See also Douglas L. Bland, *The Military Committee of the North*

Atlantic Alliance: A Study of Structure and Strategy (New York: Praeger, 1991), 35. For a theoretical examination of command systems, see Martin Van Creveld, *Command in War* (Cambridge, MA: Harvard University Press, 1985).

30 Maloney, *Securing Command of the Sea*, 12; John Ehrman, *Grand Strategy*, vol. 5: *August 1943–September 1944* (London: HMSO, 1956), 20. Quote from latter.
31 British Joint Staff Mission, "Higher Direction and Command in War," November 19, 1941, DHH 72/145.
32 Ibid.
33 Ibid.
34 Quoted in Forrest C. Pogue, *The Supreme Command: The European Theater of Operations, United States Army in World War II* (Washington, DC: Office of the Chief of Military History, Department of the Army, 1954), 37.
35 Pogue, *Supreme Command*, 37; Richard M. Leighton, "Allied Unity of Command in the Second World War: A Study in Regional Military Organization," *Political Science Quarterly* 67, 3 (September 1952): 412. Kenneth Allard has called this "the tactical autonomy of the field commander" in the "classic era of command and control." Allard, *Command, Control, and the Common Defense*, 41, 43.
36 British Joint Staff Mission, "Higher Direction and Command in War," November 19, 1941, DHH 72/145.
37 Ibid. Although the Allies adopted the American unity of command system for *operational* command and control for coalition warfare (see next section), they used the British joint committee system for *strategic* management and direction of the war. This included the establishment in 1942 of both the US Joint Chiefs of Staff and the British–American Combined Chiefs of Staff. Col. Ian Hope, "Unity of Command in Afghanistan: A Forsaken Principle of War," Carlisle Paper in Security Strategy, Strategic Studies Institute, US Army War College, Carlisle, PA, 2008, 3. For the formation of the Combined Chiefs of Staff and the Joint Chiefs of Staff, see Ray S. Cline, *United States Army in World War II, The War Department, Washington Command Post: The Operations Division* (Washington, DC: Office of the Chief of Military History, Department of the Army, 1951), 98–106.
38 Ehrman, *Grand Strategy*, vol. 5: 20; Maloney, *Securing Command of the Sea*, 12. Quote from latter.
39 Hope, "Unity of Command in Afghanistan," 1. Emphasis in original. See also Leighton, "Allied Unity of Command," 402. Canadian doctrine defines unity of command as follows: "A single clearly identified commander must be appointed for each operation. The commander has the authority to plan and direct operations and will be held responsible for an operation's success or failure." *Canadian Forces Aerospace Command Doctrine*, B-GA-0401-000/FP-001, 3.
40 Allard, *Command, Control, and the Common Defense*, 31. Allard notes that Washington was heavily influenced by his inspector general and drillmaster of the Continental Army, Baron Friedrich von Steuben, a Prussian who was a product of the military system of Frederick the Great.
41 Ibid., 1–2.
42 See for instance the 1914 *Field Service Regulations*. War Department, Office of the Chief of Staff, *Field Service Regulations, US Army* (Washington, DC: US GPO, 1914), 67–68.
43 Bland, *The Military Committee of the North Atlantic Alliance*, 29–31, 48–49; Hope, "Unity of Command in Afghanistan," 2. Quote from latter. On Foch's practice of unity of command, see Elizabeth Greenhalgh, "General Ferdinand Foch and Unified Allied Command in 1918," *Journal of Military History* 79, 4 (October 2015): 997–1023.

44 Allard, *Command, Control, and the Common Defense*, 2–4, 8, 32, 64, 96–97, 116–17, 125. Allard details the army/navy cultural "disparity in perspective regarding command relationships" as follows: "The Army approach was built around the principles of mass, subordination, and concentration of force. It stressed a strictly hierarchical organization that distributed the commander's authority through the ranks and echelons in a pyramiding control structure that enabled him to intervene personally and directly as the tactical situation required. The Navy approach was, from the outset, far more federal in character, not only because of the relatively small number of ships to be controlled, but also because of the extremely limited means of controlling them. Equally significant was the relative inability of naval commanders to intervene with their subordinates as often or as effectively as their ground force counterparts did." Allard, *Command, Control, and the Common Defense*, 45–46.

45 Ibid., 90, 47. US service autonomy also meant that each service retained responsibility for discipline and administration, both aspects of full or national command. Maurer, *Coalition Command and Control*, 19.

46 Allard, *Command, Control, and the Common Defense*, 63.

47 Maurer, *Coalition Command and Control*, 19; Allard, *Command, Control, and the Common Defense*, 32. Allard remarks that these officers saw unity of command as "a concept so threatening to traditional service autonomy in the operational sphere that it acquired an almost pejorative meaning as it was thrashed out in Joint Board and Joint Chiefs of Staff proceedings for the next generation." Ibid., 96.

48 Maurer, *Coalition Command and Control*, 20; *Joint Action of the Army and the Navy FTP-155*, Prepared and Revised by the Joint [USN-US Army] Board, November 15, 1935, US National Archives and Records Administration, College Park, Maryland (NARA), Record Group (RG) 165, file WPD 2917–35. Quote from former.

49 *Joint Action of the Army and the Navy FTP-155*, Prepared and Revised by the Joint [USN-US Army] Board, November 15, 1935, NARA, RG 165, file WPD 2917–35. As Chapter 3 demonstrates, the command and control arrangement in Joint Action formed the basis of the ABC-22 Canada–US defence plan.

50 Ibid.

51 The similarities between the two command and control terms are as follows (they are numbered and italicized for easier comparison):

Current RCAF definition of "Operational Command":
The *authority* [2] granted to a *commander* [1] to *assign missions or tasks* [3] to subordinate commanders, to deploy units, to reassign forces and *to retain or delegate operational control (OPCON) and/or tactical control (TACON) as may be deemed necessary* [4]. It *does not of itself include responsibility for administration* [5].
Source: *Canadian Forces Aerospace Command Doctrine*, B-GA-0401–000/FP-001, 7.

Joint Action of the Army and the Navy definition of "Unity of Command":
... unity of command in an operation vests in *one commander* [1] the *responsibility and authority* [2] to coordinate the operations of the participating forces of both services by the organization of task forces, the *assignment of missions* [3], the designation of objectives, and *the exercise of such coordinating control as he deems necessary* [4] to insure the success of the operation.... Unity of command *does not authorize the commander exercising it to control the administration and discipline* [5] of the forces of the service to which he does not belong, nor to issue any instructions to such forces beyond those necessary for effective coordination.

52 Maurer, *Coalition Command and Control*, 19; Leighton, "Allied Unity of Command," 402, 416; Cline, *Washington Command Post*, 2–3.
53 This was especially the case for the US Navy, which was a staunch advocate of mutual cooperation. Allard describes the arrangement as "little more than a nonaggression pact concluded between the Army and Navy of the United States." Allard, *Command, Control, and the Common Defense*, 97.
54 Foulkes, "The Complications of Continental Defence," 107.
55 Order-in-Council, P.C. 2049, August 4, 1914, reproduced in C.P. Stacey, ed., *Historical Documents of Canada*, vol. 5: *1914–1945* (Toronto: Macmillan of Canada, 1972), 552.
56 Order in Council by the Canadian Government, November 17, 1939, reproduced in David R. Murray, ed., *Documents on Canadian External Relations*, vol. 7: *1939–1941*, pt. 1 (Ottawa: Information Canada, 1974), 842. See also: W.G.D. Lund, "Command Relationships in the North West Atlantic, 1939–1943," MA thesis, History, Queen's University, 1972, 6–7; W.A.B. Douglas, Roger Sarty, and Michael Whitby, with Robert H. Caldwell, William Johnston, and William G.P. Rawling, *No Higher Purpose: The Official Operational History of the Royal Canadian Navy in the Second World War, 1939–1943*, vol. 2, pt. 1 (St Catharines: Vanwell, 2002), 56–57.
57 Stacey, *Canada and the Age of Conflict*, vol. 2, 129–35; Adrian W. Preston, "Canada and the Higher Direction of the Second World War 1939–1945," in *Canada's Defence: Perspectives on Policy in the Twentieth Century*, ed. B.D. Hunt and R.G. Haycock (Toronto: Copp Clark Pitman, 1993), 99–100, 104; Stacey, *Canada and the Age of Conflict*, vol. 2, 270. Quote from latter.
58 See Foulkes, "The Complications of Continental Defence," 107, for discussion about how Canada's "struggle for autonomy" during the First World War and the ill-feelings of Canadians towards their British commanders during this conflict "keep alive the prejudice to any form of bilateral or multilateral command of Canadian troops."
59 Roger Sarty, *The Maritime Defence of Canada* (Toronto: Canadian Institute of Strategic Studies, 1996), 206.
60 Douglas, *The Creation of a National Air Force*, vol. 2, 382.
61 Lt.-Col. H.D.G. Crerar, "The Difficulties of Unified Control of Allied Operations," *Canadian Defence Quarterly* 3, 1 (October 1925): 71–74; Lt.-Col. E.L.M. Burns, "Theory of Military Organization," *Canadian Defence Quarterly* 14, 3 (April 1937): 326–31.
62 Capt. M.V. Bezeau, "The Role and Organization of Canadian Military Staffs," MA thesis, War Studies, Royal Military College of Canada, 1978, 68.
63 Stacey, *Arms, Men, and Governments*, 130–32; Maloney, *Securing Command of the Sea*, 26; Bezeau, "Role and Organization," 121.
64 British Joint Staff Mission, "Higher Direction and Command in War," November 1941, 19, DHH 72/145.
65 Richard Goette, "The British Joint Area Combined Headquarters Scheme and the Command and Control of Maritime Air Power," *Royal Air Force Air Power Review* 14, 3 (Autumn–Winter 2011): 119–35. This arrangement also included the assignment of army liaison officers as required.
66 Coastal Command – Memorandum by the First Lord of the Admiralty, November 22, 1940, United Kingdom National Archives [TNA], Public Record Office [PRO], War Cabinet [CAB] file 66/13 280.
67 DHH 79/599, Capt. D.V. Peyton-Ward, *The RAF in the Maritime War*, vol. 2: *The Atlantic and Home Waters: September 1939–June 1940* (RAF Air Historical Branch Narrative, n.d.), 280, 285.
68 Ibid., Appendix VII, 394.

69 DHH 79/599, Peyton-Ward, *The RAF in the Maritime War*, vol. 2, 28.
70 Committee on Coastal Command Report, March 19, 1941, TNA, PRO, Air Ministry (Air) file 15/338; DHH 79/599, Peyton-Ward, *The RAF in the Maritime War*, vol. 2, 275, 286. Quote from latter.
71 AM Sir John C. Slessor, AOCinC Coastal Command, to AVM N.R. Anderson, RCAF AMAS, June 24, 1943, DHH 181.009 (D6734).
72 Operational direction is defined here as "the authority to issue directives as to the objectives to be pursued (i.e., the effect that one wants achieved) in operations. It does not include the planning and issuing of detailed instructions for the actual execution of operations." Richard Evan Goette, "The Struggle for a Joint Command and Control System in the Northwest Atlantic Theatre of Operations: A Study of the RCAF and RCN Trade Defence Efforts during the Battle of the Atlantic," MA thesis, Queen's University, 2002, vii. For a more detailed discussion of "operational direction," see ibid., Chapter 3 and 124-26. See also Allan English, *Command and Control of Canadian Aerospace Forces: Conceptual Foundations* (Ottawa: Minister of National Defence, 2008), 3-7, for a discussion of the historical evolution of command and control terminology.
73 CinC USN to US Navy Commands, Admiralty, Air Ministry, and CoS Army, February 11, 1944, TNA, PRO, Air 15/339.
74 The modern definition of operational control is: "The authority delegated to a commander to direct assigned forces to accomplish specific missions or tasks that are usually limited by function, time or location, to deploy units concerned, and to retain or assign tactical control of those units. Operational control does not include authority to assign separate employment of components of the units concerned, neither does it, of itself, include administrative or logistic control." *Canadian Forces Aerospace Command Doctrine*, B-GA-0401-000/FP-001, 50.
75 Maurer, *Coalition Command and Control*, 21; Cohen and Gooch, *Military Misfortunes*, 47-56.
76 Pogue, *Supreme Command*, 41-43; Ehrman, *Grand Strategy*, V, 20; Leighton, "Allied Unity of Command," 402-4; Mitchell, *Network Centric Warfare and Coalition Operations*, 83.
77 Joint Chiefs of Staff Directive, Unified Command for US Joint Operations, April 20, 1943 (JCS 263/2/D), reproduced as Appendix L to Morton, *Strategy and Command*, 642-43. See also Morton, *Strategy and Command*, 474-75.
78 Ibid. This document also used the term "component" commanders for the first time in US military terminology to refer to the operational commanders under the overall joint commander.
79 On the principles of mission command and centralized command and decentralized execution, see Lt.-Col. Pux Barnes, "Mission Command and the RCAF: Considerations for the Employment of Air Power in Joint Operations," Article #4 in a series on command and control and the Royal Canadian Air Force, Canadian Forces Aerospace Warfare Centre, 2014, http://www.rcaf-arc.forces.gc.ca/en/cf-aerospace-warfare-centre/c2-article-4-mission-command-and-the-rcaf.page; and Allan English, "Rethinking 'Centralized Command and Decentralized Execution,'" in *Air Force Command and Control*, ed. Douglas L. Erlandson and Allan English (Toronto: Canadian Forces College, 2002), 71-82.
80 Ronald H. Cole, Walter S. Poole, James F. Schnabel, Robert B. Watson and Williard J. Webb, *The History of the Unified Command Plan 1946-1993*, Joint Chiefs of Staff Committee Narrative, Joint History Office, Office of the Chairman of the Joint Chiefs of Staff, Washington, DC, 1995, 11-14. See also Lt.-Col. Marcus Fielding, "The United States Unified Command Plan," *Canadian Military Journal* 7, 3 (Autumn 2006): 35-40.
81 Quoted in Cole et al., *History of the Unified Command Plan*, 14.

82 Hope, "Unity of Command in Afghanistan," 5.
83 Cole et al., *History of the Unified Command* Plan, 14; Hope, "Unity of Command in Afghanistan," 5.
84 Cole et al., *History of the Unified Command Plan*, 137, 14; Robert J. Watson, *History of the Joint Chiefs of Staff*, vol. 5: *The Joint Chiefs of Staff and National Policy 1953–1954* (Washington, DC: Historical Division, Joint Chiefs of Staff, 1986), 137; James F. Schnabel, *History of the Joint Chiefs of Staff, Volume I, The Joint Chiefs of Staff and National Policy 1945–1947* (Washington, DC: Office of Joint History and Office of the Chairman of the Joint Chiefs of Staff, 1996), 85.
85 General Marshal quoted in Leighton, "Allied Unity of Command," 403. In forwarding the unity of command system for Allied coalition command, Marshall had the full support of President Franklin Roosevelt. Reminiscing about the war, he admitted: "I was very fortunate that the president didn't interfere with me on command and allowed me more or less complete freedom of action." Larry I. Bland, ed., *George C. Marshall Interviews and Reminiscences for Forrest C. Pogue*, rev. ed. (Lexington: George C. Marshall Research Foundation, 1991), 452.
86 Pogue, *Supreme Command*, 41–42; Leighton, "Allied Unity of Command," 403–5, 410–11; Maloney, *Securing Command of the Sea*, 13–15, 21–23; Maurice Matloff and Edwin M. Snell, *United States Army in World War II, The War Department: Strategic Planning for Coalition Warfare 1941–1942* (Washington, DC: Office of the Chief of Military History, Department of the Army, 1953), 123, 196.
87 Leighton, "Allied Unity of Command," 411, 400–3, 421; Morton, *Strategy and Command*, 250. Quote from former.
88 Leighton, "Allied Unity of Command," 404–6, 425; Ehrman, *Grand Strategy*, vol. 5, 21; Matloff and Snell, *Strategic Planning for Coalition Warfare 1941–1942*, 123–25. Quote from Leighton.
89 Leighton, "Allied Unity of Command," 403–4, 421; Ehrman, *Grand Strategy*, vol. V, 21. Quote from former.
90 Bland, *The Military Committee*, 44, 47–49, 52; Pogue, *Supreme Command*, 42; Leighton, "Allied Unity of Command," 405–6, 420–21.
91 Bland, *The Military Committee*, 44. This is similar to the "Red Card" holder in modern-day Targeting.
92 Ibid., 44, 52.
93 Maloney, *Securing Command of the Sea*, 83, 85. The formation of NATO and its command organization is also dealt with in further detail in the following: James Eayrs, *In Defence of Canada*, vol. 4: *Growing Up Allied* (Toronto: University of Toronto Press, 1980); Don Cook, *Forging the Alliance: NATO, 1945–1950* (New York: Arbor House, 1989); Escott Reid, *Time of Fear and Hope: The Making of the North Atlantic Treaty 1947–1949* (Toronto: McClelland and Stewart, 1977).
94 Bland, *The Military Committee*, 44, 186–94; CUSRPG Document CUS-1, "Organization for Canada United States Regional Planning," October 31, 1949, NARA, RG 59, State Department, PJBD-MCC, entry E1184, box 25, file "Military Co-operation Committee III." Quote from Bland. There was an evolution of the NATO command structure between 1951 and 1953 reflecting both operational requirements and, more importantly, political compromises, especially over the appointment of a USN Flag officer as Supreme Allied Commander Atlantic (SACLANT) and French preferences. The consolation to the United Kingdom was Commander-in-Chief Channel Command (CINCHAN). Below SACEUR were CINCS, Central, South, North and Mediterranean (another consolation prize to the UK as CINCAFMED did not have the USN 6th Fleet, which belonged to CINCSOUTH, a USN flag officer). The first CINCAFMED was Louis Mountbatten.

On the development of the NATO command structure, see Maloney, *Securing Command of the Sea*, Chapters 2 to 5.
95 CUSRPG Document CUS-1, "Organization for Canada United States Regional Planning," October 31, 1949, NARA, RG 59, State Department, PJBD-MCC, entry E1184, box 25, file "Military Co-operation Committee III"; DHH 87/47, Col R.L. Raymont, *The Evolution of the Structure*, Appendix A, "The Organization of Higher Control and Coordination in the Formulation of Defence Policy, 1945–1964," 27.
96 Jockel, *No Boundaries Upstairs*, 97. See also Canadian Ambassador to the United States to the Secretary of State for External Affairs, July 18, 1949, Joint Staff Fonds, Chiefs of Staff Committee Papers, July 1949, DHH 2002/17, box 75, file 5.
97 William R. Willoughby, *The Joint Organizations of Canada and the United States* (Toronto: University of Toronto Press, 1979), 132–33; Jockel, *No Boundaries Upstairs*, 96; Foulkes, "The Complications of Continental Defence," 118. The Americans were careful to ensure that their military and political dominance of NATO, and the US nuclear deterrence on which it rested, would guarantee that multilateralism would be an instrument of, and not a restraint on, American actions.
98 Willoughby, *The Joint Organizations*, 133; Foulkes, "The Complications of Continental Defence," 118; Jockel, *No Boundaries Upstairs*, 97. Quote from first source.

Chapter 3: Wartime Planning for Command and Control
Epigraph: Stetson Conn and Byron Fairchild, *The Western Hemisphere: The Framework of Hemisphere Defense* (Washington, DC: Office of the Chief of Military History, Department of the Army, 1960), 381.
1 Wise, *Canadian Airmen and the First World War*, 42, 93–95; 604–5; J.D.F. Kealy and E.C. Russell, *A History of Canadian Naval Aviation* (Ottawa: Queen's Printer, 1967), 3–5.
2 During the time of austerity of the interwar period, the Canadian military was required to devise such defensive plans (and, notably, ones that did not entail an overseas expeditionary commitment) such as "Defence Scheme Number 1" in order to justify funding. The Americans also had a plan for war with Britain involving the invasion of Canada, the incorporation of its provinces into the Union, and the abolishment of the Dominion government. Stacey, *Canada and the Age of Conflict*, vol. 2, 155–57; Richard A. Preston, *The Defence of the Undefended Border: Planning for War in North America 1867–1939* (Montreal and Kingston: McGill-Queen's University Press, 1977), Chapter 8; Galen Roger Perras, *Franklin Roosevelt and the Origins of the Canadian-American Security Alliance 1933–1945: Necessary, but Not Necessary Enough* (London: Praeger, 1998), 16, 29n.
3 Preston, *Defence of the Undefended Border*, 225.
4 Perras, *Stepping Stones to Nowhere*, 24.
5 Stacey, *Canada and the Age of Conflict*, vol. 2, 226; Eayrs, *In Defence of Canada*, vol. 2 (Toronto: University of Toronto Press, 1967), 183; Perras, *Franklin Roosevelt and the Origins*, 43–44; Preston, *Defence of the Undefended Border*, 228–29. Quotes from Stacey.
6 Fergusson, *Canada and Ballistic Missile Defence*, 12; David Haglund and Michel Fortmann, "Canada and the Issue of Homeland Security: Does the Kingston Dispensation Still Hold?," *Canadian Military Journal* 3, 1 (Spring 2002): 17–22.
7 On the Kingston dispensation and "defence against help," see Lackenbauer, "From 'Defence against Help' to 'a Piece of the Action,'" 4.
8 Douglas, *The Creation of a National Air Force*, vol. 2, 348–49; Raoul Dandurand, Speech to the League of Nations, October 2, 1924, quoted in Stacey, *Canada and the Age of Conflict*, vol. 2, 61.
9 Roger Sarty, "Silent Sentry: A Military and Political History of Canadian Coast Defence 1860–1945," PhD diss., History, University of Toronto, 1982, 611, 617. On early wartime

Canadian air defence measures, see Richard Goette, "Plugging the Holes in the Roof: The Origins of a Northern Aerial Threat to North America and the Defence of Central Canada," proceedings of the 7th Annual Air Force Historical Conference: Canada in NORAD, Colorado Springs, June 4–8, 2001 (Office of Air Force Heritage and History, 2001), 27–40; and Douglas, *The Creation of a National Air Force*, 373–91.

10 Grimshaw, "On Guard," 43–44; Sarty, "Silent Sentry," 616; "The System of Command across Canada Since Confederation," Historical Section, Army Headquarters, December 4, 1957, DHH 943.009 (D164).

11 C.P. Stacey, *Arms, Men, and Governments*, 131–32; Maloney, *Securing Command of the Sea*, 29; Goette, "The Struggle for a Joint Command and Control System," 35–37, 73–84.

12 Chiefs of Staff Committee, "Defence of Canada Plan, August, 1940," DHH 112.3M2 (D497).

13 Rear-Adm. L.W. Murray, CB, CBE, RCN (Ret.d), Recollections of His Naval Career, recorded on tape at the Directorate of History, CFHQ, Ottawa, May 1970, DHH Biographical File, Murray, L.W., Rear-Admiral.

14 Memorandum by DCNS, "Explanation and Précis of Reports Submitted by the Three Officers Who Recently Visited Washington, DC," [mid-July] 1940, reproduced in Murray, ed., *DCER*, 8, pt. 2, 156–59. The CGS and CNS had visited Washington in January 1938, also incognito, but this meeting between the Canadian and American militaries resulted in only a basic exchange of information. Stacey, *Canada and the Age of Conflict*, vol. 2, 226.

15 Memorandum by DCGS, DCNS, and AMAS, [n.d.–mid-July] 1940, reproduced in Murray, ed., *DCER*, 8, pt. 2, 160–61.

16 Stacey, *Official History of the Canadian Army in the Second World War*, vol. 1, 161; George F.G. Stanley, *Canada's Soldiers: The Military History of an Unmilitary People*, 3rd ed. (Toronto: Macmillan of Canada, 1974), 408.

17 Perras, *Franklin Roosevelt and the Origins*, 69, 71.

18 Joint Canadian–United States Basic Defense Plan–1940, October 10, 1940, PJBD Journals, vol. 1, DHH 82/196, file 1b.

19 Stanley W. Dziuban, *United States Army in World War II Special Studies: Military Relations between the United States and Canada, 1939–1945* (Washington, DC: Office of the Chief of Military History, Department of the Army, 1959), 111.

20 Joint Canadian–United States Basic Defense Plan (First Joint Draft), 1–2, September 11, 1940, DHH 112.11 (D1A), vol. 1.

21 Biggar to Mackenzie King, October 7, 1940, reproduced in David R. Murray, ed., *DCER*, vol. 8, 1939–1941, pt. 2 (Ottawa: Department of Foreign Affairs and International Trade, 1976), 170; Biggar to Canadian PJBD Section members, October 12, 1940, attached to PJBD Journals, vol. 1, DHH 82/196, file 1b.

22 Perras, *Franklin Roosevelt and the Origins*, 79.

23 Dziuban, *Military Relations between the United States and Canada, 1939–1945*, 111.

24 Memorandum from Commissioner for Justice and Defence to Commission of Government of Newfoundland, November 19, 1940, reproduced in *Documents on Relations Between Canada and Newfoundland* (hereafter *DRBCN*), ed. Paul Bridle (Ottawa: Information Canada, 1974), 884–87. I will use the term "Newfoundland" to describe the current Canadian province of Newfoundland and Labrador since that was the usage at the time examined. Due to financial issues in the 1930s, Newfoundland had been forced to abandon responsible government and revert back to colonial status, so the term "colony" will be used.

25 Douglas, *The Creation of a National Air Force*, 381–83; Samuel Elliot Morrison, *History of the United States Naval Operations in World War II*, vol. 2: *The Battle of the Atlantic, September 1939–May 1943* (Boston: Little, Brown, 1947), 68–69, 85; David MacKenzie,

Inside the Atlantic Triangle: Canada and the Entrance of Newfoundland into Confederation, 1939–1949 (Toronto: University of Toronto Press, 1986), 79, 81.

26. W.A.B. Douglas, "Alliance Warfare, 1939–1945: Canada's Maritime Forces," *Revue Internationale d'Histoire Militaire* 54 (1982): 159, 162; Wesley Frank Craven and James Lea Cate, eds., *The Army Air Forces In World War II*, vol. 1: *Plans and Early Operations, January 1939–August 1942* (Chicago: University of Chicago Press, 1948), 156; Dziuban, *Military Relations between the United States and Canada, 1939–1945*, 110. Quotes from Douglas.

27. Ibid., 110.

28. Stuart Memorandum, "J.D.B. Plans for 1941," January 16, 1941, DHH 112.11 (D1A) vol. 1; Appendix, "Report from Service Members on Progress Made in Carrying out Recommendations of the Board," attached to Minutes of the 14th Meeting of the PJBD, Montreal, January 20, 1941, PJBD Journals, vol. 1, DHH 82/196, file 1b.

29. Memorandum from CSC to Ministers of National Defence, April 22, 1941, LAC, RG 25, vol. 5695, file 1-AL(s); Extract from Minutes of the Cabinet War Committee, April 23, 1941, LAC, RG 25, vol. 2724, file HQS 5199-W-1; Lt.-Gen. Maurice Pope, *Soldiers and Politicians: The Memoirs of Lt.-Gen. Maurice A. Pope* (Toronto: University of Toronto Press, 1962), 163–64.

30. Pope Memo, "Definition – Strategical Direction," April 21, 1941, DHH 112.11 (D1A), vol. 2.

31. "Permanent Joint Board on Defense – Second Report," n.d. [May 29, 1941], NARA, RG 59, State Department files, PJBD, box 8, file correspondence of PJBD, January to June 1941.

32. Chiefs of Staff Committee Memo to Ministers of National Defence, "Brief Appreciation of the Situation as of 24th February 1941," February 25, 1941, Joint Staff Fonds, Chiefs of Staff Committee Papers, DHH 2002/17, box 65, file 3.

33. Capt. H.E. Reid, DCNS, to German, March 7, 1941, LAC, RG 25, vol. 5695, file 1-AL(s); Memorandum from CCS to Ministers of National Defence, April 22, 1941, LAC, RG 25, vol. 5695, file 1-AL(s). See also 15th and 16th PJBD Meetings, January 21 and February 27, 1941, PJBD Journals, DHH 82/196, vol. 1, file 1. The Canadian "observer" was Cdr. P.B. German from the RCN.

34. Report on United States–British Staff Conversations (Short Title ABC-1), March 27, 1941, DHH 112.3M2.009 (D70); German to CNS, March 20, 1941, LAC, RG 25, vol. 5695, file 1-AL(s). The acronym ABC stood for "American–British Conversations," not "American–British–Canadian" as is incorrectly stated in some sources.

35. Annex II, "Responsibility for the Strategic Direction of Military Forces," of ABC-1, DHH 112.3M2.009 (D70). See also German to CNS, March 13, 1941, LAC, RG 25, vol. 5695, file 1-AL(s).

36. Dziuban, *Military Relations between the United States and Canada, 1939–1945*, 112; Stacey, *Arms, Men, and Governments*, 349. Quote from latter.

37. Bissel to Marshall, April 21, 1941, NARA, RG 165, WPD 4330-23; Memorandum by USN PJBD Member, "Strategic direction and command features for incorporation in Plan No. 2," April 17, 1941, and Lt.-Col. Clayton Bissel, US Army, Memorandum, "Control of United States–Canadian Military Forces," April 19, 1941, Queen's University Archives, C.G. Power Papers, A.Arch 2150, box 69, file D2020; Pope to Crerar, April 18, 1941, LAC, RG 24, vol. 2427, vol. HQS 5199-W-B. Quote from last document. See also Pope, *Soldiers and Politicians*, 164.

38. CSC to Ministers of National Defence, April 22, 1941, CWC Documents, vol. 5, LAC, RG 2, 7C, vol. 4, microfilm # C-11789, copy at CFC IRC, Toronto; Crerar to Pope, May 5, 1941, DHH 112.11 (D1A) vol. 2. See also Perras, *Franklin Roosevelt and the Origins*, 80.

39 CSC to Ministers of National Defence, April 22, 1941, CWC Documents, vol. 5, LAC, RG 2, 7C, vol. 4, microfilm # C-11789; Reid Memorandum to Nelles, "Strategic Direction of Canadian and United States Naval Forces Operating in Coastal Waters of Canada and Newfoundland," April 19, 1941, LAC, RG 24, vol. 2724, file HQS 5199-W-1-B, pt. 1. Quote from latter.

40 CSC to Ministers of National Defence, April 22, 1941, and Minutes of the 78th Meeting of the Cabinet War Committee, April 23, 1941, CWC Documents, vol. 5, LAC, RG 2, 7C, vol. 4, microfilm # C-11789.

41 Biggar to La Guardia, April 29, 1941, NARA, RG 59, State Department files, PJBD, box 8, file correspondence of PJBD, January to June 1941; Biggar to La Guardia, May 3, 1941, NARA, RG 165, WPD 4330-25. Quote from latter. See also Stacey, *Arms, Men, and Governments*, 351.

42 La Guardia to Biggar, May 2, 1941, NARA, RG 165, WPD 4330-25.

43 Stacey, *Arms, Men, and Governments*, 351; Memorandum for Embick by Brig.-Gen. Harry J. Malony, Acting Assistant Chief of Staff, War Plans Division, May 7, 1941, NARA, RG 165, WPD 4330-25; Stuart to Crerar, May 14, 1941, DHH 314.009 (D116); Memorandum from ACGS for CGS, "Meeting of Permanent Joint Board on Defence, Washington, 28-29 May 1941," May 31, 1941, DHH 112.11 (D1A) vol. 2. Quote from first source.

44 La Guardia Memorandum for the President, n.d. [May 7, 1941], and Roosevelt to La Guardia, May 16, 1941, F.D. Roosevelt Diplomatic Correspondence, Subject: PJBD August 1940-May 1941, microfilm reel 3, copy at ML RMC, Kingston.

45 Stacey, *Arms, Men, and Governments*, 351.

46 Pope to Biggar, May 14, 1941, DHH 112.11 (D1A) vol. 2.

47 Pope, *Soldiers and Politicians*, 164.

48 19th Meeting of the PJBD, Washington, May 28-29, 1941, PJBD Journals, vol. 2, DHH 82/196, file 2.

49 Pope, *Soldiers and Politicians*, 165; Pope to Stacey, March 25, 1954, LAC, RG 24, vol. 20, 317, file 951.001 (D1).

50 This provision was consistent with the British-American agreements in ABC-1, which granted US strategic direction for forces operating in the Western Atlantic and Eastern Pacific Areas. (Draft) Permanent Joint Board on Defense – Second Report, n.d. [May 29, 1941], NARA, RG 59, State Department files, PJBD, box 8, file correspondence of PJBD, January to June 1941.

51 Pope, *Soldiers and Politicians*, 165-66; Memorandum from ACGS for CGS, "Meeting of Permanent Joint Board on Defence, Washington, 28-29 May 1941," May 31, 1941, DHH 112.11 (D1A), vol. 2.

52 William Lyon Mackenzie King, Diary Entry, June 22, 1941, LAC, MG 26-J13, Diaries of William Lyon Mackenzie King (Online Resource), http://www.bac-lac.gc.ca/eng/discover/politics-government/prime-ministers/william-lyon-mackenzie-king/pages/item.aspx?IdNumber=22849. On the impact of Nazi Germany's invasion of the Soviet Union on Canadian policy, see Stacey, *Canada and the Age of Conflict*, vol. 2, 319.

53 Pope, *Soldiers and Politicians*, 166; 19th Meeting of the PJBD, Washington, May 28-29, 1941, PJBD Journals, vol. 22, DHH 82/196, file 2. Quote from latter.

54 Biggar Memorandum, "Permanent Joint Board on Defence," June 1, 1941, DHH 112.11 (D1A), vol. 2.

55 Joint Canadian-United States Basic Defence Plan no. 2 (Short Title ABC-22), July 28, 1941, DHH 355.009 (D20).

56 ABC-22, July 28, 1941, DHH 355.009 (D20).

57 Ibid.

58 Brig.-Gen. L.T. Gerow, Acting ACoS WPD, to Senior US Army Member, PJBD, June 17, 1941, NARA, RG 165, WPD 4330–21.
59 Embick Memorandum to Acting ACoS WPD, "Joint Canadian–United States Plan No. 2 (ABC-22)," June 7, 1941, NARA, RG 165, WPD 4330–21.
60 Embick to Pope, October 10, 1941, NARA, RG 59, State Department files, PJBD, box 8, file correspondence of PJBD, July to October 1941; 112th CWC Meeting, October 15, 1941, CWC Minutes, vol. 6, LAC, RG 2, 7C, vol. 5, Microfilm # C-4654; Pope to Embick, October 16, 1941, NARA, RG 165, WPD 4330–22.
61 Wing Cdr. J.G. Bryans, Assistant Air Attaché, Canadian Legation, Washington, to Chief of the Air Staff, RCAF Headquarters, Ottawa, September 25, 1941, DHH 314.009 (D116); DMO&I to General Officer Commanding-in-Chief (GOCinC), Atlantic Command, October 24, 1941, LAC, RG 24, vol. 2725, HQC 5199-W-1-B pt. 2; Biggar to Ralston, October 8, 1941, LAC, MG 27 III B11, J.L. Ralston Fonds, file "Joint Defence board – Gen. (Secret)."
62 Pope to Crerar, May 16, 1941, DHH 112.11 (D1A), vol. 2.

Chapter 4: Wartime Operational Level Command and Control

1 Governor of Newfoundland to Secretary of State for External Affairs, Ottawa, November 22, 1940, and Memorandum of Meeting – St. John's, Newfoundland, November 28 to December 1, 1940, LAC, RG 25, vol. 1991, file 1156-C; Memorandum from Commissioner for Justice and Defence to Commission of Government of Newfoundland, November 19, 1940, Secretary of State for External Affairs to Governor of Newfoundland, November 29, 1940, and same to same, December 16, 1940, reproduced in Bridle, ed., *DRBCN*, 886–88.
2 Appendix, "Report from Service Members of Progress Made in Carrying Out Recommendations of the Board," attached to Sixth Meeting of the PJBD, Boston, October 2, 1940, PJBD Journals, vol. 1, DHH 82/196, file 1b; Secretary of State for External Affairs to Governor of Newfoundland, November 29, 1940, reproduced in Bridle, ed., *DRBCN*, 887.
3 Douglas, *The Creation of a National Air Force*, 387, 652–53; Stacey, *Arms, Men, and Governments*, 132; Air Officer Commanding Eastern Air Command to Minister of National Defence for Air, National Defence Headquarters, August 6, 1941, DHH 181.002 (D173); Minutes of a Meeting of the CSC, June 25, 1941, reproduced in Bridle, ed., *DRBCN*, 892.
4 12th Meeting of the PJBD, New York, December 16, 1940, PJBD Journals, vol. 1, DHH 82/196 file 1b; Earnshaw to Elkins, February 3, 1941, DHH 355.009 (D29); Craven and Cate, eds., *The Army Air Forces in World War II*, vol. 1, 156; Morrison, *History of the United States Naval Operations in World War II*, vol. 2, 68–69, 85; Stacey, *Official History of the Canadian Army in the Second World War*, vol. 2, 179; Douglas, *The Creation of a National Air Force*, 383.
5 Elkins to Pope, December 30 and 31, 1941, ML, RMC, Maj.-Gen. W.H.P. Elkins Papers, file 19, "Semi-Official Correspondence vol. # 3 – November 1941 to May 1942"; AMAS to CAS, March 5, 1942, LAC, RG 24, vol. 5290, file HQS 15–73–4, pt. 1; Memorandum from Joint Planning Sub-Committee to CSC, August 8, 1942, LAC, RG 24, vol. 5209, file HQS 15–73–4, pt. 2, copy obtained through DND ATIP, A-2006–00357.
6 Memorandum for the Chief of Staff, US Army, Gen. George Marshall and Commanding General, US First Army, Lt.-Gen. Hugh A. Drum, n.d. [December 26, 1940], DHH 82/970, Extracts from General H. "Hap" Arnold Papers, box 111, file 10.
7 See the correspondence between Brig.-Gen. Earnshaw and his superiors in Halifax and Ottawa in early to mid-1941 in DHH 355.009 (D29), DHH 72/145, and Elkins Papers, ML, RMC, file 3, "Semi-official Correspondence vol. 1 – General Officer Commanding Canadian Troops in Newfoundland."

8 Stacey, *Arms, Men, and Governments*, 363; DHH 73/770, "US Air Defense in the Northeast 1940–1957," Continental Air Defense Command Historical Reference Paper Number One, Colorado Springs, Headquarters Continental Air Defense Command, Ent Air Force Base, April 1, 1957, 31.
9 W.A.B. Douglas, "Democratic Spirit and Purpose: Problems in Canadian–American Relations, 1934–1945," in *Fifty Years of Canada-United States Defense Cooperation: The Road From Ogdensburg*, ed. Joel Sokolsky and Joseph T. Jockel (Lewiston: Edwin Mellen Press, 1992), 43. Newfoundland Governor Vice-Adm. (Ret.) Sir Humphrey Walwyn once described Brant as an "impossible character ... a better sportsman than a soldier." Quoted in Peter Neary, *Newfoundland and the North Atlantic World, 1929–1949* (Kingston and Montreal: McGill-Queen's University Press, 1988), 205.
10 "Major General Gerald Clark Brant," Biography, United States Air Force, http://www.af.mil/DesktopModules/ArticleCS/Print.aspx?PortalId=1&ModuleId=858&Article=108024; Douglas C. Waller, *A Question of Loyalty: Gen. Billy Mitchell and the Court-Martial That Gripped the Nation* (New York: HarperCollins, 2004), 187–90.
11 Pope Memo to Crerar, "Permanent Joint Board on Defence Montreal, November 10–11, 1941," November 12, 1941, DHH 112.11 (D1A), vol. 3.
12 Ibid. Pope to Crerar, September 11, 1941, LAC, MG 27 III B11, Ralston fonds, file "Joint Defence Board – Gen. (Secret)."
13 Earnshaw to Elkins, December 1, 1941, DHH 355.009 (D29); Earnshaw to Elkins, December 3, 1941, ML, RMC, Elkins Papers, file 3 – "Semi-official correspondence, vol. 1 – General Officer Commanding Canadian Troops in Newfoundland"; Cdre Murray to the Naval Secretary, Naval Service Headquarters Ottawa, December 2, 1941, LAC, RG 24, vol. 11,505, file 1550-146/36-1; Stacey, *Arms, Men, and Governments*, 363.
14 Order-in-Council P.C. 9592, December 7, 1941, and CWC 125th Meeting Minutes, December 7, 1941, LAC, RG 2, vol. 6, microfilm reel # C-4654, copy at CFC IRC; Chiefs of Staff Committee to Ministers of National Defence, December 8, 1941, LAC, RG 24, vol. 2725, file HQC 5199-W-1-B, pt. 2; Atlantic Command to Officer Commanding Canadian Troops in Newfoundland, December 8, 1941, DHH 355.009 (D20).
15 John F. Shiner, "The Air Corps, the Navy, and Coast Defense, 1919–1941," *Military Affairs* 45, 3 (October 1981): 119.
16 Perras, *Franklin Roosevelt and the Origins*, 84.
17 Douglas, *The Creation of a National Air Force*, 353–54.
18 American Minister, Ottawa, "Memorandum of conversation with Adm. Nelles, Chief of the Naval Staff," to the Secretary of State, Washington, December 11, 1941, NARA, RG 337, Headquarters, Army Field Forces, General Correspondence, 1940–42, entry 57, file 334.8 "Joint Boards Canada-United States"; "Unity of Joint Defence Command Not Essential, Ottawa Declares," *Gazette* (Montreal) December 11, 1941; Pope Memorandum, "Note on Question of United States–Canada Unity of Command," December 18, 1941, DHH 112.3M2 (D495).
19 La Guardia to Biggar, January 2, 1942, NARA, RG 59, State Department files, PJBD, box 9, file correspondence of PJBD, January to March 1942.
20 Ibid.
21 "Notes for Meeting Canadian Section, Permanent Joint Board on Defence," January 3, 1941, LAC, RG 24, vol. 5209, file HQS 15-73-4, pt. 1. Large-scale Japanese operations against the Aleutians "as a Japanese defensive measure against future bombing of Japan" from the Alaskan islands, however, was "in a different category." For Canadian–American operations in the Aleutians during the war, see Perras, *Stepping Stones to Nowhere*.
22 CSC 135th Meeting Minutes, January 7, 1942, Joint Staff Fonds, DHH 2002/17, box 69, file 14.

23 CNO memorandum, "Unity of command over joint operations," to CoS US Army, n.d. [December 2, 1941], NARA, RG 165, Files of the War Department, file WPD 2917-35; Perras, *Franklin Roosevelt and the Origins*, 84.
24 CSC 135th Meeting Minutes, January 7, 1942, Joint Staff fonds, DHH 2002/17, box 69, file 14.
25 Allan E. Gotlieb, "Canada-US Relations: The Rules of the Game," *SAIS Review* 2 (Summer 1982): 172-87. See also Brian Bow, *Politics of Linkage: Power, Interdependence, and Ideas in Canada-US Relations* (Vancouver: UBC Press, 2010), 7-15, 23. A more recent example of Canada capitalizing on the American political system and culture in this way is the Mulroney government's ability to secure an agreement with the United States in 1991 on how to combat the acid rain problem. In this and other instances, the "human element" aspect of a personal approach or "personal diplomacy" in addition to official institutional and diplomatic channels has been vital for Canada's ability to play Washington politics. On the acid rain example, see Brian Mulroney, "Acid Rain: A Case Study in Canada-US relations," *Policy Options* (April 1, 2012); "Leveraging Canada-US Relations 'to Get Big Things Done (Interview with Brian Mulroney)," *Policy Options* (March 1, 2011); Allan Gotlieb, *The Washington Diaries, 1981-1989* (Toronto: McClelland and Stewart, 2006). I thank an anonymous reviewer for pointing out this factor and Adam Chapnick for his suggestions of examples.
26 CSC 135th Meeting Minutes, January 7, 1942, Joint Staff fonds, DHH 2002/17, box 69, file 14; Pope Memorandum to Mackenzie King, January 13, 1942, and Biggar to Mackenzie King, January 13, 1942, reproduced in Hilliker, ed., *Documents on Canadian External Relations* [hereafter *DCER*], 9, 1942-43 (Ottawa: Department of External Affairs and International Trade Canada, 1990), 1151-53; Embick to Biggar, January 14, 1942, NARA, RG 59, State Department files, PJBD, box 9, file correspondence of PJBD, January to March 1942; Pope Memorandum, "Southern British Columbia-Puget Sound US Request for Institution of Unity of Command" to Stuart, January 16, 1942, DHH 112.11 (D1A), vol. 3; Pope to Embick, January 19, 1942, DHH 112.11 (D1A), vol. 3; CSC 138th Meeting Minutes, January 17, 1942, Joint Staff fonds, DHH 2002/17, box 17, file 14; Minutes of the 25th Meeting of the PJBD, Montreal, January 20, 1942, DHH 82/196, file 2, PJBD Journals, vol. 2.
27 CSC 151st Meeting Minutes, March 18, 1942, LAC, RG 2, 7C, vol. 8, microfilm reel # C-4874, copy at CFC IRC; Memorandum from CSC to Ministers of National Defence, March 10, 1942, reproduced in Bridle, ed., *DRBCN*, 926-27; Dziuban, *Military Relations between the United States and Canada, 1939-1945*, 119; Douglas, *The Creation of a National Air Force*, 393. Quotes from latter.
28 Ibid., 393.
29 SSEA to High Commissioner in Newfoundland, December 12, 1941, LAC, RG 25, Vol. 1991, File 1156-C; Stacey, *Arms, Men, and Governments*, 363; Dziuban, *Military Relations between the United States and Canada, 1939-1945*, 117-18. Murray's title changed from Commodore Commanding Newfoundland Escort Force to Flag Officer Newfoundland Force (FONF). Jones's title remained Commanding Officer Atlantic Coast.
30 Extract from Gen. Page's Diary, January 15, 1942, DHH 321.009 (D201). Page, "Memorandum of a Conference between Major General G.C. Brant Commanding US Forces in Newfoundland and Major General L.F. Page Commanding Canadian Forces in Newfoundland," January 5, 1942, DHH 355.009 (D29); Page to Elkins and Elkins to Page, January 20, 1942, DHH 355.009 (D29).
31 Waller, *A Question of Loyalty*, 190; Douglas, *The Creation of a National Air Force*, 496.
32 Extract from General Page's Diary, June 18, 1942, DHH 321.009 (D201); DRAFT Memorandum by Page to Murray and McEwen, n.d. [June 25, 1942], LAC, RG 24, vol. 11,927,

file 1550-146/36-2. Despite incidents like this, the United States awarded McEwen with the Legion of Merit for his services in Newfoundland "during the critical days of the Battle of the Atlantic." Burchnell to McEwen, October 21, 1946, Air Vice-Marshal C.E. McEwen Papers, Canadian War Museum Archives, file 58A1 95.4, 1946-1956, 1900192.

33 AOC No. 1 Group to AOC EAC, n.d. [18 June 1942], quoted in AOC EAC to Air Force Headquarters, Ottawa (AFHQ), June 19, 1941, LAC, RG 25, vol. 1991, file 1156-C.
34 DRAFT Memorandum by Page to Murray and McEwen, n.d. [June 25, 1942], and Murray to Page, June 25, 1942, LAC, RG 24, vol. 11,927, file 1550-146/36-2.
35 Drum to Elkins, August 21, 1942, CSC Memoranda, Joint Staff fonds, DHH 2002/17, box 65, file 11; CSC 177th and 178th Meeting Minutes, August 25 and September 1, 1942, Joint Staff fonds, DHH 2002/17, box 70, file 1; Secretary, CSC, to GOCinC Atlantic Command, September 3, 1942, LAC, RG 25, vol. 1991, file 1556-C.
36 CWC 199th Meeting Minutes, October 14, 1942, LAC, RG 2, 7C, vol. 11, microfilm reel # C-4874, copy at CFC IRC; La Guardia to Drum, October 1, 1942, copy in DHH 99/36, W.A.B. Douglas fonds, box 14, file 1.
37 Brant to Drum, n.d., quoted in Drum to Marshall, May 13, 1942, G.C. Marshall Papers, microfilm reel 5, Subject: 1942 Unity of Command in Newfoundland, copy at ML, RMC; La Guardia to Drum, October 1, 1942, copy in DHH 99/36, W.A.B. Douglas fonds, box 14, file 1. Quote from latter.
38 Stacey, *Arms, Men, and Governments*, 366.
39 Douglas, *The Creation of a National Air Force*, 378; Douglas et al., *No Higher Purpose*, vol. 2, pt. 1, 183.
40 Morrison, *History of the United States Naval Operations in World War II*, vol. 2, 69; Douglas, *The Creation of a National Air Force*, 387, 187; Douglas, "Alliance Warfare," 166; Admiralty to BAD Washington, July 11, 1941, DHH 75/191.
41 Marc Milner, *North Atlantic Run: The Royal Canadian Navy and the Battle for the Convoys* (Toronto: University of Toronto Press, 1986), 58-59; COMINCH to Commander Task Force Four, Rear Admiral A.L. Bristol, USN, October 10, 1941, LAC, RG 24, vol. 11,505, file 1550-146/36-1. Quote from latter.
42 Douglas, *The Creation of a National Air Force*, 387-88.
43 AOC EAC to AFHQ, September 21, 1941, and G/C Heakes to CAS and AOC EAC, September 23, 1941, LAC, RG 24, vol. 5177, S. 15-1-350, pt. 1.
44 Stacey, *Arms, Men, and Governments*, 362; USN CNO to CAS RCAF, October 2, 1941, LAC, RG 24, vol. 5177, S. 15-1-350, pt. I. Quote from latter.
45 Air Vice-Marshal G. Johnson, DCAS, to C.G. Power, MND for Air, October 9, 1941, LAC, RG 24, vol. 5177, S. 15-1-350, pt. 1; Joint Canadian-United States Basic Defence Plan No. 2 (Short Title ABC-22), July 28, 1941, DHH 355.009 (D20).
46 CWC 111th Meeting Minutes, October 9, 1941, LAC, RG 2 7C, vol. 5, microfilm reel # C-4654, copy at CFC IRC.
47 RCAF CAS to USN CNO, October 15, 1941, LAC, RG, 24, vol. 5177, S. 15-1-350, I; Minutes of the 112th Meeting of the Cabinet War Committee, October 15, 1941, LAC, RG 2, 7C, vol. 5, microfilm reel # C-4654, copy at CFC IRC; Douglas, "Democratic Spirit," 40.
48 USN CNO to RCAF CAS, October 27, 1941, LAC, RG 24, vol. 5177, S. 15-1-350, I; CWC 115th Meeting Minutes, November 6, 1941, LAC, RG 2, 7C, vol. 6, microfilm # C-4654, copy at CFC IRC; minute by CAS to Power on Admiral Stark's letter, November 3, 1941, LAC, RG 24, vol. 5177, S. 15-1-350, I. Quote from latter.
49 Douglas, *The Creation of a National Air Force*, 389; Dziuban, *Military Relations between the United States and Canada, 1939-1945*, 124, 118.

50 Pope to Ralston, January 2, 1942, DHH 112.11 (D1A), vol. 3; 22nd PJBD Recommendation, December 20, 1941, DHH 79/35. President Roosevelt approved the recommendation on December 24, 1941, followed by the Canadian government on January 14, 1942.
51 Dziuban, *Military Relations between the United States and Canada, 1939-1945*, 124, 117; PJBD Journal, December 20, 1941, reproduced in Bridle, ed., *DRBCN*, 910.
52 AOC EAC to AFHQ, December 20, 1941, DHH 79/237.
53 *Joint Action of the Army and the Navy*, FTP-155, Joint Board, November 15, 1935, NARA, RG 165, WPD 2917-35.
54 Craven and Cate, eds., *Army Air Forces in World War II*, I, 521-24; Timothy A. Warnock, *Air Power versus U-boats: Confronting Hitler's Submarine Menace in the European Theater* (Washington, DC: Office of Air Force History, 1999), 2; Kenneth Schaffel, *The Emerging Shield: The Air Force and the Evolution of Continental Air Defense, 1945-1960* (Washington, DC: Office of Air Force History, United States Air Force, 1991), 3-4; Meilinger, "The Historiography of Airpower," 476.
55 The concept of the indivisibility of air power dictates that all military air assets of a nation – including maritime air power – should be under a separate service, the air force, to ensure the proper concentration and specialized use of air power in the hands of those best trained for it: air force officers. James A. Winnefeld and Dana J. Johnson, *Joint Air Operations: Pursuit of Unity in Command and Control, 1942-1991* (Annapolis: Naval Institute Press, 1993), 7; David MacIsaac, "Voices from the Central Blue: The Air Power Theorists," in *Makers of Modern Strategy from Machiavelli to the Nuclear Age*, ed. Peter Paret (Princeton: Princeton University Press, 1986), 631.
56 AOC EAC to AFHQ, December 21, 1941, DHH 79/237; Murray to Secretary of the Naval Board, May 16, 1942, DHH 355.009 (D20).
57 Douglas, *The Creation of a National Air Force*, 389-90; Cohen and Gooch, *Military Misfortunes*, 74. The USAAF's 49th Bombardment Squadron is a good example. It was scheduled to arrive in Newfoundland in early December 1941 but had been delayed at Mitchell Field in New York due to poor weather. After Pearl Harbor, instead of proceeding to Gander, the squadron was diverted to Hawaii. See Air Commodore Cuffe for CAS to AOC EAC, December 23, 1941, LAC, RG 24, vol. 5174, file HQS 15-1-204.
58 AOC EAC to AFHQ, December 22, 1941, DHH 79/237.
59 Ibid.; Stacey, *Arms, Men, and Governments*, 363-64; Douglas, *The Creation of a National Air Force*, 389-90.
60 Dziuban, *Military Relations between the United States and Canada, 1939-1945*, 124.
61 "Air Coverage," Notes on Convoy Organization Western North Atlantic, March 20, 1942, DHH 181.002 (D156).
62 See correspondence between the Canadian and American air forces in Newfoundland from December 20, 1941, to March 1942, DHH 181.002 (D173). See also Douglas et al., *No Higher Purpose*, 595.
63 See Chapter 4, "Failure to Learn: American Antisubmarine Warfare in 1942," in Cohen and Gooch, *Military Misfortunes*.
64 On navy complaints of scanty air coverage, see "Review of Conditions by Local Escorts (March to September, 1942) and Suggestions for Increasing the Effectiveness," Memorandum by Commander J.M. Rowland, RN, captain of HMS *Walker*, September 21, 1942, DHH 181.002 (D121).
65 The best source on Coastal Command's efforts during the Second World War remains John Buckley, *The RAF and Trade Defence, 1919-1945: Constant Endeavour* (Keele: Ryburn Publishing, Keele University Press, 1995).
66 Report on Visit of EAC Halifax and RCAF Station, Dartmouth, N.S., by Wing Cdr S.R. Gibbs, RAF Coastal Command, n.d. [July 1942], and Report by Commander P.B.

Martineau, R.N. (from staff of RAF Coastal Command) on Visit to Eastern Air Command and Western Air Command, October 31 and November 8, 1942, TNA, PRO, Air 15/217.
67 Martineau, Memorandum to Staff, October 31, 1942, TNA, PRO, Air 15/217; Douglas, *The Creation of a National Air Force*, 530.
68 Ibid., 531. See also Chapters 2 and 3 of W.G.D. Lund, "Command Relationships in the North West Atlantic, 1939–1943," MA thesis, History, Queen's University, 1972; and Chapter 11 of Douglas et al., *No Higher Purpose*.
69 Douglas, "Alliance Warfare," 168; Douglas et al., *No Higher Purpose*, 595; Douglas, *The Creation of a National Air Force*, Appendix E.
70 Canning to Chief of Staff to the Commander Eastern Sea Frontier, Liaison Officer, Eastern Theatre of Operations and 1st Air Force, and Executive Officer, I Bomber Command, February 17, 1942, TNA, PRO, Air 15/217; Craven and Cate, eds., *Army Air Forces in World War II*, vol. 1, 541, and vol. 2, 393; Douglas, *The Creation of a National Air Force*, 546.
71 AFHQ to AOC EAC, February 3, 1943, and CNS to COAC, February 4, 1943, LAC RG 24, vol. 5270, S.28-1-2; DCAS Memorandum to Power, February 4, 1943, DHH 77/528; RCAF HQ Ottawa to Air Force, Combined Staff, Washington, February 5, 1943, DHH 77/528; Douglas, *The Creation of a National Air Force*, 547. It was also consistent with the fundamental air power tenet of centralized control and decentralized execution, which entails the most efficient use of limited air assets. Canada, RCAF, *Royal Canadian Air Force Doctrine*, B-GA-400-000/FP-001, November 2016 (Trenton: Canadian Forces Aerospace Warfare Centre, 2016), 16–17.
72 "Report of Sub-Committee on Command Relations," March 9, 1943, Atlantic Convoy Conference Minutes, 181.003 (D5027); Lund, "Command Relationships," 51; Goette, "Joint Command and Control," 92–93; Douglas et al., *No Higher Purpose*, 624–30; W.A.B. Douglas, Roger Sarty, Michael Whitby, with Robert H. Caldwell, William Johnston, and William G.P. Rawling, *Blue Water Navy: The Official Operational History of the Royal Canadian Navy in the Second World War, 1939-1943, Volume II, Part 2* (St. Catharines: Vanwell Publishing, 2007), 23–24.
73 Report of Sub-Committee on Command Relations, March 9, 1943, Atlantic Convoy Conference Minutes, DHH 181.003 (D5027).
74 No. 1 Group to EAC HQ, March 30, 1943, DHH 181.002 (D124).
75 Douglas, *The Creation of a National Air Force*, 549.
76 Goette, "Joint Command and Control," 71; Douglas et al., *Blue Water Navy*, 23.
77 Dziuban, *Military Relations between the United States and Canada, 1939–1945*, 126.

Chapter 5: Replacing ABC-22
Epigraphs: US Ambassador to Canada to Secretary of State, August 28, 1946, NARA, RG 59, State Department Files, PJBD, box 10, file Correspondence of PJBD, January to December 1946; quoted in Grant, *Sovereignty or Security*, 344.
1 Maj.-Gen. Maurice Pope to Col. J.H. Jenkins, April 4, 1944, LAC, RG 25, vol. 5749, file 52C(s), pt. 1. He added: "Should then, the United States go to war with Russia they would look to us to make common cause with them, and, as I judge their public opinion, they would brook no delay."
2 Ibid. "If," Pope added, "we do enough to assure the United States we shall have done a good deal more than a cold assessment of the risk would indicate to be necessary." See also Memorandum from First Secretary to Under-Secretary of State for External Affairs, February 29, 1944, reproduced in Hilliker, ed., *DCER*, vol. 11, 1944–1945, pt. 2, 1400–5. Pope also elaborated on this subject further a couple of months later. See Chairman,

Canadian Joint Staff Mission, Washington, to Department of National Defence, June 27, 1944, in ibid., 1540–41.
3 Quoted in Eayrs, *In Defence of Canada*, vol. 3, 320.
4 Jockel, *No Boundaries Upstairs*, 6–7; Robert Frank Futrell, *Ideas, Concepts, Doctrine: Basic Thinking in the United States Air Force 1907–1960* (Montgomery: Air University Press, 1989), 222–23. As this USAF official history notes, USAF Chief of Staff General Carl Spaatz argued that "the only way to prevent bombs from falling on the United States 'is to get them at the place they start from,' and that is primarily our mission." Ibid., 222.
5 Scott Sagan notes that "at the highest levels of the government ... U.S. military and political leaders quickly realized the revolutionary character of atomic power." Scott D. Sagan, *Moving Targets: Nuclear Strategy and National Security* (Princeton: Princeton University Press, 1989), 14.
6 Bernard Brodie, ed., *The Absolute Weapon: Atomic Power and World Order* (New York: Harcourt Brace, 1946), 75–76; Richter, *Avoiding Armageddon*, 3–4. Quote from former. Brodie's writings also appeared in the RCAF's professional military journal, the *RCAF Staff College Journal*, during the early Cold War period. See Bernard Brodie, "Implications of Nuclear Weapons in Total War," *The R.C.A.F. Staff College Journal* (1957): 12–22.
7 Eayrs, *In Defence of Canada*, vol. 3, 274; John Hilliker, *Canada's Department of External Affairs*, vol. 1: *The Early Years, 1909–1946* (Kingston: Institute of Public Administration of Canada and McGill-Queen's University Press, 1990), 307; Richter, *Avoiding Armageddon*, 19. Quote from former. Howe was the Minister of Munitions and Supply and Mackenzie was head of the National Research Council. A few Canadian scientists were also involved in the Manhattan Project.
8 J.W. Pickersgill and D.F. Foster, *The Mackenzie King Record*, vol. 2: *1944–1945* (Toronto: University of Toronto Press, 1968), 447–51; Eayrs, *In Defence of Canada*, vol. 3, 276.
9 Richter, *Avoiding Armageddon*, 19–22; Eayrs, *In Defence of Canada*, vol. 3, Chapter 5; Denis Smith, *Diplomacy of Fear: Canada and the Cold War 1941–1948* (Toronto: University of Toronto Press, 1988), 92–93; Hilliker, *Canada's Department of External Affairs*, vol. 1, 307–8. Former CGS and MND Lt.-Gen. (Ret.) Andy McNaughton was the first president of the commission. It could be argued that the Atomic Energy Commission was an early example of Canada exercising the diplomatic functionalism principle.
10 Quoted in Smith, *Diplomacy of Fear*, 93. Andrew Burtch nicely captures the implications of the revolutionary weapon: "The Second World War ended with mushroom clouds over the Japanese cities of Hiroshima and Nagasaki. The atomic bombing of Japan killed more than 100,000 and injured a further 90,000 people. The intense light and heat released by the blasts created horrific burn injuries, and thousands of people displayed devastating symptoms of radiation poisoning. The atomic weapon cast a long shadow over the world. Its terrible potential threatened the future survival of civilization itself. Atomic bombs, and their massively more destructive thermonuclear cousins, reshaped international relations, military strategy, and the manner in which the world imagined its future. As military professionals around the world struggled to incorporate the weapon into their plans, governments looked on at the ruins of Hiroshima and Nagasaki and were forced to grapple with the prospect of the destruction of civilian targets on a massive scale." Andrew Burtch, *Give Me Shelter: The Failure of Canada's Cold War Civil Defence* (Vancouver: UBC Press, 2012), 15.
11 J.I. Jackson, "An Article on Air Power," in *Readings in Air Power: RCAF Officers' Examinations Study Material*, ed. Clare Annis (Ottawa: Training Command, 1955), 11–12, copy available at the Keith Hodson Memorial Library, IRC, CFC.

12 On the Gouzenko Affair, see Reg Whitaker and Gary Marcuse, *Cold War Canada: The Making of a National Insecurity State, 1945–1957* (Toronto: University of Toronto Press, 1995).
13 Yefim Gordon and Vladimir Rigmant, *Tupolev Tu-4: Soviet Superfortress* (Hinkley: Midland Publishing, 2002). See esp. Chapter 1, "Presents from America." See also Andrew B. Godefroy, *In Peace Prepared: Innovation and Adaptation in Canada's Cold War Army* (Vancouver: UBC Press, 2014), 80, 241–42.
14 Report of the Advisory Committee on Post-Hostilities Problems, "Post-War Canadian Defence with the United States: General Considerations," January 23, 1945, *DCER*, vol. 11, 1567–73.
15 David Cox, *Canada and NORAD, 1958–1978: A Cautionary Retrospective*, Aurora Papers no. 1 (Ottawa: Canadian Centre for Arms Control and Disarmament, 1985), 7; Trudgen, "The Search for Continental Security," 21, 102–3, 132, 141–42, 250; Schaffel, *The Emerging Shield*, 54–55, 64, 83; Richter, *Avoiding Armageddon*, 34; Sagan, *Moving Targets*, 15–18.
16 Richter, *Avoiding Armageddon*, 17; Souffer, *Swords, Clunks, and Widowmakers*, 13; Goette, "A Snapshot of Early Cold War RCAF Writing," 52; Barnes, "'Fighters First.'"
17 On Canadian strategic interest and continental air defence, see Richter, *Avoiding Armageddon*, 46–47, 57. I thank an anonymous reader for linking national strategic culture to Canada's functional approach to protecting sovereignty.
18 Canada–United States outline security plan CA-1, January 17, 1946, NARA, RG 59, State Department files, PJBD, box 2, file Basic Security Plan II.
19 Jockel, *No Boundaries Upstairs*, 15. To be fair, at least one JCS planner defended US PJBD members, stating that they, and Maj.-Gen. Henry in particular, had "not been provided with adequate help" and that due to this lack of guidance had resorted to "approaching the problems in only very broad terms." 241st Meeting of the Joint Planning Staff, March 13, 1946, NARA, RG 218, Records of the JCS, entry 943011, box 018, file CCS 092 (1–10–45) section 2.
20 Maj.-Gen. Pope Memorandum, "Combined Chiefs of Staff," May 30, 1944, DHH 2002/17, box 68, file 4; Adrian W. Preston, "Canada and the Higher Direction of the Second World War 1939–1945," in Hunt and Haycock, eds., *Canada's Defence*, 110–13; Stacey, *Arms, Men, and Governments*, 164–65.
21 Jockel, *No Boundaries Upstairs*, 16; Director of Military Operations and Planning (DMO&P) Memorandum "Canada–US Collaboration" to CGS, January 20, 1946, DHH 112.3M2.009 (D151), vol. 1.
22 Jockel, *No Boundaries Upstairs*, 15–16; 241st Meeting of the Joint Planning Staff, March 13, 1946, NARA, RG 218, Records of the JCS, entry 943011, box 018, file CCS 092 (1–10–45), section 2; JSP 788/4, Report by the Joint Staff Planners, "Canadian-US Military Cooperation," March 16, 1946, NARA, RG 218, Records of the JCS, Entry 943011, box 018, file CCS 092 (1–10–45), section 3; JCS 1541/5, "Canadian–US Military Cooperation," March 21, 1946, NARA, RG 59, State Department files, PJBD, box 2, file Basic Security Plan II; Adm. Leahy Memorandum for the Service Members, Permanent Joint Board of [sic] Defense, Canada–US, March 30, 1946, NARA, RG 59, RG 218, JCS, entry 940311, box 018, file CCS 092 (1–10–45), section 3.
23 The most famous example of the United States' desire to avoid alliances that would limit its strategic freedom is, of course, President George Washington's farewell address of 1796. Another is Thomas Jefferson's inaugural pledge of "peace, commerce, and honest friendship with all nations – entangling alliances with none." "Washington's Farewell Address 1796," The Avalon Project: Documents in Law, History and Diplomacy, Yale Law

School, Lillian Goldman Law Library, http://avalon.law.yale.edu/18th_century/washing.asp; David Fromkin, "Entangling Alliances," *Foreign Affairs* (July 1970). Quote from latter.
24 Robert Endicott Osgood, *NATO: The Entangling Alliance* (Chicago: University of Chicago Press, 1962). See also Sean Maloney, *Learning to Love the Bomb: Canada's Nuclear Weapons during the Cold War* (Washington, DC: Potomac Books, 2007), Chapters 1–3.
25 Pearson to Robertson, January 19, 1946, LAC, RG 25, vol. 5749, file 52-C(s).
26 The DEA representative reported directly to A.D.P. Heeney, Secretary to the Cabinet. DHH 87/47, Raymont, *The Evolution of the Structure*, Appendix A, 10, 27–28; US PJBD Section to Secretary of War, January 21, 1946, NARA, RG 59, State Department files, PJBD, box 5, file Military Cooperation Committee I; Grant, *Sovereignty or Security*, 344; Jeff Noakes, "Air Force Architect: Air Marshal Wilfred Curtis, Chief of the Air Staff, 1947–1953," in *Warrior Chiefs: Perspectives on Senior Canadian Military Leaders*, ed. Lt.-Col. Bernd Horn and Stephen Harris (Toronto: Dundurn Press, 2001), 243.
27 Memorandum from the Senior US Army Member, PJBD, to the Secretary of the Canadian Section, PJBD, March 18, 1946, LAC, RG 25, vol. 5750, file 52-C(s), pt. 2.2; Memorandum by Secretary, Canadian Section, PJBD, January 18, 1946, *DCER*, vol. 12, 1600–1.
28 JWPC 433/2, "Matchpoint," Report by the Joint War Plans Committee, May 6, 1946, NARA, RG 218, Records of the JCS, entry 943011, box 018, file CCS 092 (1–10–45), section 3. Emphasis in original.
29 Ibid.
30 JWPC 433/2, Appendix "A," May 6, 1946, NARA, RG 218, Records of the JCS, Entry 943011, box 018, file CCS 092 (1–10–45), section 3.
31 Ibid.
32 JWPC 433/2, Appendix "B," "Basic Outline of Joint Canadian–US Basic Defence Plan" (Draft), May 6, 1946, NARA, RG 218, Records of the JCS, entry 943011, box 018, file CCS 092 (1–10–45), section 3.
33 MCC paper, "An Appreciation of the Requirements for Canadian–US Security," May 23, 1946, reproduced in *DCER*, vol. 12, 1946, 1617–23; Jockel, *No Boundaries Upstairs*, 17–18; Eayrs, *In Defence of Canada*, vol. 3, 337. The appreciation also did not discount the use of biological weapons.
34 Ibid.
35 Joint Canadian–United States Basic Security Plan (Draft), June 5, 1946, LAC, RG 25, vol. 5750, file 52-C(s) pt. 2.2. Newfoundland did not enter Confederation until 1949, and the US retained its base rights in the colony after the war as per its wartime agreement with the British government.
36 Other priorities had to do with familiarizing and preparing Canadian and American forces to operate in the still relatively little known areas of the Far North. Specifically, they included a program of air photography, mapping and charting; familiarization of operations in and testing of equipment and clothing in Arctic conditions; the collection of scientific data in the Arctic; and the "collection of strategic information necessary for military operations in Canada, Newfoundland and Alaska." Section VII, BSP, June 5, 1946, LAC, RG 25, vol. 5750, file 52-C(s), pt. 2.2. See also Memorandum by Air Vice-Marshal W.A. Curtis, Senior Canadian Member, MCC, to Chiefs of Staff Committee, June 18, 1946, DHH 112.3M2 (D116).
37 Section IX, BSP, June 5, 1946, LAC, RG 25, vol. 5750, file 52-C(s). pt. 2.2.
38 Ibid., Section VIII, BSP.
39 Leighton, "Allied Unity of Command," 407.
40 JCS, "Memorandum for the Senior US Army and Navy Members, Canadian–US Military Cooperation Committee," July 2, 1946, NARA, RG 333, files of International Military Agencies – PJBD, entry 17-A, box 1, file "Top Secret General Correspondence,

1941–1956," folder 2; Minutes of the 357th Meeting of the Chiefs of Staff Committee, July 11, 1946, Chiefs of Staff Committee Minutes, DHH 193.009 (D53); Chiefs of Staff Committee Memorandum to the Cabinet Defence Committee, "Canada–United States Joint Appreciation and Basic Security Plan; comments thereon by the Chiefs of Staff," July 15, 1946, DHH 112.1 (D178).

41 The implementation programs were based on the capability of them being put "in full operation at the time required in accordance with the current strategic appreciation which is reviewed not less than annually." DHH 87/47, Raymont, *Evolution of the Structure*, Appendix A, 10; Mann, "Situation Report on the Canada–United States Basic Security Plan," December 11, 1947, DHH 327.009 (D201); MCC Memorandum, "Procedures Followed in Canada in Preparation of Canada–US Basic Security Plan and Implementation Programmes," May 5, 1948, DHH 112.3M2.009 (D106).

42 MCC Memorandum, "Submission of Appendices of the Basic Security Plan to Higher Authority," December 11, 1946, NARA, RG 218, JCS, entry 943011, box 019, file CCS 092 (1–10–45), section 6.

43 DMO&P to Vice-Chief of the General Staff (VCGS), June 17, 1946, DHH 112.3M2 (D116).

44 Haydon, *The 1962 Cuban Missile Crisis*, 69.

45 As of December 1947, there were twelve appendices either already approved or in preparation by the MCC. Once approved, they were attached to the BSP and subject to further revisions as changes in the strategic situation, military technological advances, and so on, dictated. In the words of Maj.-Gen. C.C. Mann, the Vice-Chief of the General Staff and also the senior Canadian Army member of the MCC, "the Basic [Security] Plan states the problems, [and] the Appendices evolve the methods by which they are to be met." Maj.-Gen. C.C. Mann, "A Situation Report on the Canada–United States Basic Security Plan," December 11, 1947, DHH 327.009 (D201).

46 Jockel, *No Boundaries Upstairs*, 19.

47 (Draft) RCAF Study, "Command Relations – Canada–US," August 25, 1947, JPC, Minutes to Meetings and Correspondence vol. 7, DHH 2002/17, box 55, file 1.

48 Minutes to the 106th Meeting of the Joint Planning Committee, September 9, 1947, LAC, RG 24, vol. 8083, file 2172–10–10, pt. 3.

49 Planning Group of the Military Cooperation Committee, Appendix A to the Canada–United States Basic Security Plan, (Draft) Air Interceptor and Air Warning Plan, December 9, 1946, DHH 112.3M2.009 (D106), vol. 1. Again, the Soviet Union was not specifically named as the enemy in the AIAW plan, but there was no other probable adversary. The AIAW appendix did not cover ground anti-aircraft artillery defences, as they were included in Appendix "G," Anti-Aircraft Ground Defence.

50 AIAW Plan, December 9, 1946, DHH 112.3M2.009 (D106), vol. 1; Memorandum by the MCC, "Air Interceptor and Air Warning Appendix to the Joint Canadian–United States Basic Security Plan," December 10, 1946, NARA, RG 59, State Department files, MCC, box 25, file MC Committee I.

51 AIAW Plan, December 9, 1946, DHH 112.3M2.009 (D106), vol. 1. See also Mann, "Situation Report on the Canada–United States Basic Security Plan," December 11, 1947, DHH 327.009 (D201).

52 AIAW Plan, December 9, 1946, DHH 112.3M2.009 (D106), vol. 1.

53 Memorandum by the MCC, "Air Interceptor and Air Warning Appendix to the Joint Canadian–United States Basic Security Plan," December 10, 1946, NARA, RG 59, State Department files, MCC, box 25, file MC Committee I.

54 The AIAW Plan was approved "as a basis for long range planning, subject to continued revision in the light of improved equipment and weapons." Minutes of the 400th Meeting

of the Chiefs of Staff Committee, September 4, 1947, Chiefs of Staff Committee Minutes 1947, Raymont Collection, DHH 73/1223/1302; Jockel, *No Boundaries Upstairs*, 27–29; Stouffer, *Swords, Clunks, and Widowmakers*, 17.

55 This will be discussed in further detail in the next chapter. MCC Report to the Canadian Chiefs of Staff Committee and the JCS on Implementation Measures for the Air Interceptor and Air Warning Appendix to the Canada–United States Basic Security Plan for the Period from April 1, 1949, to June 30, 1950, March 18, 1948, NARA, RG 59, State Department records, PJBD-MCC, entry E1184, box 25, file "Military Cooperation Committee II."

56 MCC Memorandum, "Programme for Implementation Measures for the Period from 1st April 1948 to 30th June 1949," July 25, 1947, DHH 112.1 (D178).

Chapter 6: Organizing and Coordinating Canada–US Air Defences

1 Ann and John Tusa, *The Berlin Blockade* (Toronto: Hodder and Stoughton, 1988); Memorandum for the Senior US MCC Members, May 19, 1948, NARA, RG 218, JCS, entry 943011, box 035, file CCS 092 (1–10–45), section 11; Jockel, *No Boundaries Upstairs*, 32.

2 MCC Memorandum for the Chairman of MCC Sub-Committees, August 9, 1948, JPC Minutes, vol. 9, DHH 2002/17, box 55, file 3; Draft MCC 200, Canada–United Sates Basic Security Program, December 10, 1948, DHH 112.3M2.009 (D106); MCC Memorandum for the Chairman of MCC Sub-Committees, December 13, 1948, NARA, RG 341, Records of HQ USAF, Deputy Chiefs of Staff Operations, Director of Communications – Electronics, Reserve Force Group, general file 1946–1949, #28 and #29, box 18, file MCM, April 14, 1949 (supersedes MCCM-60, February 7, 1949); MCC 100/4, Canada–United States Basic Security Plan, August 5, 1948, NARA, RG 341, Records of HQ USAF, Deputy Chiefs of Staff Operations, Director of Communications – Electronics, Reserve Force Group, general file 1946–1949, box 17, #27 to #28, bk. 2, Canadian–US MCC Subcommittee on Air Interceptor and Air Warning System, December 1946 to October 1948 (MCCM papers). I thank Matt Trudgen for sharing copies of NARA files with me.

3 MCC 300/1, Canada–United States Emergency Defense Plan, March 25, 1949, NARA, RG 218, JCS, entry 943011, file CCS 092 (1–10–45), section 16.

4 On the development of the Soviet Navy during the early Cold War, see Jürgen Rohwer and Mikhail S. Monakov, *Stalin's Ocean-Going Fleet: Soviet Naval Strategy and Shipbuilding Programmes, 1935–1953* (London: Frank Cass, 2001), 216–17, Chapter 11; Donald W. Mitchell, *A History of Russian and Soviet Sea Power* (New York: MacMillan, 1974), Chapter 21.

5 MCC 300/1, Canada–United States Emergency Defense Plan, March 25, 1949, NARA, RG 218, JCS, entry 943011, file CCS 092 (1–10–45), section 16.

6 Draft Appendix "E" Command to Canada–US Emergency Defence plan, n.d. [November 1948], JPC Minutes, DHH 2002/17, box 55, file 4.

7 Richter, *Avoiding Armageddon*, 23; Trudgen, "The Search for Continental Security," 131; Schaffel, *The Emerging Shield*, 54–55.

8 JPC Memo, "Background: Canada–United States Emergency Defence Plan (MCC 300/2)," May 22, 1951, and Extract from the Minutes of the CSC 446th Meeting, April 26, 1949, DHH 112.3M2.009 (D106), vol. 3; 161st JPC Meeting Minutes, May 5, 1949, JPC Minutes, DHH 2002/17, box 52, file 5; Secretary JPC to Secretary CSC, August 29, 1949, and Brigadier General Kitching, BGS Plans, to CGS, September 12, 1949, CSC Papers, November 1948, DHH 2002/17, box 75, file 8.

9 Appendix "E" Command (MCC 305), to MCC 300/1, EDP, March 25, 1948, NARA, RG 218, JCS, entry 943011, file CCS 092 (1–10–45), section 16. See also Memorandum

from Acting Head, Defence Liaison Division to Under-Secretary of State for External Affairs, April 11, 1949, *DCER*, 15, 1560-61.
10 Appendix "E" Command (MCC 305), to MCC 300/1, EDP, March 25, 1948, NARA, RG 218, JCS, entry 943011, file CCS 092 (1-10-45), section 16.
11 US Navy Department Memo by W. Miller to Rear-Adm., C.D. Glover, Jr., USN, Rear-Adm. W.F. Boone, USN, Brig.-Gen. C.V.R. Schuyler, US Army, and Col. J.S. Cary, USAF, "Canada-United States Basic Security Plan - Report of Status," May 23, 1949, NARA, RG 333, Files of International Military Agencies - PJBD, entry 17-A, box 2, file "Top Secret Correspondence, 1941-1956," folder 11; Extract from Minutes of the 482nd Chiefs of Staff Committee Meeting, October 6, 1949, Chiefs of Staff Committee Papers, November 1948, DHH 2002/17, box 75, file 8.
12 The best account of the radar air defence system in the 1950s is Chapters 3 and 4 in Jockel, *No Boundaries Upstairs*; and C.L. Grant, *The Development of Continental Air Defense to 1 September 1954*, USAF Historical Study no. 126 (Montgomery: USAF Historical Division, Research Studies Institute, Air University [1954]).
13 EDP, MCC 300/2, May 23, 1950, NARA, RG 218, JCS, entry 943011, file CCS 092 (1-10-45), section 22; MCC to US JCS and CSC, May 25, 1950, NARA, RG 341, entry 335, box 723, Air Force, Plans, Project Decimal file 1942-54, file Canada 600.2 (May 12, 1948), section 3; DEA Defence Liaison Memorandum, "Revised Emergency Defence Plan ("MCC 300/2") approved by MCC on May 25," June 29, 1950, LAC, MG 30, E133, McNaughton Papers, file PJBD, Misc. Papers, Mitchell Field, Langley, Ft. Monroe Mtgs, October 2, 1950; JPC Report 17-14, "Military Cooperation Committee Planning," to CSC, September 12, 1950, NARA, RG 59, State Department Records, PJBD-MCC, entry E1184, box 25, file "Military Co-operation Committee III." Quote from last document. See also Godefroy, *In Peace Prepared* (80, 242n17-18), regarding Soviet atomic bomber capabilities and intentions during the early 1950s.
14 MCC 300/1 was subsequently cancelled in December 1950. JPC Memo, "Background: Canada-United States Emergency Defence Plan (MCC 300/2)," May 22, 1951, DHH 112.3 M2.009 (D106), vol. 3; JCS Minute of August 14, 1950, NARA, RG 341, entry 335, box 723, Air Force, Plans, Project Decimal file 1942-1954, file Canada 600.2 (May 12, 1948), section 3.
15 See MCC 300/3, June 1, 1951, NARA, RG 218, JCS, entry 943011, file CCS 092 (1-10-45), section B.P. [Bulky Package - section 28].
16 DHH 73/1501, *Nineteen Years of Air Defense*, NORAD Historical Reference Paper no. 11 (Colorado Springs: North American Air Defense Command, Ent Air Force Base, 1965), 1; Schaffel, *The Emerging Shield*, 53-54; *Chronology of JCS Involvement in North American Air Defense*, 3. These Second World War air power culture and doctrinal lessons dictated that "the normal composition of an air force included a strategic air force, a tactical air force, and air defense command, and various supporting commands." Grant, *The Development of Continental Air Defense*, 3.
17 DHH 73/1501, *Nineteen Years of Air Defense*, 1; Schaffel, *The Emerging Shield*, 53-54; *Chronology of JCS Involvement in North American Air Defense*, 10; Jockel, *No Boundaries Upstairs*, 30.
18 Grant, *The Development of Continental Air Defense*, 4-6. 12. Quote from page 4.
19 Jockel, *No Boundaries Upstairs*, 6; Schaffel, *The Emerging Shield*, 50, 54, 61, 64; DHH 73/1501, *Nineteen Years of Air Defense*, 1.
20 Grant, *The Development of Continental Air Defense*, 3.
21 *Chronology of JCS Involvement in North American Air Defense*, 22, 24; Grant, *The Development of Continental Air Defense*, 6; Schaffel, *The Emerging Shield*, 60.

22 *Chronology of JCS Involvement in North American Air Defense*, 12–13, 17; DHH 73/1501, *Nineteen Years of Air Defense*, 29; Grant, *The Development of Continental Air Defense*, 17; Schaffel, *The Emerging Shield*, 79. Quote from first source.
23 Grant, *The Development of Continental Air Defense*, 17; DHH 73/1501, *Nineteen Years of Air Defense*, 30; *Chronology of JCS Involvement in North American Air Defense*, 22–27.
24 ADC was replaced with a Western Air Defense Force and an Eastern Air Defence Force, which reported directly to ConAC. DHH 73/1501, *Nineteen Years of Air Defense*, 5–6; Schaffel, *The Emerging Shield*, 57; Jockel, *No Boundaries Upstairs*, 9, 36; *Chronology of JCS Involvement in North American Air Defense*, 16; Grant, *The Development of Continental Air Defense*, 29, 34; Bill Green, *The First Line: Air Defense in the Northeast 1952–1960* (Fairview: Wonderhorse, 1994), 9.
25 Anti-Aircraft Command began with twenty-three gun battalions in April 1951, increasing to forty-five by the end of the year (ten from the National Guard). DHH 73/1501, *Nineteen Years of Air Defense*, 5–6; Schaffel, *The Emerging Shield*, 57; Jockel, *No Boundaries Upstairs*, 9, 36; *Chronology of JCS Involvement in North American Air Defense*, 16; Grant, *Development of Continental Air Defense to 1 September 1954*, 26, 30, 33–34.
26 *Chronology of JCS Involvement in North American Air Defense*, 30; DHH 73/1501, *Nineteen Years of Air Defense*, 26, 30; Grant, *The Development of Continental Air Defense*, 36.
27 Extract from Minutes of the 304th CSC Meeting, November 14, 1944, and CGS to Secretary, Chiefs of Staff Committee, November 8, 1944, CSC Memoranda, vol. 29, February to April 1944, DHH 2002/17, box 68, file 3 [formerly 193.009 (D30)]; Minutes of the 171st Meeting of the Joint Service Committee Atlantic Coast, October 12, 1945, DHH 2002/17, box 69, file 8a; "Terms of Reference, Joint Planners," n.d., attached to CGS to MND, January 27, 1947, DHH 112.1 (D178). Canadian Army study, "Central Organization for Defence," n.d. [February 4, 1947], JPC Minutes to Meetings and Correspondence, vol. 5, DHH 2002/17, box 54, file 4; Chiefs of Staff Committee, "Report on Arrangements for Defence Cooperation with the United Kingdom and United States," n.d. [July 1947], Chiefs of Staff Committee Papers, July 1947, DHH 2002/17, box 73, file 1.
28 Report by the Joint Planning Committee to the Chiefs of Staff Committee, "A Proposed Canadian System of Unified Operational Commands" (CSC 5-11-22 [JPC]), May 4, 1951, DHH 2002/17, box 77, file 24.
29 Extract from CSC 499th Meeting Minutes, July 6, 1951, DHH 2002/17, box 77, file 24. The Chairman, Chiefs of Staff Committee position was created in February 1951 to coordinate the three Canadian services.
30 JPC Report, 25-13, "Planning and Control of Joint Operations in Defence of Canada," May 15, 1950, Chiefs of Staff Committee, Organization, Defences against Enemy Lodgements, DHH 2002/17, box 111, file 6.
31 LAC, MG B5, Claxton fonds, vol. 221, Unpublished Memoirs of Brooke Claxton, 862–63; Wing Cdr. J.H. Roberts, AFC, "The RCAF's Functional Command Organization," *The Roundel* 4, 10 (November 1952): 20–23.
32 Kostenuk and Griffin, *RCAF Squadrons and Aircraft*, 144; K.J. Goodspeed, *The Armed Forces of Canada: A Century of Achievement* (Ottawa: Directorate of History, Canadian Forces Headquarters, 1967), 219; Larry Milberry, *Sixty Years: The RCAF and CF Air Command 1924–1984* (Toronto: CANAV Books, 1984), 196, 203, 212–15; Maj. Mat Joost, "The RCAF Auxiliary and the Air Defence of North America, 1948 to 1960," in *Proceedings*, 7th Annual Air Force Historical Conference: Canada in NORAD, Colorado Springs, June 4–8, 2001 (Winnipeg: Office of Air Force Heritage and History, 2001), 27;

English and Westrop, *Canadian Air Force Leadership and Command*, 24, 26. Quote from latter.

33 Appendix "A," Air Defence [hereafter Air Defence Appendix] to JPC Study, "Summary of Joint Defence Capabilities," September 16, 1948, JPC Minutes to Meetings and Correspondence, vol. 10, DHH 2002/17, box 55, file 4. For a stark assessment of the poor state of Canada's air defence capabilities, see JPC Report, "Canadian Air Defence Requirements," July 14, 1948, JPC Minutes to Meetings and Correspondence, vol. 9, March to August 1948, DHH 2002/17, box 55, file 3.

34 Minutes of a Conference of Air Officers' Commanding and Group Commanders Held at Air Force Headquarters, Ottawa, June 27–29, 1949, Air Officers Commanding Conferences, Raymont Collection, DHH 73/1223/2000; Air Vice-Marshal C.R. Slemon, Air Member for Air Planning (AMP), Memorandum, "Change in Name of Model Interceptor Plan," to RCAF Commands, December 2, 1948, Group Capt. W.R. MacBrien, AOC No. 1 Air Defence Group, Memorandum, "Planning Function No. 1 Air Defence Group," to Air Member for Operations and Training (AMOT), October 21, 1948, and Air Marshal W.A. Curtis, CAS, Memorandum, "Air Defence Group," to MND, June 21, 1950, LAC, RG 24, acc. 1983–84/216, box 3108, file HQS-895–100–69/14, pt. 1, copy obtained through DND ATIP, A-2007–00189. Quote from last document.

35 Joint Organization Order 14, May 23,1951, LAC, RG 24, acc. 1983–84/216, box 3108, file HQS-895–100–69/14, pt. 1; Minutes of Conference of Air Officers' Commanding and Group Commanders, March 20–21 and December 6–7, 1950, Air Officers' Commanding Conferences, Raymont Collection, DHH 73/1223/2000; Don Nicks, John Bradley, and Chris Charland, *A History of the Air Defence of Canada 1948–1997* (Ottawa: Canadian Fighter Group, 1997), 9–10.

36 Claxton Memorandum, "Acceleration of RCAF Programme," to Cabinet, July 19, 1950, LAC, MG 32, B5, Claxton fonds, vol. 94, file Accelerated Defence Program; Kostenuk and Griffin, *RCAF Squadrons and Aircraft*, 146, 208; English and Westrop, *Canadian Air Force Leadership and Command*, 26. It is also worth mentioning that in addition to RCAF ADC, the RCN's carrier-based Sea Fury and Banshee fighter squadrons also made an important contribution to the defence of the Atlantic seaboard during the early Cold War period, and in many cases were the only air defence forces present in the area. Michael Whitby, "Letter to the Editor," *Airforce* 38, 4 (2015): 81.

37 The AOC ADC also had "administrative control of RCAF units" in his command. This authority had previously been held by the AOC Training Command. Joint Organization Order 14, May 23, 1951, LAC, RG 24, acc. 1983–84/216, box 3108, file HQS-895–100–69/14, pt. 1; RCAF ADC history "for background information purposes," titled "Air Defence Command," n.d. [August 1953], LAC, RG 24, acc. 1983–84/216, box 3108, file HQS-895–100–69/14, pt. 5. Quote from former.

38 As per Canadian practice, administration remained a service prerogative as part of the army's exercise of command over the Anti-Aircraft Command. RCAF ADC history "for background information purposes," titled "Air Defence Command," n.d. [August 1953], LAC, RG 24, acc. 1983–84/216, box 3108, file HQS-895–100–69/14, pt. 5. On the Canadian Army's anti-aircraft capabilities during the late 1940s and early 1950s, see Godefroy, *In Peace Prepared*, 79–83.

39 Sean Maloney, *Learning to Love the Bomb*, 100; Jockel, *No Boundaries Upstairs*, 93.

40 G/C K.L.B. Hodson, DAPS, to D/AMAP/P, April 19, 1951, LAC, RG 24, vol. 6172, file 15–73–3. G/C Hodson would later go on to become the first Deputy Chief of Operations for NORAD at the rank of Air Vice-Marshal. Tragically, he was killed in a training flight in 1960. The library at the RCAF Staff College, now the Information Resource

Centre at Canadian Forces College, was named the Air Vice-Marshal K.L.B. Hodson Memorial Library in his memory.

41 Although the EADP is still classified and not available to researchers, a description of it is available in a Chiefs of Staff Committee document of December 9, 1950.

42 DMO&P Memorandum, "Canada–United States Emergency Air Defence Plan," to CGS, December 9, 1950, Chiefs of Staff Committee Papers, DHH 112.3M2 (D121).

43 Secretary, US Section MCC to Canadian Section MCC, February 9, 1951, DHH 112.3M2.009 (D106), vol. 3. Although the specific CSC decision on the EADP remains classified, I have been able to glean this information from a number of documents that make reference to this document. See Agreed Policy Set Forth in the Canada–US Emergency Air Defence Plan (CANUSEADP), Appendix "D" to (Draft) RCAF DAPS Study, "A Staff Study on Desireable [sic] Control Arrangements for the Integrated Canada–US Air Defence System," February 21, 1951, A/C H.B. Godwin, C/Plans, to V/CAS, September 7, 1951, and D/CAS to AOC ADC, September 15, 1951, LAC, RG 24, vol. 6172, file 15-73-3. These documents also indicate that the RCAF and USAF approved a revised Canada–US EADP that contained agreed-upon common technical and doctrinal air defence practices (see the first document cited in this note for a summary of the EADP) but excluded command and control provisions. As with the EDP, the EADP was revised annually. See documents regarding the EADP, August to November 1951, in NARA, RG 341, entry 335, box 723, Air Force, Plans, Project Decimal file 1942–1954, file Canada 660.2 (May 12, 1948), section 5; and NARA, RG 218, JCS, entry 943011, box 31, file CCS 092 (9–10–45), section 29.

44 USAF–RCAF Plan for the Extension of the Presently Authorized Air Defence Radar Systems of the Continental United States and Canada, n.d. [February 1, 1951], Appendix "A" to Minutes of the 74th Meeting of the PJBD, January 29–30 and February 1, 1951, PJBD Journals, DHH 82/196, vol. 9; *Chronology of JCS Involvement in North American Air Defense*, 29; Jockel, *No Boundaries Upstairs*, 44–46.

45 USAF–RCAF Plan for the Extension of the Presently Authorized Air Defence Radar Systems of the Continental United States and Canada, n.d. [February 1, 1951], Appendix "A" to Minutes of the 74th PJBD Meeting, January 29–30 and February 1, 1951, January 10–11, 1951, PJBD Journals, DHH 82/196, vol. 9.

46 The RCAF's current definition of mission command is: "The CF philosophy of mission command, which emphasizes that only the requisite amount of control should be imposed on subordinates, argues in general for a greater decentralization of execution." *Canadian Forces Aerospace Command Doctrine*, B-GA-0401-000/FP-001, 22.

47 By this time, the PJBD had renumbered its recommendations to reflect the year in which each one was made. Minutes of the 74th PJBD Meeting, January 29–30 and February 1, 1951, January 10–11, 1951, PJBD Journals, DHH 82/196, vol. 9; PJBD Recommendation 51/1, January 31, 1951, DHH 79/35. The Canadian government approved Recommendation 51/1 on February 20, 1951; this was followed by President Truman's assent on April 14. Minutes of the 71st Cabinet Defence Committee (CDC) Meeting, February 20, 1950, CDC Meetings 1951, LAC, RG 2, 18, box 244, file C-10-9M; *Chronology of JCS Involvement in North American Air Defense*, 33.

48 (Draft) "RCAF-USAF Agreement Concerning Air Defence of Canada and the United States," April 24, 1951, LAC, RG 24, vol. 6172, file 15-73-3. Arrangements for command and control of air defence forces in the northeast were dealt with in an attachment to this document. It is discussed in the next chapter.

49 G/C K.L.B. Hodson, DAPS, to D/AMAP/P, April 19, 1951, AMAP/P to AMAP, April 24, 1951, and AVM F.R. Miller, AMOT, to AMAP, May 14, 1951, LAC, RG 24, vol. 6172, file 15-73-3; Minutes of the September 21–26, 1952 PJBD Meeting, PJBD Journal, copy in

LAC, MG 33, B12, Paul Martin Sr. Papers, file 70-5 – National Defence, Permanent Joint Board on Defence, August 1951 to October 1953; Maj.-Gen. R.L. Walsh, USAF PJBD Member, to Air Vice-Marshal F.R. Miller, RCAF PJBD Member, January 8, 1953, NARA, RG 333, Files of International Military Agencies – PJBD, entry 17-A, box 4, file "Top Secret Correspondence, 1941–1956," folder 21; Minutes of the January 26–28, 1953, PJBD Meeting, PJBD Journal, copy in LAC, MG 33, B12, Martin Papers, file 70-5 – National Defence, Permanent Joint Board on Defence, August 1951 to October 1953.

50 Foulkes, "The Complications of Continental Defence," 112.

51 RCAF PJBD Member to USAF PJBD Member, December 4, 1950, Enclosure "B" to JCS 1541/69, NARA, RG 218, JCS, entry 943011, box 36, file CS 092 (9-10-45), section 24; Richter, *Avoiding Armageddon*, 40–41; Goette, "Air Defence Leadership," 53–54. Focusing on the air defence mission also served service institutional objectives of securing greater government funding. See also the RCAF's service magazine, *The Roundel*, which in the early 1950s began to dedicate an increasing number of articles and features to air defence topics.

52 DHH 73/1501, *Nineteen Years of Air Defense*, 11–12; Richter, *Avoiding Armageddon*, 41. Quote from former. The concept is also called "forward air defence," and James Fergusson explains that it "increased the probability of the successful defence of North American cities and of industrial and transportation centres. Forward air defence gave the defender many chances to intercept bombers, and successful intercepts would down the bombers with their nuclear payloads far away from any major and minor targets of value." Fergusson, *Canada and Ballistic Missile Defence*, 12.

53 JCS 2084/19, Chairman, JCS, Memorandum, "Interception and Engagement of Identified Hostile Aircraft," to Secretary of Defense, August 29, 1950, NARA, RG 218, JCS, Geographic file, 1951–53, box 56, file CCS 373.24 US (9-8-49), section 3; Jockel, *No Boundaries Upstairs*, 50.

54 A/C Clare Annis, "The Role of the R.C.A.F.," Address delivered before the Trenton Chamber of Commerce, March 26, 1952, *Airpower 1952: Three Speeches by Air Commodore Clare L. Annis*, Canadian Forces College Collection, Toronto. Also see Goette, "A Snapshot of Early Cold War RCAF Writing," 50–61.

55 This book addresses the purely command and control aspects of the mutual reinforcement and cross-border interception issue. For a more detailed examination of the subject, see Jockel, *No Boundaries Upstairs*, 46–57; and Trudgen, "The Search for Continental Security," 159–90.

56 RCAF PJBD Member to USAF PJBD Member, December 4, 1950, enclosure "B" to JCS 1541/69, NARA, RG 218, JCS, entry 943011, box 36, file CS 092 (9-10-45), section 24; *Chronology of JCS Involvement in North American Air Defense*, 30.

57 Memorandum by the Chief of Staff, USAF, "Canada–U.S. Air Defense Exercises and Operations," to the JCS, December 27, 1950, and JCS to Chairman, U.S. Section PJBD, January 5, 1951, enclosure "A" to JCS 1541/69, NARA, RG 218, JCS, entry 943011, box 36, file CS 092 (9-10-45), section 24; Secretary, US Section MCC, to PJBD, January 7, 1951, NARA, RG 333, PJBD, entry 17-A, box 3, file "Top Secret Correspondence, 1941–1956," folder 16.

58 Curtis to Claxton, December 15, 1950, quoted in Jockel, *No Boundaries Upstairs*, 53. See also the notes on Curtis's meeting with the USAF Chief of Staff, General Hoyt Vandenberg. Agenda – Conference with Air Marshal W.A. Curtis, December 19, 1950, and Memorandum, Maj.-Gen. Truman H. Landon, USAF Director of Plans, to Vandenberg, December 19, 1950, NARA, RG 333, PJBD, entry 17-A, box 2, file "Top Secret Correspondence, 1941–1956," folder 12. Quote from former.

59 Watson, *History of the Joint Chiefs of Staff*, vol. 5, 112.

60 Jockel, *No Boundaries Upstairs*, 54; Grant, *The Development of Continental Air Defense*, 54; Nicks et al., *History of the Air Defence of Canada*, 9–10; USAF Directorate of Requirements, "General Operational Requirement for an Aircraft Control and Warning System for Air Defense 1952–1958," December 27, 1951, NARA, RG 341, Records of Headquarters USAF (Air Staff), Directorate of Operational Requirements Air Defense Division Control and Warning Branch Files, 1951–1961, box 1, entry E-1040, file SOR-3, General Operational Requirements for an Aircraft Control and Warning System for Air Defense, 1952–58; Watson, *History of the Joint Chiefs of Staff*, vol. 5, 112. Academic work on the history of air defence doctrine such as the ground control centre–interceptor relationship is very minimal and therefore offers an interesting area of future study for air force scholars.

61 Jockel, *No Boundaries Upstairs*, 50. According to one RCAF directive, an ADCC commander's major responsibilities included "control and direction of the operations of the flying squadrons allocated to him by the Command HQ" and also "exercise of operational control over the [Army] anti-aircraft forces employed in his area through the appropriate" army anti-aircraft commander. This is a good example of the evolutionary nature of command and control terminology at this time. AOC ADC Memorandum, "Establishment – Upgrading Officer Position, All ADCC Establishments – Air Defence Command," December 11, 1951, LAC, RG 24, acc. 1983–84/216, box 3108, file HQS-895-100-69/14, pt. 3.

62 AVM F.R. Miller, AMOT, to AMAP, May 14, 1951, LAC, RG 24, vol. 6172, file 15-73-3.

63 (Draft) "RCAF-USAF Agreement Concerning Air Defence of Canada and the United States," April 24, 1951, LAC, RG 24, vol. 6172, file 15-73-3. For delineation of functions between air defence headquarters, Air Defence Control Centres, and Ground Control Intercept stations, see "Functions of AC&W Units of an Air Defence Group," n.d. [September 1950], LAC, RG 24, acc. 1983–84/216, box 3108, file HQS-895-100-69/14, pt. 1.

64 As a revised version of the draft RCAF-USAF agreement explained it, the Canadian air defence commander "will exercise tactical control of aircraft airborne under control of Canadian GCI's [Ground Control Intercept station, i.e., an ADCC]," while the USAF air defence commander "will exercise tactical control of all aircraft airborne under control of United States GCI's." Revised page three to (Draft) "RCAF–USAF Agreement Concerning Air Defence of Canada and the United States," attached to G/C K.L.B. Hodson, DAPS, to AMAP, May 21, 1951, LAC, RG 24, vol. 6172, file 15-73-3. The issue of command and control of air defence forces in Newfoundland is explored in the next chapter.

65 (Draft) RCAF-USAF Agreement Concerning Air Defense of Canada and the US, n.d. [early July 1951], NARA, RG 59, State Department, PJBD, entry 1177, box 5, file "Northeast Air Command – Canadian Participation, 1952 – PJBD"; (Draft) Curtis Memorandum, "Canada–United States Air Defence Mutual Re-Inforcement," n.d. [late September 1951], Appendix "A" to Curtis to Secretary, Chiefs of Staff Committee, September 29, 1951, Cabinet Defence Committee fonds, DHH 2002/03, series I, file 59. Again, the "designated by" wording was a functional exercise of sovereignty.

66 Extract from Minutes of the 509th Meeting of the Chiefs of Staff Committee, October 18, 1951, Cabinet Defence Committee fonds, DHH 2002/03, series I, file 59.

67 Jockel, *No Boundaries Upstairs*, 54; Minutes of the 77th PJBD Meeting, November 11–21, 1951, PJBD Journals, DHH 82/196, file 9; PJBD Recommendation 51/6, November 12, 1951, PJBD Recommendations, DHH 79/35.

68 Claxton Memorandum, "Canada–United States Air Defence Mutual Re-Inforcement," to CDC, November 5, 1954, LAC, RG 2, vol. 2751, CDC Documents, vol. 10; Jockel, *No Boundaries Upstairs*, 54. Quote from former.

69 On the west coast, the AOC ADC would delegate his responsibility to the AOC No. 12 Air Defence Group at Comox, British Columbia. No. 12 Air Defence Group was an ADC unit and in 1955 was renamed No. 5 Air Division. Kostenuk and Griffin, *RCAF Squadrons and Aircraft*, 211. As the next chapter will detail, the arrangements for Newfoundland changed in 1953 when the AOC ADC took over operational control of USAF forces operating in the province's airspace.

70 Claxton Memorandum, "Canada–United States Air Defence: Mutual Reinforcement," December 3, 1951, LAC, RG 2, vol. 2751, CDC Documents, vol. 11; Decision on JCS 1541/76, "Movement of Service Aircraft across the Canada–US Border," January 18, 1952, NARA, RG 218, JCS, entry 943011, box 32, file CCS 092 (9–10–45), section 30; Secretary, US Section PJBD to Chairman, US Section PJBD, April 23, 1952, NARA, RG 333, PJBD, entry 17-A, box 4, file "Top Secret Correspondence, 1941–1956," folder 23.

71 Jockel, *No Boundaries Upstairs*, 54–55.

72 Chief of Staff, USAF, Memorandum for Secretary of Defense, n.d. [August 1950], NARA, RG 218, JCS, Geographic file, 1951–53, box 56, file CCS 373.24 US (9–8–49), section 3. This document was approved by both the Secretary of Defense and the President on August 24, 1950. Minute by Secretary of Defense Louis Johnson and minute by President Harry S. Truman in ibid., both dated August 24, 1950. In July, the US had established Air Defence Identification Zones (ADIZ) in certain "vital areas" where "all military aircraft were required, and civil aircraft requested, to file flight plans as an aid to identification." Grant, *The Development of Continental Air Defense*, 32.

73 Jockel, *No Boundaries Upstairs*, 50.

74 PJBD Recommendation 51/4, 9 May 1951, PJBD Recommendations, DHH 79/35.

75 Ibid.; Minutes of the 72nd PJBD Meeting, October 2–5, 1950, PJBD Journals, DHH 82/196, file 8; Claxton Memorandum for the CDC, "United States Air Operations over Canadian Territory – Interception of Unidentified Aircraft," November 27, 1950, LAC, RG 2, vol. 2751, CDC Documents, vol. 9; Minutes of the 68th CDC Meeting, December 1, 1950, LAC, RG 18, vol. 244, file C-10–9M, "Cabinet Defence Committee Meetings."

76 Memorandum by A.D.P. Heeney, Secretary of State for External Affairs, "United States Air Operations over Canadian Territory – Interception of Unidentified Aircraft," to CDC, May 25, 1951, LAC, RG 2, vol. 2751, CDC Documents, vol. 10; extract from Minutes of the 74th CDC Meeting, May 29, 1951, CDC fonds, DHH 2002/03, series I, file 59; Secretary, Canadian Section PJBD, to Secretary, US Section, PJBD, May 31, 1951, LAC, MG 30, E133, McNaughton Papers, vol. 293, file PJBD – Interceptor Flight Rec 51/4.

77 Maj.-Gen. R.M. Ramey, USAF Director of Operations, to Maj.-Gen. R.L. Walsh, USAF PJBD Member, June 1951, Walsh to Ramey, August 10, 1951, and Ramey to Walsh, August 24, 1951, NARA, RG 333, PJBD, entry 17-A, file "Top Secret Correspondence, 1941–1956," folder 17; Minutes of the 76th Meeting of the PJBD, August 20–25, 1951, and Minutes of the 77th PJBD Meeting, November 11–21, 1951, PJBD Journals, DHH 82/196, file 9; Jockel, *No Boundaries Upstairs*, 36, 52.

78 RCAF Directive, "Authority to Intercept and Engage Hostile Aircraft," November 22, 1951; Claxton Memorandum, "Authority to Intercept and Engage Unidentified Aircraft," November 22, 1951, CDC fonds, DHH 2002/03, series I, file 59; Minutes of the 81st CDC Meeting, December 12, 1951, LAC, RG 18, vol. 244, file C-10–9M, "Cabinet Defence Committee Meetings"; Jockel, *No Boundaries Upstairs*, 54.

79 Maj.-Gen. Robert W. Burns, USAF Acting Deputy Chief of Staff, Operations, to Maj.-Gen. R.L. Walsh, USAF PJBD Member, April 14, 1952, NARA, RG 333, PJBD, entry 17-A, box 4, file "Top Secret Correspondence, 1941–1956," folder 24; Jockel, *No Boundaries Upstairs*, 55. Quote from former.

80 Ibid. The USAF proposal did not define "hostile act" and "manifestly hostile intent" because it was "considered to be impracticable to attempt to list or otherwise define all of the many ways in which an aircraft can commit a hostile act or evidence hostile intent" and because "any agreement specifically spelling out such methods would be too restrictive upon the Air Defence Commander concerned." As we will see shortly, the RCAF disagreed with this viewpoint.
81 The only difference was the clause in the USAF proposal that allowed the pilot to determine whether an unidentified aircraft had "manifestly hostile intent."
82 M.H. Wershof, Defence Liaison (1) Division, Memorandum, "Interceptor Flights by the United States in Canada," July 28, 1952, reproduced in *DCER*, vol. 18, 1952, 1126–29. Emphasis in original.
83 RCAF PJBD Member Memorandum, "Aircraft Interception – Modification of Agreement," to Secretary, Canadian Section PJBD, September 8, 1952, LAC, MG 30, E133, McNaughton Papers, file US Interceptor Flights over Canada 1951–52. Again, the provisions were reciprocal, depending on whose airspace the intercept took place. It was therefore a functional exercise of sovereignty.
84 R.A. MacKay, DEA Member PJBD, to Air Vice-Marshal F.R. Miller, RCAF PJBD Member, November 12, 1952, LAC, MG 30, E133, McNaughton Papers, vol. 293, file Proposed Extension Rec. 51/4, Interceptor Flights; Claxton Memorandum, "Aircraft Interception Modification of Agreement," to Cabinet Defence Committee, March 23, 1953, CSC Papers, DHH 2002/17, box 78, file 30.
85 Minutes of the September 28–October 1, 1953, PJBD Meeting, PJBD Journals, copy in LAC, MG 33, B12, Martin Papers, file 70-5 – National Defence, Permanent Joint Board on Defence, August 1951 to October 1953; PJBD Recommendation 53/1, October 1, 1953, PJBD Recommendations, DHH 79/35.
86 PJBD Recommendation 53/1, October 1, 1953, PJBD Recommendations, DHH 79/35. At the time, Canadian ADCC commanders had to be at the Wing Commander rank, although the RCAF was hoping to raise this to Group Captain. See A/C W.I. Clements to AOC ADC, March 29, 1952, LAC, RG 24, acc. 1983–84/216, box 3108, file HQS-895–100–69/14, pt. 3.
87 Jockel, *No Boundaries Upstairs*, 57.
88 Senior USAF PJBD Member to JCS, October 6, 1953, quoted in ibid.; R.A. MacKay, DEA Member PJBD, to Air Vice-Marshal F.R. Miller, RCAF PJBD Member, November 12, 1952, LAC, MG 30, E133, McNaughton Papers, vol. 293, file Proposed Extension Rec. 51/4, Interceptor Flights; Claxton Memorandum, "Aircraft Interception Modification of Agreement," to CDC, March 25, 1953, CSC Papers to 533rd Meeting, DHH 2002/17, box 78, file 30.
89 Jockel, *No Boundaries Upstairs*, 57–58; Claxton Memorandum, "Principles Governing the Interception of Unidentified Aircraft in Peace Time," to CDC, November 2, 1953, CSC Papers to 537th Meeting, DHH 2002/17, box 78, file 34; Extract from Minutes of CDC Meeting, November 3, 1953, reproduced in *DCER*, vol. 19, 1953, 1023; Walter D. Smith, Secretary of Defense, Memorandum for the President, "Recommendation 53/1 of the Permanent Joint Board on Defense, Canada–United States, concerning the principles governing the interception of unidentified aircraft in peacetime," December 4, 1953, and minute "Approved" by President Dwight D. Eisenhower in ibid., December 9, 1953, NARA, RG 59, State Department, Central Decimal File 1950–54, box 3186, file 711.56342/2–1853; Minutes of the January 10–14, 1954, PJBD Meeting, PJBD Journals, copy in LAC, MG 30, E133, McNaughton Papers, vol. 286, file PJBD Meetings, June 1953 to January 1955. The recommendation was also attached to the Command Appendix of the EDP.
90 Jockel, *No Boundaries Upstairs*, 55.

Chapter 7: The US Northeast Command

Epigraph: A/C H.B. Godwin, C/Plans, to V/CAS, September 7, 1951, LAC, RG 24 Vol 6172, File 15-73-3.

1 Committee on Post-Hostilities Problems Report, "Text of Canadian Position on Post-War Defence of Newfoundland and Labrador," January 12, 1945, reproduced in Paul Bridle, ed., *Documents on Relations between Canada and Newfoundland, Volume I: 1935–1949* (Ottawa: Information Canada, 1974), 949–55.
2 Extract from Minutes of the 1096th Meeting of Commission of Government of Newfoundland, March 28, 1949 and Ceremonies at St. John's and Ottawa on April 1, 1949, reproduced in Bridle, ed., *DRBCN*, vol. 2, 1633, 1675–88. See also Neary, *Newfoundland and the North Atlantic World*, which gives the best account of the circumstances surrounding Newfoundland's entry into Confederation.
3 JPC Study 19-7, "Federation of Newfoundland with Canada – Military Considerations with Regard to United States," for Chiefs of Staff Committee, February 3, 1949, JPC Correspondence, vol. 11, 1948–49, DHH 2002/17, box 56, file 1. See also H.H. Wrong, Canadian Ambassador to the United States, to the Honourable Robert A. Lovett, Acting Secretary of State, Department of State, November 19, 1949, LAC, RG 24, vol. 5185, S-15-9-56, pt. 3.
4 *Chronology of JCS Involvement in North American Air Defense 1946–1975* (Washington, DC: Historical Division, Joint Secretariat, Joint Chiefs of Staff, March 20, 1976), 5, 15. Quote from page 15.
5 Cole et al., *The History of the Unified Command Plan*, 17.
6 JCS, Decision on JCS 1259/136, Memorandum by the Chief of Staff USAF on Establishment of the Northeast Command, April 11, 1949, Records of the JCS, Strategic Issues, reel 9 (section 2), pt. 2, 1946–53, microfilm copy at ML, RMC; Memorandum by the Secretary of Defense for the US Section, PJBD, "Establishment of the Northeast Command," April 20, 1949 and Memorandum by US Section PJBD for Secretary Johnson, July 29, 1949, NARA, RG 333, PJBD, entry 17-A, box 2, file "Top Secret General Correspondence, 1941–1956," folder 9.
7 Henry to McNaughton, May 2, 1949, NARA, RG 333, PJBD, entry 17-A, box 2, file "Top Secret General Correspondence, 1941–1956," folder 9.
8 Memoranda for the Director of Plans and Operations, CSUSA by Secretary, US Section PJBD, May 3 and 5, 1949, and William P. Snow, Secretary, US Section PJBD, to Maj.-Gen. Robert L. Walsh, USAF Member, PJBD, June 8, 1949, NARA, RG 333, PJBD, entry 17-A, box 2, file "Top Secret General Correspondence, 1941–1956," folder 9; Defence Liaison Division to USSEA, May 21, 1949, LAC, RG 25, vol. 5961, file 50221-40, copy obtained through DND ATIP, A-2007-00353. Quote from first source. These special rights largely had to do with civilian issues such as customs duties and tax exemptions, postal facilities, and legal jurisdiction, not military matters related to the operation of the US bases. However, it was feared that incidents between American servicemen and Newfoundlanders, now Canadian citizens, might have an overall negative effect on the Canada–US defence relationship in the new province. In light of Newfoundland's entry into Confederation, the Canadian government was seeking to renegotiate several of these rights, which had been granted to the United States in the original Destroyers-for-Bases Deal in 1941. See Bercuson, "SAC vs. Sovereignty," 210.
9 McNaughton to McKay, August 17, 1949, LAC, MG 30, E133, McNaughton Papers, file PJBD, US Government, Northeast Command, 1-4-2; Briefing for Secretary of Defense for his visit to Ottawa – August 11, 49, n.d. [likely early August 1949], and NARA, RG 333, PJBD, entry 17-A, box 2, file "Top Secret General Correspondence, 1941–1956," folder 10. Quote from first source. The Canadian military in particular liked the idea of

dealing with a single American commander in Newfoundland. See Brig.-Gen. Kitching, BGS (Plans), to CGS, May 20, 1949, and Minute by Vice-Chief of the General Staff, n.d. [May 21 or 22, 1949], Chiefs of Staff Committee Papers, April 1950, DHH 2002/17, box 76, file 3.

10 Secretary JCS, Memorandum to Senior US Army Member, PJBD, "Establishment of the United States Northeast Command," September 23, 1949, Records of the JCS, Strategic Issues, reel 9 (section 2), pt. 2, 1946–53, microfilm copy at ML, RMC.

11 Extract from Minutes of the 452nd CSC Meeting, October 6, 1949, CSC Papers, April 1950, DHH 2002/17, box 76, file 3; (Unofficial) Memorandum for the JCS from the US Section PJBD, October 20, 1949 and USAF PJBD Member to Commanding General, Newfoundland Base Command, October 21, 1949, NARA, RG 333, PJBD, entry 17-A, box 2, file "Top Secret General Correspondence, 1941–1956," folder 9; R.A. MacKay, DEA, to A.D.P. Heeney, USSEA, October 5, 1949, LAC, MG 20, E133, McNaughton Papers, file PJBD, US Government, Northeast Command, 1-4-2; Secretary, Canadian Section PJBD, to Secretary CSC, October 26, 1949, DHH 112.3M2.009 (D114). Quote from last document.

12 Secretary JCS, to Senior US Army PJBD Member, "Establishment of the United States Northeast Command," December 2, 1949, Records of the JCS, Strategic Issues, reel 9 (section 2), pt. 2, 1946–53, microfilm copy at ML, RMC.

13 Report to the JPC by the Joint Planning Staff, "Proposed United States Northeast Command," March 15, 1950, DHH 112.3M2.009 (D114); Minutes of the 182nd JPC Meeting, March 23, 1950, JPC Minutes, DHH 2002/17, box 52, file 5; Extract from 460th CSC Meeting, April 18, 1950, LAC, RG 25, vol. 5961, file 50221–40.

14 SSEA to MND, May 6, 1950, LAC, RG 25, vol. 5961, file 50221–40.

15 Extract from 463rd CSC Meeting, May 16, 1950, LAC, RG 25, vol. 5961, file 50221–40; Secretary, Canadian Section PJBD, to Canadian Section PJBD, May 24, 1950, LAC, MG 20, E133, McNaughton Papers, file PJBD, US Government, Northeast Command, 1-4-2; Memorandum from MND, "Proposed US Northeast Command," to Cabinet, May 16, 1950, reproduced in *DCER*, vol. 16, 1950, 1505–6; Minutes of the 71st Meeting of the PJBD, May 27–31, 1950, PJBD Journals, DHH 82/196, vol. 8; Extract from May 18, 1950, Cabinet Meeting and Canadian Section PJBD to US Section PJBD, May 24, 1950, LAC, RG 25, vol. 5961, file 50221–40. Quote from last document.

16 Henry to JCS, June 13, 1950, JCS to Commander in Chief, US Northeast Command and Commander in Chief, Atlantic, August 10, 1950, and Decision on JCS 1259/189, Memorandum by the Chief of Staff US Air Force on Establishment of the US Northeast Command, August 29, 1950, Records of the JCS, Strategic Issues, reel 9 (section 2), pt. 2, 1946–53, microfilm copy at ML, RMC; Canadian Ambassador to the United States to SSEA, August 21, 1950, LAC, RG 25, vol. 5961, file 50221–40. Maj.-Gen. Whitten had previously been the commander of the Newfoundland Base Command, which was inactivated with the standing up of US Northeast Command.

17 US Northeast Command was what later was called a "specific command" under the Unified Command Plan. DHH 73/1501, *Nineteen Years of Air Defense*, NORAD Historical Reference Paper No. 11, 8; DHH 73/770, Lydus H. Buss, *US Air Defense in the Northeast 1940–1957*, Historical Reference Paper No. 1 (Colorado Springs: Directorate of Command History, Office of Information Services, Headquarters Continental Air Defense Command, 1957), 7–8, 18; Webb et al., *The History of the Unified Command Plan*, 14; "Northeast Command Area Briefing," January 17, 1951, DHH 112.3M2.009 (D115).

18 D/AMAP/P Study, "RCAF Viewpoint on Control Arrangements for the Integrated Canada–US Air Defence System with Particular Reference to the Position of the US

Northeast Command," March 21, 1951, LAC, RG 24, vol. 6172, file 15-73-3; Foulkes Memorandum, "Command Arrangements in Newfoundland," to Claxton, November 6, 1952, LAC, RG 25, vol. 5961, file 50221-40. The USAF also used Goose Bay from time to time for SAC exercises throughout the early 1950s before the lease was formalized in late 1952. On the Goose Bay agreement see Bercuson, "SAC vs. Sovereignty," 202-22; and John Clearwater, *US Nuclear Weapons in Canada* (Toronto: Dundurn, 1999), Chapter 5.

19 Canadian air force plans in 1950 called for an establishment of five fighter squadrons to be located at St-Hubert, Bagotville, Chatham, Trenton (interim), and Toronto to cover vital points in the Great Lakes-St. Lawrence area. This also included radar units at Chatham, Lac-St-Joseph, the Bagotville area, the Pembroke area, and the Toronto area. In 1954, an additional CF-100 squadron was to be assigned to Vancouver to defend the west coast, but no further interceptor aircraft were available for Newfoundland. Minutes of a Conference of Air Officers' Commanding and Group Commanders Held at Air Force Headquarters, Ottawa, March 20-21, 1950, Air Officers Commanding Conferences, Raymont Collection, DHH 73/1223/2000; Jockel, *No Boundaries Upstairs*, 92.

20 Memorandum from Head, Defence Liaison (1) Division, to USSEA, April 17, 1951, reproduced in Donaghy, ed., *DCER*, vol. 17, 1951, 1455-56; M.H. Wershof, Defence Liaison, to R.A. MacKay, USSEA, March 27, 1951, LAC, RG 25, vol. 5961, file 50221-40. Quote from latter.

21 See, for example, Eugene Griffin, "US Air Force Protects Weak Canadian Areas," *Chicago Daily Tribune*, November 20, 1950; and Ross Munroe, "US 'Ghost Command' in Canada," *Ottawa Citizen*, November 23, 1950, clippings in LAC, RG 25, vol. 5961, file 50221-40.

22 Memorandum from Chairman, Joint Services Committee East Coast, "(Draft) Basic Provisions for Canada-US Collaboration on Defence in the Northeastern Areas of Canada," to Chairman CSC, February 22, 1951, LAC, RG 25, vol. 5961, file 50221-40. The Joint Services Committee East Coast and US Northeast Command were the authorized Canadian and American planning authorities as per the EDP (see previous chapter).

23 Minutes of the 487th CSC Meeting, March 14, 1951, Raymont Collection, DHH 73/1223, box 60, file 1306.

24 D/AMAP/P Study, March 21, 1951, LAC, RG 24, vol. 6172, file 15-73-3.

25 Ibid. The Canadian Army was also scheduled to have AAA batteries posted to St. John's. The study also called for "Canadian representation in the appropriate operations rooms."

26 Ibid.; "Agreed Policy Set Forth in the Canada-US Emergency Air Defence Plan (CANUSEADP)," Appendix "B" to (Draft) D/AMAP Study, "A Staff Study on Desireable [sic] Control Arrangements for the Integrated Canada-US Air Defence System," February 21, 1951, LAC, RG 24, vol. 6172, file 15-73-3.

27 M.H. Wershof, Defence Liaison, to R.A. MacKay, USSEA, March 27, 1951, LAC, RG 25, vol. 5961, file 50221-40. McKay was also the DEA Member of the PJBD.

28 US Northeast Command Headquarters Memorandum to AFHQ, "NEC View on Command Arrangements," April 5, 1951, LAC, RG 24, vol. 6172, file 15-73-3.

29 Certain staff positions in Northeast Air Command were also to be earmarked for Canadian personnel. RCAF Study, "An Agreement on the Control Arrangements Pertaining to the Air Defence Forces of US Northeast Command and those of the RCAF," April 24, 1951, LAC, RG 24, vol. 6172, file 15-73-3; Group Capt. K.L.B. Hodson, Director of Air Plans (DAPS), to Secretary, JPC, September 4, 1951, JPC Minutes to Meetings and Correspondence, September 1951-January 1952, DHH 2002/17, box 57, file 1. Quotes from former.

30 AVM F.R. Miller, AMOT, Minute to AMAP, May 14, 1951, LAC, RG 24, vol. 6172, file 15-73-3.

31 G/C K.L.B. Hodson, DAPS, to AMAP, May 21, 1951, LAC, RG 24, vol. 6172, file 15-73-3.
32 DMO&P to CGS, October 31, 1951, JPC Minutes to Meetings and Correspondence September 1951 to January 1952, DHH 2002/17, box 57, file 1.
33 *Chronology of JCS Involvement in North American Air Defense*, 37; Webb et al., *The History of the Unified Command Plan*, 37; Bercuson, "SAC vs. Sovereignty," 218; Maj.-Gen. Joseph Smith, USAF Director of Plans, to HQ USAF Operational Plans Division, September 20, 1951, NARA, RG 333, Files of International Military Agencies – PJBD, entry 17-A, box 3, file "Top Secret Correspondence, 1941–1956," folder 15; Memorandum from Assistant Secretary to the Cabinet to Secretary to the Cabinet, "Appointment of R.C.A.F. Officer as Deputy C-IN-C, US Northeast Command," October 29, 1951, reproduced in *DCER*, vol. 17, 1951, 1461–62.
34 Memorandum by the Chief of Staff USAF to the JCS on Canadian Participation in the US Northeast Command, October 23, 1951, NARA, RG 333, Files of International Military Agencies – PJBD, entry 17-A, box 3, file "Top Secret Correspondence, 1941–1956," folder 15; Maj.-Gen. Guy Henry, Chairman, US Section, to Canadian PJBD Section, November 9, 1951, DHH 112.3M2.009 (D114); Memorandum from Defence Liaison (2) Division to Head, Defence Liaison (1) Division, August 28, 1951, reproduced in *DCER*, vol. 17, 1951, 1456–62; Jockel, *No Boundaries Upstairs*, 58. Quote from Jockel.
35 Minutes of the 253rd JPC Meeting, December 4, 1951, JPC Minutes to Meetings and Correspondence, September 1951 to June 1952, DHH 2002/17, box 57, file 1; Secretary JPC, Memorandum to JPC, "Command Relationship between CINCNE and Canadian Commanders," January 28, 1952, and same to same, February 8, 1952, LAC, RG 25, vol. 5961, file 50221–40.
36 See USSEA to Secretary CSC, March 1, 1952, JPC Minutes to Meetings and Correspondence, vol. 11, January to March 1952, DHH 2002/17, box 57, file 2. In addition, DEA felt that the CinC US Northeast Command "appears to consider his responsibilities greater than the Canadian Government anticipated or accepted when the establishment of the Command was approved." J.M. Cook, Defence Liaison (Draft) "Note for File: The US Northeast Command," April 2, 1952, LAC, RG 25, vol. 5961, file 50221–40.
37 Foulkes to Secretary CSC, March 12, 1952, JPC Minutes to Meetings and Correspondence, vol. 12, March to June 1952, DHH 2002/17, box 57, file 3.
38 G/C S.W. Coleman, DAPs, Memorandum "Command and Control – Newfoundland and Labrador" to C Plans I, April 17, 1952, LAC, RG 24, vol. 6172, file 15-73-3.
39 Minutes of the March 4–5, 1952, Meeting of the PJBD, PJBD Journal, copy in LAC, MG 33, B12, Martin Papers, file 70–5 – National Defence, Permanent Joint Board on Defence, August 1951 to October 1953.
40 Minutes of the 11/52 Meeting of the JPC, March 25, 1952, JPC Minutes to Meetings and Correspondence, vol. 12, DHH 202/17, box 57, file 3. The numbering of the JPC meetings changed as of January 1952.
41 For a discussion of the decision to send RCAF fighter aircraft to Europe, see Carl A. Christie, "Canada Commits to NATO: No. 1 (Fighter) Wing, RCAF Station North Luffenham, Rutland, England," paper presented at "Canada's Air Forces @ Eighty: An Historical Symposium," March 31–April 4, 2004, Ottawa, ON; and Stouffer, *Swords, Clunks, and Widowmakers*, 34–39.
42 Foulkes Memorandum, "Command Arrangements in Newfoundland," to Claxton, November 6, 1952, LAC, RG 25, vol. 5961, file 50221–40; extract from Minutes of a Meeting of the CDC, October 9, 1952, reproduced in *DCER*, vol. 18, 1952, 1148–49. See also Pearson to Claxton, August 15, 1952, reproduced in ibid., 1140–41. Critics of the 1 Canadian

Air Division deployment decision included Progressive Conservative MP George Pearkes, who felt that the fighters would be best used to defend Canada. Pearkes would later become Minister of National Defence in John Diefenbaker's government in 1957 and play a key role in the formation of NORAD (see next chapter). Maloney, *Learning to Live with the Bomb*, 102.

43 Claxton to Pearson, August 8, 1952, reproduced in *DCER*, vol. 18, 1952, 1140.
44 On the establishment of 1 Canadian Air Division, see Stouffer, *Swords, Clunks, and Widowmakers*, Chapter 4.
45 Foulkes Memorandum, "Command Arrangements in Newfoundland," to Claxton, November 6, 1952, LAC, RG 25, vol. 5961, file 50221-40. Foulkes unfortunately overlooked the fact that the CF-100 was not yet available at that time. The "Clunk," as the new all-Canadian designed and built CF-100 interceptor was affectionately known, did not deploy to frontline RCAF squadrons until April 1953. Kostenuk and Griffin, *RCAF Squadron Histories and Aircraft*, 145; T.F.J. Leversedge, *Canadian Combat and Support Aircraft: A Military Compendium* (St. Catharines: Vanwell, 2007), 63. On the development of the Canadair F-86 Sabre and Avro CF-100 Canuck fighters, see Wakelam, *Cold War Fighters*.
46 Secretary CSC to CAS, June 3, 1952, DHH 112.3M2.009 (D114). The bases were valuable because of their role in American offensive operations. SAC bombers were based there, and the US – and by extension NATO – was relying on offensive strategic nuclear deterrence to defend the United States rather than strategic defence. Therefore, the primary role of continental air defence became the protection of the SAC deterrent. Maloney, *Learning to Love the Bomb*, 29, 37; Goette, "A Snapshot of Early Cold War RCAF Writing," 52-53.
47 JPC Report, "Review of the Factors Involved in the Canada-US Relationship in the Northeast Areas of Canada," May 7, 1952, JPC Minutes to Meetings and Correspondence, vol. 12, March to June 1952, DHH 2002/17, box 57, file 3.
48 Secretary CSC to CAS, June 3, 1952, and extract from 523rd CSC Meeting, May 30, 1952, DHH 112.3M2.009 (D114); Secretary, Canadian Section PJBD, Memorandum "Canadian Participation in and Command Relations with US North East Command," to Canadian Section PJBD, June 3, 1952, LAC, RG 25, vol. 5961, file 50221-40; JPC Report, "Review of the Factors Involved in the Canada-US Relationship in the Northeast Areas of Canada," May 7, 1952, JPC Minutes to Meetings and Correspondence, vol. 12, March to June 1952, DHH 2002/17, box 57, file 3.
49 The two relevant clauses in the Command Appendix to the EDP read as follows: "The Command principles set out in paragraph 3a below do not apply in the case of those bases in Canada leased and/or occupied by United States forces for which special agreements exist" and "The Commander-in-Chief, United States Northeast Command is an exception in that while he is within the command structure of the United States Armed Forces, some of his forces are located in Canada. The principle outlined in paragraph 3a above, therefore does not apply in this case." Paragraph 3a of the Command Appendix read as follows: "Any force located in Canada and employed in execution of the tasks set forth in this plan will operate under a commander designated by Canada." Appendix "F" Command to MCC 300/3, Canada-United States Emergency Defence Plan, June 1, 1951, NARA, RG 218, JCS, entry 943011, box 37, file CCS 092 (9-10-45), section B.P. [Bulky Package – section 28]. The "special agreements" refers to the original Leased Bases Agreement of 1941.
50 DHH 73/770, Buss, *US Air Defense in the Northeast*, 20; M.H. Wershof, Defence Liaison Division, Memorandum "Basic Provisions for Canada-United States Collaboration on

Defence in the Northeastern Areas of Canada" to Heeney, July 17, 1952, LAC, RG 25, vol. 5961, file 50221-40. Once again this demonstrates the "human element" importance that individual personalities sometimes play in the Canada-US air defence command and control relationship.

51 DHH 73/770, Buss, *US Air Defense in the Northeast 1940-1957*, 20; *Chronology of JCS Involvement in North American Air Defense*, 20-21, 40; JCS 1259/253, Memorandum by the Chief of Staff USAF for the JCS on Canadian Operational Control of US Defense Forces Operating from Bases within Canada, October 15, 1952, and JCS 1259/260, Memorandum by the Chief of Staff USAF for the JCS on Canadian Operational Control of US Defense Forces Operating from Bases within Canada, November 4, 1952, Records of the JCS, Strategic Issues, reel 9 (section 2), pt. 2, 1946-53, microfilm copy at ML.

52 Minutes of the September 21-26, 1952, Meeting of the PJBD, PJBD Journal, copy in LAC, MG 33, B12, Martin Papers, file 70-5 – National Defence, Permanent Joint Board on Defence, August 1951 to October 1953; Memoranda "Goose Bay Lease" and "Posting of Fighter Squadrons to Goose Bay" from Dana Wilgress, USSEA, to Lester Pearson, SSEA, September 23 and 27, 1952, in Barry, ed., *DCER*, vol. 18, 1952, 1171-75; Clearwater, *US Nuclear Weapons in Canada*, 156.

53 Minutes of a Conference of Air Officers' Commanding and Group Commanders Held at Air Force Headquarters, Ottawa, November 5, 6, and 7, 1952, Air Officers' Commanding Conferences, 1952, Raymont Collection, DHH 73/1223/2000; Foulkes Memorandum, "Command Arrangements in Newfoundland," to Claxton, November 6, 1952, LAC, RG 25, vol. 5961, file 50221-40. Quote from latter. David Bercuson notes that this arrangement "was the same understanding on which practically all US defence operations in Canada, from arctic weather stations to continental defence radar installations, were based." Bercuson, "SAC vs. Sovereignty," 220.

54 Clearwater, *US Nuclear Weapons in Canada*, 128; DHH 73/330, Buss, *US Air Defense in the Northeast*, 17. The 61st and 59th FIS converted to the F-89D Scorpion in 1954; and later to the F-102 Dagger.

55 Vandenberg to Curtis, n.d. [approx. October 9, 1952]; and McKay to Wilgress, November 17, 1952, in Barry, ed., *DCER*, vol. 18, 1952, 1178; Minutes of a Conference of Air Officers' Commanding and Group Commanders Held at Air Force Headquarters, Ottawa, November 5, 6, and 7, 1952, Air Officers' Commanding Conferences, 1952, Raymont Collection, DHH 73/1223/2000. See also Extract from Minutes of the 531st CSC Meeting, November 12, 1952, CSC Papers, DHH 2002/17, box 78, file 28; and Minutes of the January 26-28, 1953, Meeting of the PJBD, PJBD Journal, copy in LAC, MG 33, B12, Martin Papers, file 70-5 – National Defence, Permanent Joint Board on Defence, August 1951 to October 1953.

56 Minute by M. Wershof, n.d. [likely November 7, 1952] on DEA copy of ibid., LAC, RG 25, vol. 5961, file 50221-40.

57 JCS 1259/260, Memorandum by the Chief of Staff USAF for the JCS on Canadian Operational Control of US Defense Forces Operating from Bases within Canada, November 4, 1952, Records of the JCS, Strategic Issues, reel 9 (section 2), pt. 2, 1946-53, microfilm copy at ML, RMC; Canadian Joint Staff Washington to Chairman CSC, November 7, 1952, CSC Papers, DHH 2002/17, box 78, file 28. Portions of the JCS Memorandum, notably the command and control and Canadian ROE proposals, were reproduced in the Canadian Joint Staff Washington document. This ROE arrangement did not, however, apply to USAF aircraft employed on operations in or over Greenland, which of course was part of US Northeast Command but not under Canadian jurisdiction.

58 Prime Minister Louis St. Laurent noted at the November 14 CDC Meeting that "a very satisfactory solution of the problem of US military activities in Newfoundland appeared

to be in sight," and added specifically that "General McNaughton was to be congratulated for his work on this problem in the Permanent Joint Board on Defence." Extract from Minutes of the 90th CDC Meeting, November 14, 1952, reproduced in *DCER*, vol. 18, 1952, 1155–56; Minutes of the 41/52 JPC Meeting, November 10, 1952, JPC Minutes to Meetings and Correspondence vol. 14, DHH 2002/17, box 58, file 2.

59 The JPC also made sure to make special note that Goose Bay was not included as a US base and that the Americans had to have the "express permission of the Canadian Government" to move forces to this air base. See JPC Report for Chiefs of Staff Committee, "Proposals of the US Section MCC for a Military Agreement on the Canadian Operational Control of US Defence Forces Operating from Bases within Canada," November 12, 1952, JPC Minutes to Meetings and Correspondence, vol. 14, DHH 2002/17, box 58, file 2.

60 Revision of the Command Appendix "F" Canada–United States Emergency Defense Plan (MCC 300/3), Appendix "A" to Minutes of the 5/52 Meeting of the MCC, November 21, 1952, MCC Minutes, DHH 80/540. This was very similar to both the definition that the AOCinC Eastern Air Command utilized in Halifax during the Second World War (see Chapters 2 and 4) and also the modern definition of operational control: "The authority delegated to a commander to direct assigned forces to accomplish specific missions or tasks that are usually limited by function, time or location, to deploy units concerned, and to retain or assign tactical control of those units. Operational control does not include authority to assign separate employment of components of the units concerned, neither does it, of itself, include administrative or logistic control." *Canadian Forces Aerospace Command Doctrine*, B-GA-0401-000/FP-001, 50.

61 Revision of the Command Appendix "F" Canada–United States Emergency Defense Plan (MCC 300/3), Appendix "A" to Minutes of the 5/52 Meeting of the MCC, November 21, 1952, MCC Minutes, DHH 80/540. This excluded "temporary tactical deployment" of USAF aircraft, although the AOC ADC was still to be informed of such movements. A previous paper had defined "temporary tactical deployment" as a period of no more than two weeks. The Vice-Chief of the Air Staff, Air Vice-Marshal F.R. Miller, described a hypothetical instance of how this exception would apply: "if a movement of SAC bombers were taking place through Goose Bay, CINCNE could move a fighter squadron from Harmon [AFB, Stephenville, Newfoundland] to Goose Bay for the duration of the operation without obtaining permission." (Draft) Proposed Directive on Amplification of the Can–US Agreement Concerning the Operational Control of US Air Defence Forces Operating from Bases within Canada, n.d. [likely November 12, 1952], JPC Minutes to Meetings and Correspondence, vol. 14, DHH 2002/17, box 58, file 2; AVM Miller to Chairman, Chiefs of Staff Committee, December 31, 1952, LAC, RG 24, acc. 1983–84 049, box 113, file 096-107-5, pt. 1, copy obtained through DND ATIP, A-2007-00185. Quote from latter.

62 AVM Miller to Chairman CSC, December 31, 1952, LAC, RG 24, acc. 1983–84 049, box 113, file 096-107-5, pt. 1. To the dismay of DEA officials, there were no references to the JCS's proposed interception arrangements in the revised EDP Command Appendix (see M.H. Barton, Defence Liaison Division, to USSEA, December 1, 1952, LAC, RG 25, vol. 5961, file 50221–40). However, as we shall see shortly, the provision for RCAF ROE and final firing authority would be included in the final detailed arrangements that the Canadian and American air force chiefs agreed to in April.

63 R.A. MacKay Minute, n.d. [November 14 or 15, 1952] on Wershof Memorandum "Canada–United States Military Installations in Newfoundland and Labrador" to R.A. MacKay, November 14, 1952, LAC, RG 25, vol. 5961, file 50221–40. Emphasis added.

64 Revision of the Command Appendix "F" Canada–United States Emergency Defense Plan (MCC 300/3), Appendix "A" to Minutes of the 5/52 Meeting of the MCC, November 21, 1952, MCC Minutes, DHH 80/540.

65 The MCC recommended quick approval of the new Command Appendix by the two countries' Chiefs of Staff, even "without awaiting approval of the revision of the entire" EDP, MCC Memorandum to PJBD, "Control of Air Defense Forces of the US Northeast Command Operating over Canada," November 27, 1952, LAC, RG 24, vol. 21418, file 1855:1, pt. 3, copy obtained through DND ATIP, A-2007-0209.

66 Minutes of the 5/52 Meeting of the MCC, December 10–13, 1952, and 1/53 Meeting, February 24–March 3, 1953, MCC Minutes, DHH 80/540; Secretary, US Section MCC to JCS, December 24, 1952, NARA, RG 218, JCS, entry 943011, box 32, file CCS 092 (9–10–45), section 32; "Chairman Chiefs of Staff Brief on Latest Developments Pertaining to Canada–US Defence Arrangements," n.d. [likely early March 1953], LAC, RG 24, vol. 21418, file 1855:1, pt. 3.

67 Minutes of the 91st CDC Meeting, February 10, 1953, LAC, RG 2, vol. 2749, file CDC Conclusions, vol. 6. American approval of the revised EDP Command Appendix only required authority from the JCS, not the president.

68 Bercuson, "SAC vs. Sovereignty," 222.

69 CAS Memorandum, "Directive – Canadian Operational Control of US Air Defence Forces Operating from Bases within Canada" to AOC ADC, March 5, 1953, Raymont Collection, DHH 73/1223/101. The directive also called for the AOC ADC to "take such actions as he considers necessary to ensure close liaison between" his headquarters and that of the CinC US Northeast Command "for the successful implementation of these plans."

70 Ibid.

71 The modern definition of operational command is: "the authority granted to a commander to assign missions or tasks to subordinate commanders, to deploy units, to reassign forces and to retain or delegate operational control, tactical command, and/or tactical control as necessary. It does not include responsibility for administration." *Canadian Forces Aerospace Command Doctrine*, B-GA-0401–000/FP-001, 7.

72 JCS 1259/279, Memorandum by the Chief of Staff USAF for the JCS, "Canadian Operational Control of US Defense Forces Operating from Bases within Canada," March 25, 1953, (Draft) Memorandum by the Chief of Staff USAF for the CinC US Northeast Command, "(Secret) Canadian Operational Control of US Defense Forces Operating from Bases within Canada and over Canadian Territory," March 23, 1953, and Decision on JCS 1259/279, April 2, 1953, Records of the JCS, Strategic Issues, reel 9 (section 2), pt. 2, 1946–53, microfilm copy at ML, RMC; Memorandum by Maj.-Gen. Robert M. Lee, USAF Director of Plans, "(Secret) Canadian Operational Control of US Defense Forces Operating from Bases within Canada and over Canadian Territory," to CinC US Northeast Command, April 10, 1953, DHH 2002/17, box 59, file 1, JPC Minutes to Meetings and Correspondence, vol. 16.

73 AVM A.L. James, AOC ADC, and Lt.-Gen. C.T. Myers, CinC US Northeast Command, "Agreed Control Arrangements by AOC ADC – C IN C NEC," April 21, 1953, LAC, RG 24, vol. 6172, file 15–73–3; DHH 73/770, Buss, *US Air Defense in the Northeast 1940–1957*, 20. 2 Air Defence Control Centre included the eastern Arctic, while 3 ADCC also covered Manitoba and Saskatchewan. 4 ADCC consisted of British Columbia. The remaining parts of Canada, western Ontario, and parts of the Prairies, were covered by USAF Air Divisions. Nicks et al., *A History of the Air Defence of Canada*, 11.

74 AVM A.L. James, AOC ADC, and Lt.-Gen. C.T. Myers, CinC US Northeast Command, "Agreed Control Arrangements by AOC ADC – C IN C NEC," April 21, 1953, LAC,

RG 24, vol. 6172, file 15-73-3. Specifically, "Rules of Engagement" was defined in the agreement as follows: "Rules of Engagement are written authorities and procedures by which an air defence commander [in this case the AOC ADC] determines when, where, by whom, and under what circumstances interceptions and engagements of unidentified or hostile aircraft may take place."

75 Foulkes Aide Memoire, "Integration of Operational Control of Canadian and Continental US Air Defence Forces in Peacetime," July 23, 1957, Raymont Papers, DHH 73/1223/84.

76 AVM A.L. James, AOC ADC, and Lt.-Gen. C.T. Myers, CinC US Northeast Command, "Agreed Control Arrangements by AOC ADC – C IN C NEC," April 21, 1953, LAC, RG 24, vol. 172, file 15-73-3; DHH 73/770; Buss, *US Air Defense in the Northeast*, 20. The agreement also provided for a Senior RCAF Liaison Officer, Newfoundland, to coordinate between the CinC US Northeast Command, the Canadian Civil Defence Administration, and the provincial and federal government agencies.

77 Don C. Bliss, Chargé d'affaires, US Embassy, Ottawa, Memorandum of Conversation with R. MacKay, Assistant USSEA, April 23, 1953, quoted in Jockel, *No Boundaries Upstairs*, 58-59.

78 DHH 73/770, Buss, *US Air Defense in the Northeast*, 20, 24-26.

79 Bercuson, "SAC vs. Sovereignty," 221-22.

80 As Sean Maloney has noted of the RCAF ADC-US Northeast Command arrangement, USAF "interceptor units would come under Canadian operational control in the event of war, something not done in other NATO countries." Maloney, *Learning to Love the Bomb*, 35.

Chapter 8: Integrating North American Air Defences under Operational Control

Epigraph: Memorandum by Defence Liaison (1) Division, "The Air Defence of North America," December 8, 1954, reproduced in *Documents on Canadian External Relations*, vol. 20, 1954, ed. Greg Donaghy (Ottawa: Department of Foreign Affairs and International Trade, 1997), 1060.

1 Memorandum from Defence Liaison (1) Division to USSEA, September 20, 1954, LAC, RG 25, vol. 6046, file 50309-40, pt. 1; John Hilliker and Donald Barry, *Canada's Department of External Affairs*, vol. 2: *Coming of Age, 1946-1968* (Kingston: Institute of Public Administration of Canada and McGill-Queen's University Press, 1990), 109; Richter, *Avoiding Armageddon*, 43-44.

2 Jockel, *No Boundaries Upstairs*, 81, 122; Fergusson, *Canada and Ballistic Missile Defence*, 11.

3 Maloney, *Learning to Love the Bomb*, 26-28; Trudgen, "The Search for Continental Security," 280.

4 Sagan, *Moving Targets*, 19; Cox, *Canada and NORAD*, 8; Trudgen, "The Search for Continental Security," 280-81. This transfer of funds pleased advocates for greater air defences and the USAF leadership and also was consistent with Eisenhower's policy to reduce overall military spending.

5 Sagan, *Moving Targets*, 19-26; Cox, *Canada and NORAD*, 10; Trudgen, "The Search for Continental Security," 153.

6 Cox, *Canada and NORAD*, 11; Maloney, *Learning to Love the Bomb*, 38-43. I thank an anonymous reviewer for bringing forth these connections.

7 Eisenhower's "New Look" policy therefore made a clear link between US nuclear war fighting and continental air defence. Jockel, *No Boundaries Upstairs*, 123; Haydon, *The 1962 Cuban Missile Crisis*, 72; Cox, *Canada and NORAD*, 12; Richter, *Avoiding Armageddon*, 46-47; Claxton Memorandum, "Continental Air Defence," for CDC, October 29,

1953, LAC, RG 2, vol. 2752, CDC Documents, vol. 13; 96th CDC Meeting Minutes, November 3, 1953, LAC, RG 2, vol. 2749, CDC Conclusions, vol. 6.
8   Cox, *Canada and NORAD*, 8–9; Jockel, *No Boundaries Upstairs*, 81–82.
9   On the establishment of these radar lines, see ibid., Chapter 4; and Trudgen, "The Search for Continental Security," Chapters 6 to 8.
10 The MSG was a subgroup of the PJBD. Second Draft Terms of Reference, Canada/United States Joint Military Study Group, as amended at the 539th CSC Meeting, May 28, 1953, n.d. [June 15, 1953]; and CJS Washington to Foulkes, June 15, 1953, Raymont Collection, DHH 73/1223/101; MG Robert M. Webster, USAF PJBD Member, Memorandum for Colonel Monteith, Chairman, Staff Group, JCS, October 20, 1953, NARA, RG 333, PJBD, entry 17-A, box 4, file "Top Secret Correspondence, 1941–1956," folder 20; Canadian Section PJBD Report for "Canadian eyes only," "Some Aspects of United States Air Defence Policy – As Enunciated to the PJBD by the Staff of the USAF Air Defence Command, Colorado Springs, Colorado," January 21, 1954, LAC, MG 30, E133, McNaughton Papers, vol. 286, file PJBD Meetings June 1953 to January 1955.
11 DHH 73/1501, *Nineteen Years of Air Defense*, 47–48; Jockel, *Canada in NORAD*, 12; Jockel, *No Boundaries Upstairs*, 56, 93.
12 MG Robert M. Webster, USAF PJBD Member, Memorandum, "Briefing for General Twining's Visit to Ottawa, February 14, 1954," for Deputy Chief of Staff, Operations, USAF, NARA, RG 333, PJBD, entry 17-A, box 2, file "Top Secret Correspondence, 1941–1956," folder 7.
13 Memorandum for Record by "F.P.B.," February 15, 1954, NARA, RG 333, PJBD, entry 17-A, box 2, file "Top Secret Correspondence, 1941–1956," folder 7; Canadian Section PJBD Report for "Canadian eyes only," "Some Aspects of United States Air Defence Policy – As Enunciated to the PJBD by the Staff of the USAF Air Defense Command, Colorado Springs, Colorado," January 21, 1954, LAC, MG 30, E133, McNaughton Papers, vol. 286, file PJBD Meetings June 1953 to January 1955.
14 Remarks by Rep. Cole before the 135th Anniversary Banquet of Colgate University, at the Waldorf-Astoria Hotel, New York, Thursday, April 29, 1954, at 9:30 p.m., copy in CSC Papers, DHH 2002/17, box 79, file 29. Cole was the Chairman of the US Joint Committee on Atomic Energy.
15 Cole to Wilson, May 7, 1954, NARA, RG 341, entry 335, USAF Plans, Project Decimal file 1942–1954, box 728, file OPD Canada 600.2 (September 14, 1945), section 19; Jockel, *No Boundaries Upstairs*, 95. Quote from former.
16 JCS Decision on JCS 1541/94, A Report by the Joint Strategic Plans Committee on Proposed North American Continental Defense Organization, June 11, 1954, NARA, RG 341, entry 335, USAF Plans, Project Decimal file 1942–1954, box 728-A, file OPD Canada 600.2 (September 14, 1945), section 20.
17 Claxton to Wilson, May 12, 1954, NARA, RG 333, PJBD, entry 17-A, box 1, file "Top Secret Correspondence, 1941–1956," folder "Briefing Book for the Secretary of Defense Visit to Canada." See also DEA correspondence during the spring of 1954 and in particular Claxton to Heeney, May 25, 1954, LAC, MG 30, E133, McNaughton Papers, file Continental Defence and CSC Chairman, to MND, Associate MND, Deputy MND, CGS, CNS, and CAS, May 11, 1954, DHH 112.3M2.009 (D114).
18 CSC Chairman to MND, Associate MND, Deputy MND, CGS, CNS, and CAS, May 11, 1954, DHH 112.3M2.009 (D114).
19 JPC Report CSC 1855-1 (JPC) for CSC, "Command of Continental Defence Forces," July 14, 1954, CSC Papers, September 21, 1954, DHH 2002/17, box 79, file 29.
20 Ibid. It will be recalled from Chapter 2 that Canada and the United States previously had the option of establishing a Supreme Command for the Canada–US NATO region in

1952 but chose against it. The prospect of an American NATO Supreme Commander for North America also did not sit well with Canadian planners.
21 Ibid.
22 Extract from 568th CSC Meeting, September 21, 1954, DHH 2002/17, box 79, file 29. See also DMO&P Brief on JPC Report, "Command of Continental Defence Forces," to VCGS and CGS, July 26, 1954, ibid.
23 Joseph Jockel has termed this US practice as "twinning" – giving one officer more than one command title depending on the specific role or mission he was performing/accomplishing (i.e., as head of a particular command organization) at the time. Chidlaw had been commanding general of USAF ADC since 1951. Jockel, "Canada in NORAD, 1957–2007: A History," Queen's Centre for International Relations (QCIR) Presentation, January 24, 2007. See also Jockel, *Canada in NORAD*, 6.
24 The USN maintained "picket ships" and airborne early warning aircraft patrolling off the American coasts as part of the US air defence posture. Naval Forces Continental Air Defense Command was established on September 1, 1954, at CONAD Headquarters in Colorado Springs "to provide centralized direction to the Navy effort." Capt. Joseph F. Bouchard, USN, "Guarding the Cold War Ramparts: The U.S. Navy's Role in Continental Air Defense," *Naval War College Review* 52, 3 (Summer 1999): 111–35.
25 There was also a Marine Corps representative assigned to CONAD's staff. *Chronology of JCS Involvement in North American Air Defense*, 52–56; Robert J. Watson, *History of the Joint Chiefs of Staff, Volume V: The Joint Chiefs of Staff and National Policy 1953-1954* (Washington, DC: Historical Division, Joint Chiefs of Staff, 1986), 137; Chargé d'affaires, Canadian Embassy, Washington, "Department of Defense Press Release No. 733-54 of 3 August, 1954," to SSEA, August 4, 1954, LAC, MG 30, E133, McNaughton Papers, file Continental Defence. Quote from first source.
26 Terms of Reference for CinC CONAD, September 1, 1954, quoted in DHH 73/1501, *Nineteen Years of Air Defense*, 34. CONAD did not include responsibility for the air defence of Alaska or the US bases in Newfoundland, which remained under the authority of the CinC Alaska Command and the CinC US Northeast Command, respectively. The CinC CONAD was also required to prepare plans for air defence and early warning systems and procedures, which he would then submit to the JCS for approval. *Chronology of JCS Involvement in North American Air Defense*, 56.
27 The US Army in particular was never comfortable with the CONAD operational control authority, feeling strongly that operational control should be exercised through the component commands. 73/1501, *Nineteen Years of Air Defense*, 34.
28 Maj.-Gen. James E. Briggs, USAF, Assistant Deputy Chief of Staff, Operations, to Director, Joint Staff, October 22, 1954, NARA, RG 218, JCS, entry 943011, box 29, file CCS 092 (9–10–45), section 38; AVM C.R. Dunlap, Chairman, Canadian Section MSG, to CSC Chairman, August 26, 1954, Raymont Collection, DHH 73/1223/89; CSC Chairman, to JCS Chairman, September 20, 1954, reproduced in Donaghy, ed., *DCER*, 20, 1954, 1026–28.
29 JCS 1541/102, Report by the Chief of Staff, USAF, to the JCS on A Combined Canada–United States North American Air Defense Command, December 5, 1955, NARA, RG 218, JCS, entry 943011, box 29, file CCS 092 (9–10–45), section 41; *Chronology of JCS Involvement in North American Air Defense*, 59.
30 Extract from a Special CSC Meeting, April 6, 1955, reproduced in Donaghy, ed., *DCER*, vol. 21, 1955, 709–12.
31 Goette, "A Snapshot of Early Cold War RCAF Writing," 56–58.
32 Extract from a Special CSC Meeting, April 6, 1955, reproduced in Donaghy, ed., *DCER*, vol. 21, 1955, 709–12.

33 Jockel, *No Boundaries Upstairs*, 101.
34 Slemon Brief for Meeting of Consultation Continental Defence, December 2, 1955, Raymont Collection, DHH 73/1223/89. Slemon's definition of command is consistent with what today is called full command, which is a service prerogative. *Canadian Forces Aerospace Command Doctrine*, B-GA-0401-000/FP-001, 6–7.
35 In addition to Foulkes, meeting attendees included officials from the US State Department, Canada's Department of External Affairs, the Canadian Embassy, and the Canadian Joint Staff Mission in Washington, plus JCS Chairman Admiral Arthur Radford. Minutes of Top Secret United States–Canadian Political-Military Meeting, December 5, 1955, NARA, RG 59, State Department, Deputy Assistant Secretary for Politico-Military Affairs, Subject Files of the Special Assistant for Atomic Energy and Aerospace, 1950–56, entry 5181, NN3-059-9-018, NND 989528, box 11, file Nuclear Sharing – Canada – Pol-Military Meetings, 1955–57.
36 Henceforth the title of American correspondence on this issue switched from "A Combined Canada–United States North American Air Defense Command" to "Integration of Operational Control of the Continental Air Defenses of Canada and the United States during Peacetime." Capt. Richard H. Phillips, JCS Secretary, Memorandum, "Operational Control of Canadian and United States Continental Air Defense Warning Systems and Air Forces," to JSPC, December 15, 1955, and JCS 1541/103, Report by the JSPC to the JCS, "Integration of Operational Control of the Continental Air Defenses of Canada and the United States During Peacetime," January 9, 1956, NARA, RG 218, JCS, entry 943011, box 29, file CCS 092 (9–10–45), section 41; *Chronology of JCS Involvement in North American Air Defense*, 61. Quotes from first document.
37 Jockel, *No Boundaries Upstairs*, 101.
38 (Draft) JCS Memorandum for the CSC Chairman, n.d. [January 5, 1956], Enclosure "B" to JCS 1541/103, Report by the JSPC to the JCS on Integration of Operational Control of the Continental Air Defenses of Canada and the United States during Peacetime, January 9, 1956, NARA, RG 218, JCS, entry 943011, box 29, file CCS 092 (9–10–45), section 41. USAF planners had been reviewing the definition of operational control as part of their re-examination of CONAD's terms of reference since September 1955 (see below). It appears that this JSPC definition of operational control reflected findings thus far. As we will see shortly, a slightly revised version of this January 1956 definition of operational control would be entrenched in the new terms of reference for CONAD in November 1956.
39 Exclusion of any mention of "command" has continued to be a common feature of subsequent definitions of operational control.
40 CinC USN to US Navy Commands, Admiralty, Air Ministry and Chief of Staff Army, February 11, 1944, TNA PRO, Air 15/339.
41 The USAF Chief of Staff assigned the staff of CONAD responsibility for undertaking these discussions on behalf of the United States. *Chronology of JCS Involvement in North American Air Defense*, 62; JCS Decision on JCS 1541/103, Report by the JSPC to the JCS on Integration of Operational Control of the Continental Air Defenses of Canada and the United States during Peacetime, January 18, 1956, Secretary JCS to Chief of Staff USAF, January 18, 1956, Chief of Staff USAF to CinC CONAD, February 11, 1956, and JCS Memorandum, "Integration of Operational Control of the Continental Air Defenses of Canada and the United States during Peacetime," to CSC Chairman, February 14, 1956, NARA, RG 218, JCS, entry 943011, box 29, file CCS 092 (9–10–45), section 41.
42 589th CSC Meeting Minutes, February 21, 1956, DHH 2002/17, box 80, file 13; Capt. Richard H. Phillips, JCS Secretary, Memorandum "Integration of Operational

Control of Continental Air Defences of Canada and the United States during Peacetime," to CSC Chairman, March 27, 1956, LAC, RG 24, vol. 21422, file CSC 1855:8. Quotes from former.
43 The Canadian members of the AHSG were A/C M.D. Lister (Chairman) and G/C R.M. Cox from the RCAF, plus Col. G.A. Turcot from the Canadian Army. Vice-Chief of the Air Staff (VCAS) to CAS, May 11, 1956, LAC, RG 42, acc. 1997–98/260, box 73, file TS 964-204, pt. 1; *Chronology of JCS Involvement in North American Air Defense*, 64, 68.
44 Canadian Ambassador to the United States to Secretary of State for External Affairs, March 19, 1956, reproduced in Donaghy, ed., *DCER*, vol. 23, 1956–57, pt. 2, 18–19.
45 CNS to CAS, April 9, 1956, and CGS to CAS, April 9, 1956, LAC, RG 24, vol. 21422, file CSC 1855:8; VCAS to CAS, May 11, 1956, LAC, RG 42, acc. 1997–98/260, box 73, file TS 964-204, pt. 1.
46 This new arrangement was placed in effect as of October 1, 1956. DHH 73/1501, *Nineteen Years of Air Defense*, 37–39, 81; Webb et al., *The History of the Unified Command Plan*, 25; *Chronology of JCS Involvement in North American Air Defense*, 67. It will be recalled that General Partridge was, as a major-general, the first USAF member of the Air Standardization Coordination Committee, where he had established an early strong association with and connection to the RCAF; Lt.-Col. Christopher England, "Air and Space Interoperability Council (ASIC) and the RCAF," Masters of Defence Studies Directed Research Project, Canadian Forces College Toronto, 2016, 1–2, 10–16.
47 DHH 73/1501, *Nineteen Years of Air Defense*, 38, 45–46; Webb et al., *The History of the Unified Command Plan*, 25–26; *Chronology of JCS Involvement in North American Air Defense*, 66, 69–70.
48 DHH 73/1501, *Nineteen Years of Air Defense*, 38.
49 Terms of Reference for CinC CONAD, September 4, 1956, quoted in DHH 73/1501, *Nineteen Years of Air Defense*, 38; also in *Chronology of JCS Involvement in North American Air Defense*, 67.
50 Ibid., JCS 1541/109, Report by the JSPC to the JCS on Integration of Operational Control of the Continental Air Defenses of Canada and the United States during Peacetime, August 13, 1956; Secretary JCS to Chairman, US Section, MSG, August 24, 1956, NARA, RG 218, JCS, entry 943011, box 8, file CCS 092 (9–10–45), Section 42. Quote from former.
51 (Second Draft) "Integration of Operational Control of the Continental Air Defenses of Canada and the United States in Peacetime," Ad Hoc MSG Committee on Integration of Operational Control, October 22, 1956, DHH 99/36, W.A.B. Douglas fonds, box 49, file 16.
52 Minutes of the 13th MSG Meeting, December 19, 1956, DHH 112.3M2 (D711); AM C.R. Dunlap, DCinC NORAD, to Gen. (Ret.) Charles Foulkes, April 11, 1967, LAC, MG 31, G5, Air Marshal C.R. Dunlap Papers. Special thanks to Matt Trudgen for sharing the latter document with me.
53 Col. Mayhue D. Baine, USAF Air Attaché, US Embassy, Ottawa, to MG James F. Briggs, USAF Deputy Chief of Staff, Operations, June 18, 1954, NARA, RG 341, entry 335, USAF Plans, Project Decimal file 1942–54, box 728-A, file OPD Canada 600.2 (September 14, 1945), section 20.
54 "Integration of Operational Control of the Continental Air Defenses of Canada and the United States in Peacetime," Ad Hoc MSG Committee Report on Integration of Operational Control, December 19, 1956, DHH 112.3M2 (D711); Jockel, *No Boundaries Upstairs*, 103. Quote from latter.
55 "Integration of Operational Control" MSG Report, December 19, 1956, DHH 112.3M2 (D711).

56 Ibid.; Jockel, *No Boundaries Upstairs*, 102. Quote from former.
57 Each of these commanders was also to be the "principal advisor to the CINCADCANUS on matters of his Service" related to continental air defence. "Integration of Operational Control" MSG Report, December 19, 1956, DHH 112.3M2 (D711).
58 Grimshaw, "On Guard," 151; Jockel, *Canada in NORAD*, 21; AVM C.R. Dunlap, VCAS, to Maj.-Gen. G. Kitching, VCGS, June 3, 1957, and Kitching to Dunlap, January 26, 1957, DHH 71/403.
59 Jockel, *Canada in NORAD*, 21; AM Slemon, CAS Directive to AOCADC, January 1, 1955, LAC, RG 24, acc. 1997-98/260, box 59, file TS 096-207-4, pt. 2; "Integration of Operational Control" MSG Report, December 19, 1956, DHH 112.3M2 (D711).
60 "Integration of Operational Control" MSG Report, December 19, 1956, DHH 112.3M2 (D711).
61 The records that may shed light on the reasons remain classified.
62 Ibid. The phrase "which may, or may not, be under the command of the authority exercising operational control" referred to the different American and Canadian air defence forces under the operational command of the component commanders. For instance, "which may ... be under the command of the authority exercising operational control" meant the forces under the operational command of the CINCADCANUS in his capacity as the CinC CONAD (i.e., with his CinC CONAD component commander "hat" on), while "may not be under the command of the authority exercising operational control" meant that CINCADCANUS would have operational control of forces that were under the operational command of the RCAF ADC component command. The emphasis here was that operational command remained with the component commanders, not CINCADCANUS, who would only have operational control over them.
63 The new definition also emphasized that the CINCADCANUS was only to have authority over units that were *already* deployed. Responsibility for the original composition of assigned forces was thus still the prerogative of the individual services themselves under their national command authority. "Integration of Operational Control" MSG Report, December 19, 1956, DHH 112.3M2 (D711).
64 Ibid.; Jockel, *No Boundaries Upstairs*, 102.
65 Schaffel, *Emerging Shield*, 250.
66 Eighth Report of the MSG, December 19, 1956, DHH 112.3M2 (D711); Chief of Staff USAF Memorandum, "Integration of Operational Control of Canadian and Continental United States Air Defence Forces in Peacetime," to JCS, March 8, 1957, NARA, RG 218, JCS, entry 943011, box 8, file CCS 092 (9-10-45), section 43; Reginald H. Roy, *For Most Conspicuous Bravery: A Biography of Major-General George R. Pearkes, V.C., through Two World Wars* (Vancouver: UBC Press, 1977), 288.
67 Minutes of the 604th and 605th Meetings of the Chiefs of Staff Committee, February 1 and 15, 1957, Chiefs of Staff Committee Special Meeting (July 16, 1957), DHH 2002/17, box 87, file 2; Slemon to Secretary, Chiefs of Staff Committee, February 1957, LAC, RG 24, vol. 21422, file CSC 1855:8; MND Memorandum, "Integration of Operational Control of Canadian and Continental United States Air Defence Forces in Peacetime," to Cabinet Defence Committee, March 11, 1957, LAC, RG 24, vol. 21422, file CSC 1855:8; Chief of Staff USAF Memorandum, "Integration of Operational Control of Canadian and Continental United States Air Defence Forces in Peacetime," to JCS, March 8, 1957, NARA, RG 218, JCS, entry 943011, box 8, file CCS 092 (9-10-45), section 43.
68 Col. R.L. Raymont, Executive Staff Officer to the CSC Chairman, to Ross Martin, Privy Council Office, March 11, 1957, and Captain F.W.T. Lucas, CSC Secretary, to JCS Secretary, April 26, 1957, Raymont Collection, DHH 73/1223/84; CGS to Chairman, Canadian Joint Staff Washington, DC, April 26, 1957, LAC, RG 24, vol. 21422, file CSC

1855:8; Memorandum on Steps in Development of Integration of Operational Control of Canadian and Continental United States Air Defence Forces in Peacetime, November 11, 1957, Raymont Collection, DHH 73/1223/87; Minutes of the April 29 to May 2, 1957, Meeting of the PJBD, PJBD Journals, copy in LAC, MG 30, E133, McNaughton Papers, vol. 286, file PJBD Journal of Meeting, Fort Bagg, North Carolina, April 29 to May 3, 1957; Jockel, *No Boundaries Upstairs*, 103–4; Reginald Roy interview with Gen. Charles Foulkes, March 9, 1967, quoted in Roy, *For Most Conspicuous Bravery*, 288.

69 Jockel, *No Boundaries Upstairs*, 104; Memorandum on Steps in Development of Integration of Operational Control of Canadian and Continental United States Air Defence Forces in Peacetime, November 11, 1957, Raymont Collection, DHH 73/1223/87; Extract from Cabinet Conclusions, June 13, 1957, reproduced in Donaghy, ed., *DCER*, vol. 23, 1956–57, pt. 2, 82–83; Foulkes, Note for Files, Telephone Conversation with General Sparling, Washington, June 18, 1957, Raymont Collection, DHH 73/1223/84.

70 Ibid.; House of Commons Special Committee on Defence, *Minutes of Proceedings and Evidence*, no. 15, October 22, 1963 (Ottawa: Queen's Printer, 1963), 510, 527.

71 Extract from CSC Special Meeting Minutes, July 16, 1957, DHH 2002/17, box 87, file 2; George Pearkes, MND, Memorandum "Integration of Operational Control of Canadian and Continental United States Air Defence Forces in Peacetime," to Cabinet, July 22, 1957, LAC, RG 25, vol. 6046, file 50309-40, pt. 2.

72 Pearkes note "Discussed with P.M. and approved," July 24, 1957, on Index of documents relating to the integration of operational control of Canada–US continental air defence forces in peacetime, MND confirmation of PM approval on NORAD, Raymont Collection, DHH 73/1223/84; CSC Chairman, to Acting USSEA, August 7, 1957, and Draft Memorandum from Prime Minister to SSEA, December 4, 1957, reproduced in Michael Stevenson, ed., *DCER*, vol. 25, 1957–58, pt. 2 (Ottawa: Department of Foreign Affairs and International Trade, 2004), 35–37; Roy interview with General Charles Foulkes, March 9, 1967, 22–23, quoted in Roy, *For Most Conspicuous Bravery*, 289; John G. Diefenbaker, *One Canada: Memoirs of the Right Honourable John G. Diefenbaker*, vol. 3: *The Tumultuous Years 1962–1967* (Toronto: Macmillan of Canada, 1977), 19–20.

73 Extract from Cabinet Conclusions, July 31, 1957, reproduced in Stevenson, ed., *DCER*, vol. 25, 1957–58, pt. 2, 21–32; Vice-Adm. B.L. Austin, Director US Joint Staff, to Adm. A. Radford, Chairman JCS, July 31, 1957, NARA, RG 218, JCS, entry 943011, box 8, file CCS 092 (9–10–45), section 44; Bryce to Holmes, August 1, 1957, LAC, RG 25, vol. 6046, file 50309-40, pt. 2.

74 Press Release by the Secretary of Defense of the United States and the Minister of National Defence of Canada, August 1, 1957, and Press Release by the Minister of National Defence Appointing Air Marshal C.R. Slemon Deputy Commander-in-Chief of the Canada–United States Air Defence Command, August 1, 1957, Raymont Collection, DHH 73/1223/84. At the July 31 Cabinet meeting the Diefenbaker government also confirmed the promotion of Air Vice-Marshal H.L. Campbell to Air Marshal and his replacement of Slemon as RCAF Chief of the Air Staff effective September 1, 1957. Campbell had previously been serving as Deputy Chief of Staff (Operations) on the staff of Gen. Lauris Norstad, NATO Supreme Allied Commander Allied Powers Europe. Foulkes to Norstad, July 31, 1957, Raymont Collection, DHH 73/1223/84. Quote from August 1, 1957, joint press release.

75 DHH 73/1501, *Nineteen Years of Air Defense*, 50. In one of his last acts as Chief of the Air Staff, Air Marshal Slemon reasserted that NORAD did not change the fact that the primacy for air defence in Canada remained with the RCAF: "the CAS/RCAF will continue to be the Executive Agent of the Canadian Chiefs of Staff with respect to matters dealing with North American Air Defence, and that communications between CINCNORAD

and the Canadian Chiefs of Staff will be channelled through CAS/RCAF." Slemon to Foulkes, August 19, 1957, Raymont Collection, DHH 73/1223/85. In 2002, the "in-chief" was eliminated from American command organizations and Canada followed suit, resulting in the current Commander NORAD and Deputy Commander NORAD. At the time of writing the commander of NORAD is Gen. Lori Robinson and the Deputy Commander is Lt.-Gen. Pierre St-Amand.

76 This arrangement ensured that Partridge, as CinC CONAD, remained "a national commander responsible to the United States Joint Chief of Staff for purely national matters." *Chronology of JCS Involvement in North American Air Defense*, 81–82, 87; DHH 73/1501, *Nineteen Years of Air Defense*, 53; Jockel, *Canada in NORAD*, 21; JCS, Terms of Reference for CINCONAD, January 8, 1958, NARA, RG 218, JCS, entry 943011, box 16, file CCS 092 (9–10–45), section 47; Brig.-Gen. R.D. Wentworth, Secretary JCS, Memorandum, "Terms of Reference for the Commander in Chief North American Air Defense Command (NORAD) (U)," January 10, 1958, Papers to the CSC 617th Meeting, DHH 2002/17, box 81, file 6. Quote from first source. A similar arrangement remains today, with the NORAD commander double-hatted as the Commander of Northern Command, a JCS Unified Command Plan combatant command.

77 Text of Canadian Note, His Excellency N.A. Robertson, Canadian Ambassador to the United States, to the Honourable John Foster Dulles, Secretary of State of the United States, May 12, 1958, Canada Treaty Series 1958, copy in George Randolph Pearkes Papers, University of Victoria Archives, acc. 74-1, box 26, file 26.3 (hereafter NORAD Agreement, May 12, 1958).

78 Ibid., Terms of Reference for the Commander in Chief, North American Air Defense Command, June 10, 1958, NARA, RG 218, JCS, entry 943011, box 16, file CCS 092 (9–10–45), section 51; CSC Chairman to CinC NORAD, June 10, 1958, Raymont Collection, DHH 73/1223/87; JCS to CinC NORAD, June 10, 1958, NARA, RG 218, JCS, entry 943011, box 16, file CCS 092 (9–10–45), section 50.

79 NORAD Agreement, May 12, 1958, Pearkes Papers, University of Victoria Archives, acc. 74-1, box 26, file 26.3.

80 Appendix "F," Command, MCC 300/10, Canada–United States Emergency Defense Plan, July 1, 1958, NARA, RG 218, JCS, entry 943011, box 16, file CCS 092 (9–10–45), section B.P. pt. 9. The procedures for Rules of Engagement (ROE) for air defence aircraft under NORAD operational control stayed the same in the EDP. These were consistent with the PJBD Recommendation 53/1, which stipulated that the ROE of the country in whose airspace the interception took place would apply. Annex "A," Principles Governing the Interception of Unidentified Aircraft in Peace Time, to Appendix "F" Command, ibid.

81 See Jockel, *No Boundaries Upstairs*, 107–17; Haydon, *The 1962 Cuban Missile Crisis*, 74–78; Robert Douglas Allin, "Implementing NORAD, 1956–1962: The Bureaucratic Tug of War for Access and Influence," MA thesis, Department of History, Carleton University, 1998, Chapters 1 and 2; J.L. Granatstein, *A Man of Influence: Norman A. Robertson and Canadian Statecraft* (Toronto: Deneau, 1981), 316–18; J.L. Granatstein, *Canada 1957–1967: The Years of Uncertainty and Innovation* (Toronto: McClelland and Stewart, 1986), 102–5; Roy, *For Most Conspicuous Bravery*, 288–94; Maloney, *Learning to Love the Bomb*, 101–4; Jon B. McLin, *Canada's Changing Defense Policy, 1957–1963: The Problems of a Middle Power in Alliance* (Baltimore: Johns Hopkins University Press in cooperation with the Washington Center of Foreign Policy Research, School of Advanced International Studies, Johns Hopkins University, 1967), Chapter 3.

82 Ibid. See Diefenbaker's memoir in particular for the prime minister's mistrust of the DEA. Diefenbaker, *One Canada*, vol. 3, 24–25. For a fresh and astute reassessment, see

Patricia I. McMahon, *Essence of Indecision: Diefenbaker's Nuclear Policy, 1957–1963* (Montreal and Kingston: McGill-Queen's University Press, 2009), 9–14.

83 Allin, "Implementing NORAD," 19; Adam Chapnick, *Canada's Voice: The Public Life of John Wendell Holmes* (Vancouver: UBC Press, 2009), 96–98; John W. Holmes to Reginald Roy, December 5, 1972, quoted in Roy, *For Most Conspicuous Bravery*, 291; Jules Léger, USSEA, to Gen. Charles Foulkes, CSC Chairman, September 10, 1957, Raymont Collection, DHH 73/1223/85; J.W. Holmes, Acting USSEA, to F.R. Miller, Deputy MND, August 2, 1957, LAC, RG 25, vol. 6046, file 50309-40, pt. 2; Memorandum from USSEA to SSEA, October 7, 1957, reproduced in Stevenson, ed., *DCER*, vol. 25, 1957–58, pt. 2, 46–48. See other correspondence from the summer and autumn of 1957 in this volume for further information on DEA's procedural and consultation concerns.

84 Peter Haydon has argued that an official exchange of notes unnecessarily complicated matters: "Rather than be submitted for formal political approval, the NORAD Agreement should have been signed jointly by the Canadian Chairman of the Chiefs of Staff and the US Chairman of the Joint Chiefs of Staff. Thus, the NORAD Agreement became a politically sensitive strategic concept rather than a simple operational plan for coordinating the activities of the two air forces." Haydon, *The 1962 Cuban Missile Crisis*, 77.

85 Memorandum from Assistant USSEA to Secretary of the Cabinet, July 31, 1957, and Memorandum from USSEA to SSEA, October 7, 1957, reproduced in Stevenson, ed., *DCER*, vol. 25, 1957–58, pt. 2, 32–33, 46–48; Jules Léger, USSEA, to Gen. Charles Foulkes, CSC Chairman, September 10, 1957, Raymont Collection, DHH 73/1223/85; Jockel, *No Boundaries Upstairs*, 106, 117; Maloney, *Learning to Love the Bomb*, 104. The DEA officials did not appreciate that the NATO arrangement involved operational control of European forces on their national territory.

86 For example, for the Canadian Northwest Atlantic Command, there was no "CinC Canada–US Maritime Air Forces" but simply the elevation of the Air Officer Commanding Eastern Air Command to CinC status with operational control authority over the maritime air forces from the RCAF, the USAAF, and USN in the theatre of operations.

87 Canada, Department of National Defence, *National Defence Act* (1950); Diefenbaker, *One Canada*, vol. 3, 20. Foulkes used the National Defence Act provision as justification for the way that NORAD was established during the summer of 1957. See Gen. Charles Foulkes, CSC Chairman, to J.W. Holmes, Acting USSEA, August 7, 1957, Raymont Collection, DHH 73/1223/85. In US constitutional practice, only the approval of the Secretary of Defense was required to establish a command organization.

88 For example, in May, Gen. Earle Partridge, the CinC CONAD, began describing the proposal as "the complete integration of the operational elements of the air defense forces of the United States and Canada into a single *Command structure*." Partridge to Twining, May 14, 1957, Raymont Collection, DHH 73/1223/84. Emphasis added.

89 MND Memorandum "Integration of Operational Control of Canadian and Continental United States Air Defence Forces in Peacetime," to Cabinet, July 22, 1957, Raymont Collection, DHH 73/1223/84; Extract from Cabinet Conclusions, July 31, 1957, reproduced in Stevenson, ed., *DCER*, vol. 25, 1957–58 pt. 2, 21–32; Vice Adm. B.L. Austin, Director US Joint Staff, to Adm. A. Radford, Chairman JCS, July 31, 1957, NARA, RG 218, JCS, entry 943011, box 8, file CCS 092 (9–10–45), section 44; Press Release by the Secretary of Defense of the United States and the Minister of National Defence of Canada, August 1, 1957, Raymont Collection, DHH 73/1223/84.

90 Jockel, *No Boundaries Upstairs*, 107.

91 Bert Frandsen in particular notes that Pearkes was "a proponent of a strong air force" and quotes from Reginald Roy's biography of the MND: "Despite his long association with

the army, Pearkes considered the air force to be Canada's first line of defence. In 1949 he suggested that there should be three dollars spent on the air force for every dollar spent on the army.'" Frandsen, "The Rise and Fall of Canada's Cold War Air Force," 76; Roy, *For Most Conspicuous Bravery*, 270–71, 277.

92 Roy, *For Most Conspicuous Bravery*, Chapter 11 and 290; Diefenbaker, *One Canada*, vol. 3, 17; Granatstein, *Canada 1957–1967*, 103. For Canada's role in the Aleutian campaign, see Perras, *Stepping Stones to Nowhere*. It is also worth noting that the Deputy MND (a civilian position, not an elected one) at the time was Frank Miller, formerly Vice-Chief of the Air Staff. Air Vice-Marshal Miller, it will be recalled, had been intimately involved in negotiations for the RCAF's operational control arrangement in Newfoundland with US Northeast Command (see Chapter 7). It is not clear, however, whether Pearkes consulted with Miller on the NORAD operational control issue before bringing it to Diefenbaker. Miller would also in 1964 become the Canadian military's first Chief of Defence Staff at the rank of Air Chief Marshal. See Ray Stouffer, "Air Chief Marshal Frank Miller: A Civilian and Military Leader," *Canadian Military Journal* 10, 2 (Spring 2010): 41–51.

93 CinC CONAD to CSC Chairman, August 13, 1957, Raymont Collection, DHH 73/1223/85; JCS 1541/120, Note by the Secretaries to the JCS on Integration of Air Defense, Canada–United States, August 16, 1957, NARA, RG 218, JCS, entry 943011, box 8, file CCS 092 (9-10-45), section 44; Jockel, *Canada in NORAD*, 25. Quote from first document.

94 Press Release by the Secretary of Defense of the United States and the Minister of National Defence of Canada, August 1, 1957, Raymont Collection, DHH 73/1223/84.

95 Draft Press Release regarding the appointment of Air Marshal C.R. Slemon as Deputy Commander-in-Chief of the Canada–United States Air Defence Command, n.d. [likely July 24, 1957], Raymont Collection, DHH 73/1223/84. This was one of the documents that Pearkes discussed with Diefenbaker on the afternoon of July 24 when the prime minister approved the integrated air defence operational control proposal. It will be recalled that the MND Memorandum to the Cabinet that Foulkes had prepared also referred to a "unified command" being established. MND Memorandum "Integration of Operational Control of Canadian and Continental United States Air Defence Forces in Peacetime," to Cabinet, July 22, 1957, Raymont Collection, DHH 73/1223/84.

96 Foulkes to Pearkes, July 26, 1957, and American editorial changes, n.d. [July 26, 1957], written on Draft Press Release by the Secretary of Defense of the United States and the Minister of National Defence of Canada, n.d. [July 24, 1957], Raymont Collection, DHH 73/1223/84; Vice-Adm. B.L. Austin, Director US Joint Staff, to Adm. A. Radford, Chairman JCS, July 31, 1957, and revised (by USAF Chief of Staff and Deputy Secretary of Defense) Draft Press Release by the Secretary of Defense of the United States and the Minister of National Defence of Canada, n.d. [July 26, 1957], NARA, RG 218, JCS, Entry 943011, box 8, file CCS 092 (9-10-45), section 44; USN Chief of Naval Operations Memorandum for the JCS, January 8, 1958, NARA, RG 218, JCS, entry 943011, box 16, file CCS 092 (9-10-45), section 47. It will be recalled that the correspondence on the air defence proposal and the actual title of the MSG study itself was "integration of operational control" of the two nations' air forces. Pearkes approved the changes, noting, "I agree, integrated is better than unified." Pearkes to Foulkes, July 26, 1957, Raymont Collection, DHH 73/1223/84.

97 Press Release by the Secretary of Defense of the United States and the Minister of National Defence of Canada, August 1, 1957, Raymont Collection, DHH 73/1223/84.

98 Jockel, *No Boundaries Upstairs*, 111–17; Allin, "Implementing NORAD," Chapter 2; Diefenbaker, *One Canada*, vol. 3, 20; Haydon, *The 1962 Cuban Missile Crisis*, 76–77. See also

Discussion with General Foulkes on the position of the new North American Air Defense Command [between Foulkes, and DEA officials Holmes, Bryce, Tremblay, McCardle, and Crean], September 23, 1957, LAC, RG 25, vol. 6046, file 50309-40, pt. 2; Foulkes to Léger, November 26, 1957, LAC, RG 24, vol. 21422, file CSC 1855:8; Comments by Diefenbaker in Extract from Verbatim Record of the Meeting of the [NATO] Council – Monday, December 16, 1957, and CSC Chairman to Chairman, Canadian Joint Staff, December 27, 1957, Raymont Collection, DHH 73/1223/86; Foulkes to Secretary Canadian Section PJBD, December 30, 1957, LAC, RG 25, vol. 6046, file 50309-40, pt. 4.

99 Schaffel, *Emerging Shield*, 252; Maloney, *Learning to Love the Bomb*, 124-31; Allin, "Implementing NORAD," Chapters 1 and 2; Jockel, *No Boundaries Upstairs*, 107-17; *Chronology of JCS Involvement in North American Air Defense*, 85-86; Brig.-Gen. R.D. Wentworth, Secretary JCS, Memorandum, "NORAD as a NATO Organization," to Gen. Sparling, Chairman, Canadian Joint Staff Washington, April 10, 1958, Raymont Collection, DHH 73/1223/86.

100 It will be recalled that the MSG study referred to both NATO and US UCP unified commands as the best "method" of command for North American air defence. "Integration of Operational Control" MSG Report, December 19, 1956, DHH 112.3M2 (D711).

101 As political commentator Jon McLin noted, "both the sectional [i.e., media] and partisan [i.e., political] opinion lamented the purely bilateral character of the agreement." The effort to make a direct connection to NATO therefore "responded to a long-felt Canadian preference – partly instinctive, partly intellectualized – for multilateral rather than bilateral commitments." McLin, *Canada's Changing Defense Policy*, 54-55. See also Diefenbaker, *One Canada*, vol. 3, 27-28; and Roussel, *The North American Democratic Peace*, 206-7.

102 Roussel, *The North American Democratic Peace*, 217; *Chronology of JCS Involvement in North American Air Defense*, 85-86; Allin, "Implementing NORAD," 51. Quote from Roussel. Later, Foulkes explained the bilateral nature of NORAD and its distinctiveness from multilateral commands in 1963 to the House of Commons Special Committee on Defence: "But you see, NORAD is an entirely different set-up. Instructions to NORAD are given by the chiefs of staff of both countries. Any instructions, for instance, going to the commander in chief of NORAD, whether they originate with the United States chief of staff or the Canadian chiefs of staff, are sent at the same time by both of them. They both have to agree and they both send messages at exactly the same time to the commander in NORAD, so that the control of that force is equally shared by Canada and the United States." House of Commons Special Committee on Defence, *Minutes of Proceedings and Evidence*, October 22, 1963, 537.

103 Terms of Reference for the Commander in Chief, North American Air Defense Command, June 10, 1958, NARA, RG 218, JCS, entry 943011, box 16, file CCS 092 (9-10-45), section 51.

104 Kostenuk and Griffin, *RCAF Squadron Histories and Aircraft*, 208.

105 Diefenbaker, *One Canada*, vol. 3, 22.

106 Memorandum from USSEA to SSEA, December 2, 1957, LAC, RG 25, vol. 6046, file 50309-40, pt. 4.

107 DHH 79/128, interview with Air Marshal Roy Slemon by W.A.B. Douglas and William McAndrew, October 20, 1978, 18. The context of the discussion related to tactical nuclear air defence weapons, which were included in NORAD operational control authority.

108 Ibid.

109 Slemon quoted in Schaffel, *The Emerging Shield*, 252, 253; "Organization for NORAD Headquarters," Appendix "A" to S964-104 (CAS), Campbell to Foulkes, March 8, 1958,

Raymont Collection, DHH 73/1223/86; discussion by author with Joseph Jockel at the QCIR Spring Conference, "NORAD's Half-Century: Securing and Defending North America," June 7, 2007, Queen's University, Kingston; Goodspeed, *The Armed Forces of Canada*, 226; multiple discussions between author and Lt. Colonel Pierre Viens, a longtime senior staff officer at NORAD headquarters in Colorado Springs. It will be recalled from Chapter 7 that Hodson was one of the key RCAF staff officers who assisted in the development of the operational control arrangements for US Northeast Command in the early 1950s. Quotes from first source.

110 Campbell to Foulkes, March 8, 1958, Raymont Collection, DHH 73/1223/86.
111 Sokolsky, "A Seat at the Table," 21–22; Jockel, "Canada in NORAD," QCIR National Security Seminar Presentation, January 24, 2007; Lackenbauer, "From 'Defence against Help' to 'a Piece of the Action.'"
112 Canada, DND, "Backgrounder – Canada–United States Defence Relations," BG-06.021, July 27, 2006, www.forces.gc.ca/en/news/article.page?doc=canada-united-states-defence-relations/hnocfojt; Richard Gimblett, *Operation Apollo: The Golden Age of the Canadian Navy in the War against Terrorism* (Ottawa: Magic Light and DND, 2004), 12; Mitchell, *Network Centric Warfare*, 88.
113 Diefenbaker, *One Canada*, vol. 3, 25.
114 *Duty with Honour*, 6; Canada, DND, *Leadership in the Canadian Forces: Conceptual Foundations* (Kingston: Canadian Defence Academy and the Canadian Forces Leadership Institute, 2005), 130.
115 Terms of Reference for the Commander in Chief, North American Air Defense Command, June 10, 1958, NARA, RG 218, JCS, entry 943011, box 16, file CCS 092 (9–10–45), section 51. As Peter Haydon has noted of this chain of responsibility, "overall, it was a carefully structured system of delegated authority that included stringent checks and balances to prevent the abuse of authority." Haydon, *The Cuban Missile Crisis Revisited*, 98.

Conclusion

1 English and Westrop also advocate this position in *Canadian Air Force Leadership and Command*, 29–30.
2 Fergusson, *Canada and Ballistic Missile Defence*, 12.
3 For a discussion of these events, see Jockel, *Canada in NORAD*, 54–60, 91–93.
4 Fergusson, *Canada and Ballistic Missile Defence*, 12.
5 On the former, see Gimblett, *Operation Apollo*; on the latter, see Richard Mayne, "The Canadian Experience: Operation Mobile," in *Precision and Purpose: Airpower in the Libyan Civil War*, ed. Karl P. Mueller (Santa Monica: RAND Corporation, 2015), 239–66.
6 Prime Minister of Canada Mandate Letter to the Minister of National Defence, accessed June 26, 2016, http://pm.gc.ca/eng/minister-national-defence-mandate-letter; Canada, House of Commons, *Canada and the Defence of North America: NORAD and Aerial Readiness*, Report of the Standing Committee on National Defence, September 2016, 42nd Parliament, 1st Session, 61. In its strategic vision for Canada, the new Canadian Defence Policy specifically articulates that Canada must be "*Strong at home*, its sovereignty well-defended by a Canadian Armed Forces also ready to assist in times of natural disaster, other emergencies, and search and rescue." Canada, *Strong, Secure, Engaged*, 14 (emphasis in original).

Select Bibliography

Archival Sources

Canadian War Museum Archives
Air Vice-Marshal C.E. McEwen Papers

Directorate of History and Heritage, Department of National Defence, Ottawa

Document Collection
Biographical file, Rear-Admiral L.W. Murray
Canada–US Military Study Group Papers
General Charles Foulkes Papers
General H. Arnold Fonds (extracts)
Interview with Air Marshal Roy Slemon, October 20, 1978
Joint Staff fonds
Royal Canadian Air Force fonds

Kardex Files
Military Co-operation Committee Meeting Minutes
Permanent Joint Board on Defence Journals
Robert Lewis Raymont fonds
W.A.B. Douglas, Roger Sarty, and Michael Whitby (*No Higher Purpose*) Collection
W.A.B. Douglas fonds

Keith Hodson Memorial Library, Canadian Forces College, Toronto
Annis, Clare, Air Commodore. "The Role of the R.C.A.F." Address delivered before the Trenton Chamber of Commerce, March 26, 1952. *Airpower 1952: Three Speeches by Air Commodore Clare L. Annis.*
Examinations Study Material, ed. Clare Annis. Ottawa: Training Command, 1955.
Hodson, K.L.B. "The RCAF Air Division in Europe: Address to the United Services Institute, London, ON, 15 December 1954."
Jackson, J.I. "An Article on Air Power." *Readings in Air Power: RCAF Officers' Examinations Study Material*, ed. Clare Annis. Ottawa: Training Command, 1955.

Library and Archives Canada, Ottawa

Record Groups
Department of National Defence
Department of External Affairs
Privy Council Office

Manuscript Groups
Air Marshal C.R. Dunlap Papers
Brooke Claxton fonds
James L. Ralston fonds
Lieutenant-General A.G.L. McNaughton Papers
Lieutenant-General Maurice Pope Papers
Paul Martin Sr. Papers

Public Record Office, The National Archives, London
Admiralty Files
Air Ministry Files
War Cabinet Files

Queen's University Archives, Kingston
Charles Gavan Power Papers

Royal Military College of Canada Archives, Massey Library, Kingston
F.D. Roosevelt Diplomatic Correspondence (microfilm)
General of the Army George C. Marshall Papers (microfilm)
Major-General W.H.P. Elkins Papers
Records of the Joint Chiefs of Staff (microfilm)

United States National Archives and Records Administration II, College Park, Maryland
Joint Chiefs of Staff Collection
State Department Files
United States Air Force Files
United States Army War Plans Division Files

University of Toronto Archives, Toronto
C.P. Stacey Papers

University of Victoria Archives, Victoria
Lieutenant-General George Randolph Pearkes Papers

Journals and Periodicals (for a complete accounting of specific articles, please see Notes)

American Review of Canadian Studies
Armed Forces and Society
Canadian Army Journal
Canadian Defence Quarterly
Canadian Historical Review
Canadian Military History
Canadian Military Journal
Foreign Affairs
International Journal
Journal of Canadian Studies
Journal of Military History
Journal of Peace Research
Journal of the Royal United Service Institution

Journal of Strategic Studies
Military Affairs
Naval War College Review
The Northern Mariner
Pacific Northwest Quarterly
Parameters
Policy Options
Political Science Quarterly
The RCAF Staff College Journal
Review of Constitutional Studies
Revue Internationale d'Histoire Militaire
The Roundel
Royal Air Force Air Power Review
Royal Canadian Air Force Journal
SAIS Review
Survival

Books and Articles

Allard, C. Kenneth. *Command, Control, and the Common Defense*. New Haven: Yale University Press, 1992.

Barnes, Bruce. "'Fighters First': The Transition of the Royal Canadian Air Force, 1945–1952." MA thesis, War Studies, Royal Military College of Canada, 2006.

Bezeau, M.V., Capt. "The Role and Organization of Canadian Military Staffs." MA thesis, Royal Military College of Canada, 1978.

Bland, Douglas. *The Administration of Defence Policy in Canada, 1947–1985*. Kingston: Ronald P. Frye, 1987.

Bland, Larry I. *George C. Marshall Interviews and Reminiscences for Forrest C. Pogue*, rev. ed. Lexington: George C. Marshall Research Foundation, 1991.

Bridle, Paul, ed. *Documents on Relations between Canada and Newfoundland*. 2 vols. Ottawa: Information Canada, 1974, 1984.

Brodie, Bernard, ed. *The Absolute Weapon: Atomic Power and World Order*. New York: Harcourt Brace, 1946.

Buckley, John. *The RAF and Trade Defence, 1919–1945: Constant Endeavour*. Keele: Ryburn Publishing, Keele University Press, 1995.

Burtch, Andrew. *Give Me Shelter: The Failure of Canada's Cold War Civil Defence*. Vancouver: UBC Press, 2012.

Buss, Lydus H. *U.S. Air Defense in the Northeast 1940–1957*. Historical Reference Paper no. 1. Colorado Springs: Directorate of Command History, Office of Information Services, Headquarters Continental Air Defense Command, 1957.

Byers, Michael. "Canadian Armed Forces under US Command." Report commissioned by the Simons Centre for Peace and Disarmament Studies, Liu Centre for the Study of Global Issues, University of British Columbia, May 6, 2002.

Canada. Department of National Defence. *Duty with Honour: The Profession of Arms in Canada*. Kingston: Canadian Forces Leadership Institute, 2009.

–. *Canadian Forces Aerospace Command Doctrine*, B-GA-0401-000/FP-001, March 2012. Trenton: Canadian Forces Aerospace Warfare Centre, 2012.

–. *Canadian Forces Joint Publication CFJP 3.0 – Operations Keystone*, B-GJ-005-300/FP-001, September 2011.

–. *National Defence Act*. Ottawa: King's Printer, 1950.

–. RCAF. *Royal Canadian Air Force Doctrine*, B-GA-400-000/FP-001, November 2016. Trenton: Canadian Forces Aerospace Warfare Centre, 2016.

–. *Strong, Secure, Engaged: Canada's Defence Policy*. Ottawa: Minister of National Defence, 2017.

Chapnick, Adam. *Canada's Voice: The Public Life of John Wendell Holmes*. Vancouver: UBC Press, 2009.

Chronology of JCS Involvement in North American Air Defense 1946–1975. Washington, DC: Historical Division, Joint Secretariat, Joint Chiefs of Staff, March 20, 1976.

Clausewitz, Carl von. *On War*. Translated and edited by Michael Howard and Peter Paret. Princeton: Princeton University Press, 1982.

Clearwater, John. *US Nuclear Weapons in Canada*. Toronto: Dundurn, 1999.

Coakley, Thomas P. *Command and Control for War and Peace*. Washington, DC: National Defense University Press, 1992.

Cohen, Eliot A., and John Gooch. *Military Misfortunes: The Anatomy of Failure in War*. New York: Macmillan, 1990.

Conn, Stetson, and Byron Fairchild. *The Western Hemisphere: The Framework of Hemisphere Defense*. Washington, DC: Office of the Chief of Military History, Department of the Army, 1960.

Craven, Wesley Frank, and James Lea Cate, eds. *The Army Air Forces in World War II*, vols. 1 and 2. Chicago: University of Chicago Press, 1948, 1949.

Diefenbaker, John G. *One Canada: Memoirs of the Right Honourable John G. Diefenbaker*, vol. 3: *The Tumultuous Years, 1962–1967*. Toronto: Macmillan of Canada, 1977.

Documents on Canadian External Relations. Volumes 8-25, 1939-1958. Ottawa: Global Affairs Canada, 1996–2004.

Douglas, W.A.B. *The Creation of a National Air Force: The Official History of the Royal Canadian Air Force*, vol. 2. Toronto: University of Toronto Press and Department of National Defence, 1986.

–. "Democratic Spirit and Purpose: Problems in Canadian-American Relations, 1934–1945." In *Fifty Years of Canada–United States Defense Cooperation: The Road from Ogdensburg*, edited by Joel Sokolsky and Joseph T. Jockel, 31–57. Lewiston: Edwin Mellen Press, 1992.

Douglas, W.A.B., Roger Sarty, and Michael Whitby, with Robert H. Caldwell, William Johnston, and William G.P. Rawling. *No Higher Purpose: The Official Operational History of the Royal Canadian Navy in the Second World War, 1939–1943*, vol. 2, pt. 1. St. Catharines: Vanwell, 2002.

Dziuban, Stanley W. *United States Army in World War II Special Studies: Military Relations between the United States and Canada, 1939–1945*. Washington, DC: Office of the Chief of Military History, Department of the Army, 1959.

Eayrs, James. *In Defence of Canada*, vols. 3 and 4. Toronto: University of Toronto Press, 1972, 1980.

English, Allan. *Understanding Military Culture: A Canadian Perspective*. Montreal and Kingston: McGill–Queen's University Press, 2004.

English, Allan, and Col. John Westrop (Ret.). *Canadian Air Force Leadership and Command: The Human Dimension of Expeditionary Air Force Operations*. Trenton: Canadian Forces Aerospace Warfare Centre, 2007.

Feaver, Peter. *Armed Servants: Agency, Oversight, and Civil-Military Relations*. Cambridge, MA: Harvard University Press, 2003.

Fergusson, James G. *Canada and Ballistic Missile Defence 1954–2009: Déjà Vu All Over Again*. Vancouver: UBC Press, 2010.

Foulkes, Gen. Charles. "The Complications of Continental Defence." In *Neighbors Taken for Granted: Canada and the United States*, edited by Livingston T. Merchant, 101–33. Toronto: Burns and MacEachern, 1966.

Gimblett, Richard. "The Canadian Way of War: Experience and Principles." In *Canadian Expeditionary Air Forces*, edited by Allan D. English. Bison Paper no. 5, 9–20. Winnipeg: Centre for Defence and Security Studies, University of Manitoba, 2002.

Godefroy, Andrew B. *In Peace Prepared: Innovation and Adaptation in Canada's Cold War Army*. Vancouver: UBC Press, 2014.

Goette, Richard. "Air Defence Leadership during the RCAF's 'Golden Years.'" In *Sic Itur Ad Astra: Canadian Aerospace Power Studies*, vol. 1: *Historical Aspects of Canadian Air Power Leadership*, edited by William March, 51–64. Ottawa: Minister of National Defence, 2009.

–. "The Struggle for a Joint Command and Control System in the Northwest Atlantic Theatre of Operations: A Study of the RCAF and RCN Trade Defence Efforts During the Battle of the Atlantic." MA thesis, Queen's University, 2002.

Gotlieb, Allan. *The Washington Diaries, 1981–1989*. Toronto: McClelland and Stewart, 2006.

Government of Canada. *Dishonoured Legacy: The Lessons of the Somalia Affair*. Report of the Commission of Inquiry into the Deployment of the Canadian Forces to Somalia, vol. 1. Ottawa: Queen's Printer, June 30, 1997.

Grant, C.L. *The Development of Continental Air Defense to 1 September 1954*. USAF Historical Study no. 126. Montgomery: USAF Historical Division, Research Studies Institute, Air University, 1954.

Grant, Shelagh D. *Polar Imperative: A History of Arctic Sovereignty in North America*. Toronto: Douglas and McIntyre, 2010.

Haydon, Peter. *The 1962 Cuban Missile Crisis: Canadian Involvement Reconsidered*. Toronto: Canadian Institute of Strategic Studies, 1993.

Hilliker, John, and Donald Barry. *Canada's Department of External Affairs*, 2 vols. Kingston: Institute of Public Administration of Canada and McGill–Queen's University Press, 1990.

House of Commons Special Committee on Defence. *Minutes of Proceedings and Evidence*, no. 15, October 22, 1963. Ottawa: Queen's Printer, 1963.

Huntington, Samuel P. *The Soldier and the State: The Theory and Politics of Civil–Military Relations*. Cambridge, MA: Belknap Press of Harvard University, 1957.

James, A. "Sovereignty." In *The Dictionary of World Politics: A Reference Guide to Concepts, Ideas and Institutions*, edited by Graham Evans and Jeffrey Newnham, 369–70. Toronto: Simon and Schuster, 1990.

Janowitz, Morris. *The Professional Soldier: A Social and Political Portrait*. New York: Free Press, 1960.

Jockel, Joseph T. *Canada in NORAD 1957–2007: A History*. Montreal and Kingston: McGill–Queen's University Press, in association with Queen's Centre for International Relations and Queen's Defence Management Program, 2007.

–. *No Boundaries Upstairs: Canada, the United States, and the Origins of North American Air Defence, 1945–1958*. Vancouver: UBC Press, 1987.

Kasurak, Peter. *A National Force: The Evolution of Canada's Army, 1950–2000*. Vancouver: UBC Press, 2013.

Kostenuk, Samuel, and John Griffin. *RCAF Squadron Histories and Aircraft, 1924–1968*. Toronto: A.M. Hakkert, 1977.

Krasner, Stephen D., ed. *Problematic Sovereignty: Contested Rules and Political Possibilities*. New York: Columbia University Press, 2001.

Lackenbauer, P. Whitney. "From 'Defence against Help' to 'a Piece of the Action': The Canadian Sovereignty and Security Paradox Revisited." Centre for Military and Strategic Studies (CMSS), University of Calgary. Occasional Paper no. 1, May 2000.

–. "Right and Honourable: Mackenzie King, Canadian, American Bilateral Relations, and Canadian Sovereignty in the Northwest, 1943–48." In *Mackenzie King: Citizenship* and Community: Essays Marking the 125th Anniversary of the Birth of William Lyon Mackenzie King, edited by John English, Kenneth McLaughlin, and P. Whitney Lackenbauer, 151–68. Toronto: Robin Brass Studio, 2002.

Lagassé, Philippe. "Accountability for National Defence: Ministerial Responsibility, Military Command, and Parliamentary Oversight." *IRPP Study* no. 4, March 2010.

Lerhe, Eric James. *At What Cost Sovereignty? Canada–US Military Interoperability in the War on Terror*. Halifax: Centre for Foreign Policy Studies, Dalhousie University, 2013.

MacKenzie, David. *Inside the Atlantic Triangle: Canada and the Entrance of Newfoundland into Confederation, 1939–1949*. Toronto: University of Toronto Press, 1986.

Maloney, Sean M. *Learning to Love the Bomb: Canada's Nuclear Weapons during the Cold War*. Washington, DC: Potomac Books, 2007.

–. *Securing Command of the Sea: NATO Naval Planning, 1948–1954*. Annapolis: Naval Institute Press, 1995.

Maurer, Martha. *Coalition Command and Control: Key Considerations*. Washington, DC: National Defence University and Harvard University Program on Information Resources Policy, 1994.

McMahon, Patricia I. *Essence of Indecision: Diefenbaker's Nuclear Policy, 1957–1963*. Montreal and Kingston: McGill–Queen's University Press, 2009.

Milberry, Larry. *Sixty Years: The RCAF and CF Air Command 1924–1984*. Toronto: CANAV Books, 1984.

Milner, Marc. *North Atlantic Run: The Royal Canadian Navy and the Battle for the Convoys*. Toronto: University of Toronto Press, 1986.

Mitchell, Paul. *Network Centric Warfare and Coalition Operations: The New Military Operating System*. New York: Routledge Global Security Studies, 2009.

Neary, Peter. *Newfoundland and the North Atlantic World, 1929–1949*. Montreal and Kingston: McGill–Queen's University Press, 1988.

Nicks, Don, John Bradley, and Chris Charland. *A History of the Air Defence of Canada 1948–1997*. Ottawa: Canadian Fighter Group, 1997.

Nineteen Years of Air Defense. NORAD Historical Reference Paper no. 11. Colorado Springs: North American Air Defense Command, Ent Air Force Base, 1965.

Osgood, Robert. *NATO: The Entangling Alliance*. Chicago: University of Chicago Press, 1962.

Perras, Galen Roger. *Franklin Roosevelt and the Origins of the Canadian–American Security Alliance 1933–1945: Necessary, but Not Necessary Enough*. London: Praeger, 1998.

–. *Stepping Stones to Nowhere: The Aleutian Islands, Alaska, and American Military Strategy*. Vancouver: UBC Press, 2003.

Pigeau, Ross, and Carol McCann. "What Is a Commander?" In *Generalship and the Art of the Admiral: Perspectives on Canadian Senior Military Leadership*, edited by Bernd Horn and Stephen J. Harris, 79–104. St. Catharines: Vanwell Publishing, 2001.

Pogue, Forrest C. *The Supreme Command: The European Theater of Operations, United States Army in World War II*. Washington, DC: Office of the Chief of Military History, Department of the Army, 1954.

Pope, Maurice, Lieutenant-General. *Soldiers and Politicians: The Memoirs of Lt.-Gen. Maurice A. Pope*. Toronto: University of Toronto Press, 1962.

Preston, Adrian W. "Canada and the Higher Direction of the Second World War, 1939–1945." In *Canada's Defence: Perspectives on Policy in the Twentieth Century*, edited by B.D. Hunt and R.G. Haycock, 98–118. Toronto: Copp Clark Pitman. 1993.

Richter, Andrew. *Avoiding Armageddon: Canadian Military Strategy and Nuclear Weapons, 1950–63*. Vancouver: UBC Press, 2002.

Roussel, Stéphane. *The North American Democratic Peace: Absence of War and Security Institution-Building in Canada–US Relations, 1867–1958*. Montreal and Kingston: McGill-Queen's University Press, 2004.
Roy, Reginald H. *For Most Conspicuous Bravery: A Biography of Major-General George R. Pearkes, V.C., through Two World Wars*. Vancouver: UBC Press, 1977.
Sagan, Scott D. *Moving Targets: Nuclear Strategy and National Security*. Princeton: Princeton University Press, 1989.
Sarty, Roger. *The Maritime Defence of Canada*. Toronto: Canadian Institute of Strategic Studies, 1996.
Schaffel, Kenneth. *The Emerging Shield: The Air Force and the Evolution of Continental Air Defense, 1945–1960*. Washington, DC: Office of Air Force History, United States Air Force, 1991.
Schnabel, James F. *History of the Joint Chiefs of Staff*, vol. 1: *The Joint Chiefs of Staff and National Policy 1945–1947*. Washington, DC: Office of Joint History and Office of the Chairman of the Joint Chiefs of Staff, 1996.
Slessor, Sir John. *The Central Blue: Recollections and Reflections*. London: Cassel, 1956.
Sokolsky, Joel J. "Exporting the 'Gap': The American Influence." In *The Soldier and the State in the Post Cold War Era*, edited by Albert Legault and Joel Sokolsky. Kingston: Queen's Quarterly Press, 2002.
Stacey, C.P. *Arms, Men, and Governments: The War Policies of Canada, 1939–1945*. Ottawa: Queen's Printer, 1970.
–. *Canada and the Age of Conflict: A History of Canadian External Policies*, vol. 2: *1921–1948, The Mackenzie King Era*. Toronto: University of Toronto Press, 1981.
–. *A Date with History: Memoirs of a Canadian Historian*. Ottawa: Deneau, 1983.
–., ed. *Historical Documents of Canada*, vol. 5: *1914–1945*. Toronto: Macmillan of Canada, 1972.
Stouffer, Ray. *Swords, Clunks, and Widowmakers: The Tumultuous Life of the RCAF's Original 1 Canadian Air Division*. Trenton: Canadian Forces Aerospace Warfare Centre, 2015.
Trudgen, Matthew. "Good Partners or Just Brass Intrigue: The Transnational Relationship between the USAF and RCAF with Respect to the North American Air Defence System, 1947–1960." In *Sic Itur Ad Astra: Canadian Aerospace Power Studies*, vol. 1: *Historical Aspects of Canadian Air Power Leadership*, edited by William March, 75–83. Ottawa: Minister of National Defence, 2009.
–. "In Search for Continental Security: The Development of the North American Air Defence System, 1949 to 1956." PhD diss., History, Queen's University, 2011.
United States. Joint Publication 1. *Doctrine of the Armed Forces of the United States*. Washington, DC: Joint Chief of Staff, March 25, 2013.
Van Creveld, Martin. *Command in War*. Cambridge, MA: Harvard University Press, 1985.
Visiting Forces (British Commonwealth) Act, George V c. 21, April 12, 1933.
Wakelam, Randall. *Cold War Fighters: Canadian Aircraft Procurement, 1945–54*. Vancouver: UBC Press, 2011.
Waller, Douglas C. *A Question of Loyalty: Gen. Billy Mitchell and the Court-Martial That Gripped the Nation*. New York: HarperCollins, 2004.
War Department. Office of the Chief of Staff. *Field Service Regulations, US Army*. Washington, DC: US Government Printing Office, 1914.
Warnock, Timothy A. *Air Power versus U-boats: Confronting Hitler's Submarine Menace in the European Theater*. Washington, DC: Office of Air Force History, 1999.
Watson, Robert J. *History of the Joint Chiefs of Staff*, vol. 5: *The Joint Chiefs of Staff and National Policy 1953–1954*. Washington, DC: Historical Division, Joint Chiefs of Staff, 1986.

Whitaker, Reg, and Gary Marcuse. *Cold War Canada: The Making of a National Insecurity State, 1945–1957.* Toronto: University of Toronto Press, 1995.

Willoughby, William R. *The Joint Organizations of Canada and the United States.* Toronto: University of Toronto Press, 1979.

Winnefeld, James A., and Dana J. Johnson. *Joint Air Operations: Pursuit of Unity in Command and Control, 1942–1991.* Annapolis: Naval Institute Press, 1993.

Wise, S.F. *Canadian Airmen and the First World War: The Official History of the Royal Canadian Air Force*, vol. 1. Toronto: University of Toronto Press and Department of National Defence, 1980.

Index

Note: "(i)" following a page number indicates an image; "(m)" following a page number indicates a map

ABC-1 Staff Agreement, 78–79
ABC-22 plan, 15, 17, 71, 77–78; and ABC-1, 78–79; Basic Security Plan, 17, 107, 114–16, 120; command and control provisions, 79–96, 104, 113–14, 194; defence of Newfoundland, 86–94
Ad Hoc Study Group (AHSG), 179–82
Admiralty (UK), 48, 50–51
Air Corps Tactical School (US), 88, 97
Air Defence Command (ADC) (Canada), 6; command badge, 67(i); established, 118–19, 140–41; interceptor coverage, 190(m); operational control, 139–50; unified command, 55; US Northeast Command, 151–65, 167(m), 182, 185, 189, 194, 198
Air Defense Command (ADC) (United States), 14, 118–19, 138–42, 148; US Northeast Command, 151, 157, 161, 165, 176–77, 180
Air Defence Control Centre (ADCC), 145–46, 148–49, 166, 204, 208, 252n61, 254n86
Air Defence Group (ADG) (Canada), 141
Air Defence Study Group (ADSG), 172, 176–77
Air Interceptor and Air Warning Appendix, 107, 116–20, 137, 195
Air Ministry (UK), 50–51
Air Space Interoperability Council, 14
Air Standardization Coordination Committee, 14
Air University, 14
Aircraft Detection Corps (Canada), 73
Alaska Command (US), 54
Aleutian campaign, 41, 187

Allard, Kenneth, 22, 27, 45, 228n44, 228n47
American-British-Dutch-Australian Command (ABDA), 56, 116
Anderson, Air Vice-Marshal N.R., 87, 95, 98–99
Annis, Air Commodore Clare, 126(i), 144, 177, 196
Anti-Aircraft Command (Canada), 141, 252n61
Anti-Aircraft Command (US), 139, 176, 180
Argentia, NL, 76, 87, 93–95, 99–102
Atkinson, Lieutenant-General Joseph, H., 180
Atlantic Command (Canada), 73, 87
Atlantic Command (US), 54
Atlantic Conference, 94
Atlantic Convoy Conference, 102
atomic bomb, 58, 107–10, 117, 132–37, 155, 171, 195, 242n10

B-17 aircraft, 87, 101
B-29 Superfortress bomber, 109
Basic Security Plan (BSP), 17, 107, 114–20, 131–32, 152; command and control provisions, 194–95. *See also* Air Interceptor and Air Warning Appendix
Bercuson, David, 31, 165, 168
Biggar, O.M., 80, 84
bilateral command of North American air defence evolution, 5, 9–12, 21, 58–60, 75, 77, 85, 111–15, 138, 142, 152–59, 168–70, 174–79, 181–82, 188–89, 193–94, 198–99; models contributing to bilateral command concept, 36–37. *See also* ABC-22,

Basic Security Plan (BSP), Emergency Defence Plan (EDP), Matchpoint plan, NORAD, United States Northeast Command (USNEC)
"Black Plan." See Joint Canadian–United States Basic Defence Plan–1940
Bland, Douglas, 4, 57
Bouchard, Lieutenant-General Charles, 199
Boyd, Colonel John, 24
Brainard, Rear-Admiral R.M., 93, 100–1
Brant, Major-General Gerald Charles, 88–90, 92–94, 101, 104, 156, 195
Breadner, Air Marshal L.S., 96
Bristol, Rear-Admiral A.A.L., 87, 95, 99
British Columbia, 89–90
British-American Combined Chiefs of Staff (CCS), 56
Brodie, Bernard, 107–8
Brooks, Brigadier-General John B., 94
Brown, Brigadier-General James Sutherland (Buster), 71
Bryce, R.B., 177
BSP. See Basic Security Plan (BSP)
Byers, Michael, 4–5

CA-1. See Canada–United States Security Plan (CA-1)
Cabinet Defence Committee, 113, 116, 183–85
Cabinet War Committee, 80, 92, 96
Campbell, Air Marshal Hugh, 191
Campney, Ralph, 183
Canada–United States Air Defence Command. See NORAD
Canada–United States Chiefs of Staff (CANUSA), 111–12
Canada–United States Permanent Joint Board on Defence. See Permanent Joint Board on Defence (PJBD)
Canada–United States Regional Planning Group (CUSRPG), 58–59, 156, 174–75
Canada–United States Security Plan (CA-1), 110–12
Canadian Army, 15; Ad Hoc Study Group, 179–80; air defence role, 140–41, 182; command and control in World War II, 36–41, 49, 73–74, 90; defence of Newfoundland, 87, 92; US Northeast Command, 158
Canadian Joint Operations Command, (CJOC), 4
Canadian Northwest Atlantic Command, 101–4, 103(m), 139, 186, 194
CANUSA. See Canada–United States Chief of Staff (CANUSA)
Caribbean Command (US), 54
CF-100 Canuck jet interceptor, 14, 67(i), 124(i), 130–31(i), 161, 259n45
Chidlaw, General Benjamin, 121(i), 176
Chiefs of Staff Committee (CSC): ABC-22, 90–92, 104, 192, 194; Basic Security Plan, 116; Canada–US Regional Planning Group, 58; continental air defence integration, 142–43, 146, 170, 175, 177, 181, 183; continental defence, 74; defence of Newfoundland, 87, 92–93, 104; EDP Command Appendix, 134, 136–37, 165; and national command, 5, 48–49, 74, 140; NORAD, 185, 187–89, 192; operational control concept, 170–71, 178–79; Pinetree System, 142; US Northeast Command, 154–64
Churchill, Winston S., 43, 94
civil-military relations (CMR), 17, 21; Cohen model, 32–33; definition, 31–32; Feaver model, 34; Huntington model, 32; Janowitz model, 33–34, 192
Claxton, Brooke, 36, 140, 146, 149, 160, 168, 174
Clements, Air Commodore W.I., 10
CMR. See civil-military relations (CMR)
Coakley, Thomas, 23–24
coalition supreme command system, 36, 44, 52–59, 78
Coastal Command, 39, 50–52, 97, 100–1
Coates, Major-General Christopher, 191
Cohen, Eliot, 13, 25, 32–33
Cole, Sterling, 172, 174
Coleman, Group Captain S.W., 160
command and control: Boyd model, 24; Canada and *Visiting Forces Act*, 37–38; Canada and World War II, 36–41, 48–50; Coakley model, 23; Cohen and Gooch model, 25; command and control systems, 36; definition, 4–5, 18, 21–23; Lawson model, 24; Pigeau and McCann CAR model,

Index 285

25–27, 35. *See also* coalition supreme command system, Joint Committee Cooperation System (Canada), Joint Committee System (United Kingdom), Newfoundland, Operational Control System (UK), sovereignty, Unity of Command System (US)
Commander-in-Chief, Air Defense Canada–United States (CINCADCANUS), 181–82, 187, 268n62, 268n63
Conn, Stetson, 71
Continental Air Command (ConAC) (US), 139
Continental Air Defense Command (CONAD) (US), 55, 175–76, 179–85, 188, 198, 265n26
control: definition, 18, 21–23, 26. *See also* operational control
Currie, Lieutenant-General Sir Arthur, 37
Curtis, Air Marshal Wilfred, 68(i), 111–12, 140, 144, 146, 154

Dandurand, Raoul, 73
Defence of Canada Plan 1940, 74
"Defence Scheme Number 1 (The United States)," 71–72
Denfeld, Admiral Louis, 152
Department of External Affairs, 7, 15, 49; Basic Security Plan, 116; Goose Bay lease negotiations, 162; and Military Cooperation Committee, 112; NORAD creation, 177–78, 186, 188–89; operational control of RCAF in Newfoundland, 163–64;US Northeast Command, 154–57, 164; views on atomic bomb, 108; views on US command of Canadian forces, 112, 149, 154–55, 170, 177, 179, 189
Destroyers-for-Bases Deal, 76, 162
Dewitt, Lieutenant-General John L., 90
DH.100 Vampire jet, 65–66(i), 140
Diefenbaker, John G., 184–85, 187–89, 192
Distant Early Warning (DEW) Line, 117, 172–73
Douglas, W.A.B., 8, 88
Dowding, Air Chief Marshal Hugh, 145
Drouin, Major General Christian, 189
Drum, Lieutenant-General Hugh A., 94

Drury, C.M., 177
Dunlap, Air Marshal C.R. (Larry), 112, 177, 181, 196
Duty with Honour, 33

EADP. *See* Emergency Air Defence Plan (EADP)
Earnshaw, Brigadier-General Philip, 87–88, 92
Eastern Air Command (EAC), 6; command badge, 62(i); established, 73–74; Hemisphere Defence Plan No. 4, 94–95; No. 1 Group, 87, 95–102; operational control of air forces in Newfoundland, 8, 87, 89, 95–102, 115, 186, 197; operational control and USN, 95–96, 197; USAAF-RCAF relations, 93–94
Eayrs, James, 15–16, 108
EDP. *See* Emergency Defence Plan
EDP Command Appendix, 134–37, 143–50, 161–65, 259n49, 261n60, 261n61. *See also* Emergency Defence Plan (EDP)
Eisenhower, Dwight D., 43, 56, 149, 171
Elkins, Major-General W.H.P., 87
Embick, Lieutenant-General Stanley, 84
Emergency Air Defence Plan (EADP), 142
Emergency Defence Plan (EDP), 117, 132–34, 170, 195; Air Defence Command, 141; air defence priorities, 137; Canada–US Defense Command, 134–35; NORAD, 185; US Northeast Command, 153, 170. *See also* EDP Command Appendix
English, Allan, 18, 39–40, 140–41
European Command (US), 54–55

F-86 Sabre fighter aircraft, 14, 66(i), 122(i), 160
F-89 Scorpion interceptor, 123(i), 130(i)
F-94 Starfire interceptor, 123(i), 162
F-102 Dagger aircraft, 130(i)
F-104A Starfighter jet fighter, 130–31(i)
Fairchild, Byron, 71
Far East Command (US), 54
Feaver, Peter, 34
Fergusson, James, 13, 198
Foch, Marshal Ferdinand, 45, 57, 78, 82
Fort Pepperrell, NL, 76, 87, 151

Foulkes, General Charles: continental air defence, 143, 171, 175, 178–79, 182–84, 188–89, 191; joint committee cooperation system, 140; national command, 48; views on supreme command, 59; US Northeast Command, 159–61

Gander, NL, 87
Godwin, Air Commodore H.B., 151
Gosselin, Major-General Daniel, 29, 33
Graham, Lieutenant-General H.G., 180
Grant, Ulysses S., 45
Grimshaw, Major Louis, 8–9

Haig, Field Marshal Douglas, 57
Harmon Field, NL, 151, 162, 190(m)
Haydon, Peter, 116
Heakes, Group Captain F.V., 64(i), 95, 98, 102
Hemisphere Defence Plan No. 4 (WPL-51), 94–95
Henry, Major-General Guy, 106, 111–12, 153
Hickerson, John, 81
Hodson, Air Vice-Marshal Keith, 40, 125(i), 142, 158, 191, 196, 249n40
Hope, Colonel Ian, 44
Howe, C.D., 108
Huntington, Samuel B., 13, 21, 32

Jackson, James L., 16, 108
James, Air Vice-Marshal A.L., 121(i), 162
Janowitz, Morris, 32–34, 142, 148, 157, 168, 170, 182, 192, 195
Jockel, Joseph, 7, 117, 145, 147, 149–50, 159, 177, 181, 187
Johnson, Air Vice-Marshal G.O., 62(i), 102
Joint Action of the Army and Navy doctrine: ABC-22, 83–85, 88, 91; Joint Canadian–United States Basic Defence Plan–1940, 76–77; and unity of command, 46–47, 53–54, 97, 228n51
Joint Appreciation, 114–15
Joint Board of the Army and the Navy (US), 46
Joint Canadian–United States Basic Defence Plan–1940, 75–78, 83, 194
Joint Chief of Staff (JCS), 28; ABC-22, 91–92; Basic Security Plan, 116; Canada–United States Security Plan, 111–12; Canada–US Regional Planning Group, 58; continental air defence, 139, 142, 144, 147, 174–80; EDP Command Appendix, 134, 136, 163–65; Military Cooperation Committee, 112; and NORAD, 185, 187–89, 192, 197; operational control concept, 177–79, 183; and unity of command, 53–55, 92, 138, 181; US Northeast Command, 152–55, 157, 159, 161–66, 180. *See also* Unified Command Plan (UCP)
Joint Committee Cooperation System (Canada), 48–50, 71, 75, 84, 86–87, 140
Joint Committee System (UK), 36, 41–44, 48, 50, 55–56, 227n37
Joint Planning Committee (JPC) (Canada), 112, 117, 158–61, 174–75
Joint Service Sub-Committee, Newfoundland, 87, 92
Joint Services Committee Atlantic Coast, 74, 87, 92
Joint Services Committee East Coast, 156
Joint Services Committee Pacific Coast, 74, 92
Joint Staff Committee. *See* Chiefs of Staff Committee (CSC)
Joint Strategic Plans Committee (JSPC) (US), 179–80
Jones, Rear-Admiral George C., 92, 95

Keenleyside, Hugh, 82
Key West Agreement, 138, 140
Kingston Dispensation, 72–73
Korean War, 58, 132, 147, 155
Krasner, Stephen, 27
Kuter, General Laurence S., 191, 196

La Guardia, Fiorello, 80, 84, 90
Lackenbauer, Whitney, 9–10, 30
Lagassé, Philippe, 28
Lawson, Joel S., 24
Leased Bases Agreement of 1941, 151–53, 155–56, 159, 161, 164
levels of war, 18–19
Léger, Jules, 189
Lincoln, Abraham, 45
Lockheed Hudson aircraft, 61(i)

M-4 Bison bomber, 124(i), 171
MacKay, R.A., 149, 164, 177

Mackenzie, C.J., 108
Mackenzie King, William Lyon, 48–49, 72–73, 75, 80, 82, 107–9, 116
Marshall, General George, 55, 75, 91
Matchpoint plan, 113–14
Maurer, Martha, 11
MC-48 NATO Strategic Concept, 171
MCC. *See* Military Cooperation Committee (MCC)
MCC 300. *See* Emergency Defence Plan (EDP)
McCann, Carol, 25–27, 29, 35
McEwen, Air Commodore C.M., 87, 93, 96, 98–99, 101
McNaughton, Lieutenant-General Andrew, 4, 37
Mediterranean Supreme Command, 56
Mid-Canada Line, 172, 173(m)
Military Cooperation Committee (MCC), 58, 112–13; air defence command and control in Newfoundland, 164; Air Interceptor and Air Warning Appendix, 117, 119; Basic Security Plan, 114–16, 119, 132; bilateral planning process, 132–33; EDP Command Appendix, 137–38, 164; Emergency Defence Plan, 133–35, 137
Military Study Group (MSG), 172; Air Defence Study Group report, 177; Ad Hoc Study Group report, 181; proposal on air defence integration, 181–88
Miller, Air Vice-Marshal, Frank, 145, 158, 164, 196, 272n92
Minister of National Defence, 5
mission command, 53, 96, 143, 250n46
Mitchell, Brigadier General William (Billy), 88
Mitchell Field, NY, 138–39
Monroe Doctrine, 72
Montgomery, Field Marshal Bernard, 38
Mountbatten, Admiral Louis, 5, 56
MSG. *See* Military Study Group
Murray, Rear-Admiral Leonard W., 74, 87, 92–93, 95, 98, 102
Myers, Lieutenant-General Charles T., 161–63, 166, 176

national command, 11, 28, 56–57; ABC-22, 71, 79, 83–85; and air defence integration, 118, 141, 143, 145, 165–66, 176, 179, 182–86, 189, 193–95; Canada, 5–6, 21, 38, 40–41, 47–48, 52, 193–95, 211n10; and Canadian forces in Newfoundland, 86, 104; and EDP Command Appendix, 136, 166; *Plan 1*, 78; and United States, 54, 176
National Defence Act (NDA), 5, 186
National Defence College, 14
NATO: as an alliance, 10–11, 112; doctrine, 22, 27; RCAF Air Division, 40–41, 160; command structure, 44, 55, 58, 134, 159, 174, 186; Regional Planning Groups, 58–59, 156; Canadian contribution, 41, 151, 160, 162, 168; nuclear deterrent, 169, 171; and North American air defence integration, 174–75; and Military Study Group, 181; Canada links to NORAD, 188–89; and *Operation Unified Protector*, 199
NATO Military Committee, 58
Naval Forces Continental Air Defense Command (US), 176
Newfoundland, 15–17; Air Defence Command and operational control, 168–69; American bases, 76, 151–52, 155, 197; *Basic Security Plan*, 76, 152; command and control of Canadian forces, 79–82; defence in World War II, 76, 86–93; Emergency Defence Plan, 153; RCAF post-war air defence role, 151–52; US Northeast Command, 151–55
Newfoundland Base Command (US), 87–94; replaced by US Northeast Command, 151
Newfoundland Escort Force (NEF), 87
NORAD: binational command, 11–12, 30, 169, 170–74, 185–96, 273n102; Canada–US Press Release on the Establishment of NORAD, 210; Colorado Springs Headquarters, 139; Deputy Commander-in-Chief, 158, 170, 181, 189–91, 197–98; emblem, 129(i); establishment, 3–4, 7, 15, 184–92; and NATO, 188–89, 210; operational control, 18, 52, 143, 168, 170–78, 184–86, 189, 192–99
NORAD Agreement, 170, 185–86, 271n84

North American Air Defense Command. *See* NORAD
North Atlantic Treaty Organization. *See* NATO
Northeast Air Command (US), 155, 157–59, 163–64, 168, 176
Northwest Europe Supreme Command, 56
NSC 162/2 US Strategic Concept, 171

Ogdensburg Agreement, 3, 75
Operation Unified Protector, 199
operational command: Air Interceptor and Air Warning Appendix, 118–19; Basic Security Plan, 116; and Canada, 5–6, 37–41, 48, 92, 194–95, 225*n*15, 225*n*16; Defence of Canada Plan 1940, 74; defence of Northwest Atlantic, 95, 197; definitions, 47, 52, 77, 83, 166, 212*n*12, 228*n*51, 262*n*71; Eastern Air Command, 95, 98, 197; EDP Command Appendix, 134–36, 145–46, 163–66, 193; and NORAD, 182–87, 193, 196; and United Kingdom, 43, 48, 224*n*10; unity of command, 47, 54, 77; and US Northeast Command, 156, 164, 168, 197
operational control, 4–6; Air Defence Command, 141, 144–45; air defence integration, 32, 118–19, 138–39, 177–81, 195; Basic Security Plan, 132 195; Command appendix, 134–37, 143, 150, 165, 195; Continental Air Defense Command, 176, 181, 188; defence of Northwest Atlantic, 17, 50–51, 99–5, 186, 194, 197; definitions, 51–52, 142–43, 170–71, 179–80, 185, 239*n*74; Eastern Air Command, 98–99, 102, 186, 197; NORAD, 7, 18, 171–79, 182–89, 192–99; and Pinetree System, 142–43, 195; US Northeast Command, 17, 152–70, 195, 197. *See also* Operational Control System (UK)
Operational Control System (UK), 17, 36, 42, 50–52, 87, 100, 104, 134, 188, 194
operational level of war, 18–19
Ørvik, Nils, 8
Osgood, Robert, 112

P-51 Mustang fighter aircraft, 65(i), 140
Pacific Command (Canada), 73, 187

Pacific Command (US), 54–55
Page, Major-General L.F., 92–94
Partridge, General Earle, 129(i), 180, 184–85, 187, 189, 191, 196
Pearkes, Lieutenant-General George, 184, 187, 210
Pearl Harbor, attack on, 17, 52, 86, 89–90, 96, 99, 101, 104
Pearson, Lester B., 112, 154, 186
Permanent Joint Board on Defence (PJBD), 14, 61(i); ABC-22, 78–85; Basic Security Plan, 114, 116; Canada–United States Security Plan, 110–12; command and control of forces in Newfoundland, 96–99; Joint Canadian–United States Basic Defence Plan–1940, 75–76; Matchpoint plan, 111–14; Military Cooperation Committee, 112–13; NORAD, 195; Pinetree System, 142–43; Recommendation 51/1, 143, 157; Recommendation 51/4, 147, 202; Recommendation 51/6, 146–47, 150, 182, 185, 201; Recommendation 53/1, 148–50, 209; role in bilateral defence, 88, 106, 110; unity of command, 90–92; US Northeast Command, 153–55, 157, 162, 164. *See Also* Military Cooperation Committee
Pigeau, Ross, 22, 25–27, 29, 35
Pinetree System, 70(i), 142–43, 157, 161, 173(m), 195
Plan 1, 77–81
Plan 2. *See ABC-22*
Pope, Lieutenant-General Maurice, 6, 8, 81–82, 106
Power, C.G. (Chubby), 96

Radford, Admiral Arthur, 174, 178
RAF. *See* Royal Air Force (RAF)
Ralston, James L., 81
RCAF. *See* Royal Canadian Air Force (RCAF)
Reid, Captain H.E. (Rastus), 79–80
Royal Air Force (RAF): Bomber Group (RAF), 39–40, 110; Coastal Command, 50–51, 97; relationship with RCAF, 13–14, 37–41, 119
Royal Canadian Air Force (RCAF): air power culture, 13; bilateral continental air defence, 9–10, 14–16, 31,

106, 140–41; command and control in World War II, 38–41; defence of Newfoundland in World War II, 86–105; national command, 5; NORAD, 18, 197–99; operational control, 6; RCAF NATO Air Division, 40; relationship with RAF, 13–14, 36–41; relationship with USAF, 13–16, 40, 195; unified command, 55; US Northeast Command, 17–18, 151–69. *See also* Air Defence Command (ADC), Eastern Air Command (EAC), NORAD, operational command, operational control, Western Air Command (Canada)

Royal Canadian Air Force "Authority to Intercept and Engage Hostile Aircraft" directive, 203–8

Royal Canadian Air Force Auxiliary, 140–41

Royal Canadian Air Force Rules of Engagement, 148, 163, 165–66, 168, 203–5, 262n74

Royal Canadian Air Force Staff College, 16, 67–68(i), 108–9

Royal Canadian Air Force Station Goose Bay, 88, 151, 156, 162, 168, 261n59, 261n61

Royal Canadian Navy (RCN), 15; Ad Hoc Study Group, 180; command and control in Northwest Atlantic campaign, 94–95; defence of Newfoundland, 87–88, 92; Hemisphere Defense Plan No. 4, 95; post-war defence roles, 140–41; and Royal Navy, 48–49, 199. *See also* Canadian Northwest Atlantic Command

Royal Navy (RN), 48–51, 100

Roosevelt, Franklin, D., 72–73, 75, 80–81, 94

Sharpe, Brigadier-General Joe, 18
Single Integrated Operational Plan (SIOP), 171
Skelton, O.D., 49
Slemon, Air Marshal Roy, 128–29(i), 165–66, 177–78, 181–85, 189, 191, 195–96
Slessor, Air Marshal Sir John, 51
Smith, Sidney, 188
Sokolsky, Joel, 8, 13, 191
Southeast Asia Supreme Command, 56

sovereignty: definition: 4–9, 27–31, 219n28, 220n29
Soviet threat, 3, 8–9, 31 58, 114, 137, 177, 189; atomic bomb, 58, 132, 136–37, 155, 170–71, 176, 197; bomber threat to North America, 3, 13, 17, 107, 109–10, 119–20, 132–33, 141, 144, 150, 151, 153, 155, 164, 172, 194–95; maritime threat, 133
St. Laurent, Louis, 184, 186–87
Stacey, C.P., 37, 40, 49, 71–72, 79, 81
Stark, Admiral Harold, 75, 91, 95–96
State Department (US), 112, 168, 179
Statute of Westminster, 37, 48
Stephenville, NL, 76
Stouffer, Ray, 14, 40
Strategic Air Command (SAC): air defence integration, 134, 139, 144, 151, 155–56, 161–62 168, 172, 175; bases, 190(m); and nuclear deterrence, 13, 16, 59, 107, 110, 171, 197; Unified Command Plan, 55, 138. *See also* Royal Canadian Air Force Station Goose Bay

Tactical Air Command (TAC), 138–39
Task Force 4, 87, 95
Truman, Harry S., 54, 147, 152
Tu-4 Bull bomber, 66(i), 109, 133
Twinning, General Nathan, 129(i)

U-970, 62(i)
Unified Command Plan (UCP), 52–55, 117, 138, 152, 154, 161
Unified Command for US Joint Operations directive, 53
United States Air Force, 12; and air defence, 14–18, 132–39, 147–51; in Newfoundland, 15, 147, 151–69; and offensive air power, 110, 138; operational control, 15; RCAF Air Division, 40; relationship with RCAF, 13–15, 40–41, 94, 141–51; "right to shoot," 147. *See also* Air Defense Command (ADC), Continental Air Defense Command (CONAD), Strategic Air Command (SAC), Tactical Air Command (TAC), and United States Northeast Command (USNEC)
United States Army, 28; air defence role, 139, 155, 176, 180; in defence

of Newfoundland, 87–99, 195; Joint Canadian–United States Basic Defence Plan–1940, 76–77; unity of command, 45–49, 53, 104. *See also* Anti-Aircraft Command (US)
United States Army Air Forces: in Newfoundland and Northwest Atlantic campaign, 87–88, 93–102, 197; post-war organization, 138; post-war role, 107, 114. *See also* Air Defense Command (ADC), Continental Air Defense Command (CONAD), Strategic Air Command (SAC), Tactical Air Command (TAC), and United States Northeast Command (USNEC)
United States Army War Plans Division (WPD), 76, 80, 84, 90
United States Navy (USN): Afghanistan, 199; command and control in Northwest Atlantic campaign, 75–76, 87–88, 91, 93–102, 197; leased British bases, 76; views on unity of command, 45–49, 52–55, 199; World War I cooperation with Canada, 71
United States Northeast Command (USNEC), 17, 54, 151–69; 64th Air Defence Sector, 166–67, 176; boundaries, 167(m); Canadian sovereignty, 152–56, 164–69; command and control, 151–53, 182, 186, 195, 197; disestablished, 180
United States Northern Command (NORTHCOM), 4
unity of command: ABC-22, 17, 79, 83–86, 89, 91–92, 95–96, 104, 111, 114–15, 119, 194; Basic Security Plan, 107, 111, 115, 119; British views on, 44; CA-1, 111; Canada–US forces in Newfoundland, 87–96, 194; Canadian views on, 16, 49, 71, 77, 82; continental air defence, 52, 60, 132, 150, 174, 199; definitions, 46–47, 227n39, 228n51; and Eastern Air Command, 95–98, 100; and EDP Command Appendix, 134–36; and Joint Canadian–United States Basic Defence Plan–1940, 75–77
Unity of Command System (US), 36, 44–47, 54–57, 71, 75, 79, 84; defence of Newfoundland, 86–98, 100, 104

Vandenberg, General Hoyt, 159
Visiting Forces Act, 37–39, 41, 224n8, 224n9

War Cabinet (UK), 43, 50–51
Washington, George, 45, 243n23
Wavell, Lieutenant-General Sir Archibald, 56
Welty, Colonel Maurice D., 87–88
Wershof, M.H., 161, 163
Western Air Command (Canada), 73
Westrop, John, 39–40, 140–41
Whitten, Major-General Lyman T., 155–58, 161
Willoughby, William, 59
Wilson, Charles, 174, 183, 210
Working Committee on Post-Hostilities Problems, 109
WPD. *See* United States Army War Plans Division (WPD)
WPL-51. *See* Hemisphere Defence Plan No. 4 (WPL-51)
Wray, L.E., 127(i), 189
Wrong, Hume, 108

Studies in Canadian Military History

John Griffith Armstrong, *The Halifax Explosion and the Royal Canadian Navy: Inquiry and Intrigue*
Andrew Richter, *Avoiding Armageddon: Canadian Military Strategy and Nuclear Weapons, 1950–63*
William Johnston, *A War of Patrols: Canadian Army Operations in Korea*
Julian Gwyn, *Frigates and Foremasts: The North American Squadron in Nova Scotia Waters, 1745–1815*
Jeffrey A. Keshen, *Saints, Sinners, and Soldiers: Canada's Second World War*
Desmond Morton, *Fight or Pay: Soldiers' Families in the Great War*
Douglas E. Delaney, *The Soldiers' General: Bert Hoffmeister at War*
Michael Whitby, ed., *Commanding Canadians: The Second World War Diaries of A.F.C. Layard*
Martin F. Auger, *Prisoners of the Home Front: German POWs and "Enemy Aliens" in Southern Quebec, 1940–46*
Tim Cook, *Clio's Warriors: Canadian Historians and the Writing of the World Wars*
Serge Marc Durflinger, *Fighting from Home: The Second World War in Verdun, Quebec*
Richard O. Mayne, *Betrayed: Scandal, Politics, and Canadian Naval Leadership*
P. Whitney Lackenbauer, *Battle Grounds: The Canadian Military and Aboriginal Lands*
Cynthia Toman, *An Officer and a Lady: Canadian Military Nursing and the Second World War*
Michael Petrou, *Renegades: Canadians in the Spanish Civil War*
Amy J. Shaw, *Crisis of Conscience: Conscientious Objection in Canada during the First World War*
Serge Marc Durflinger, *Veterans with a Vision: Canada's War Blinded in Peace and War*
James G. Fergusson, *Canada and Ballistic Missile Defence, 1954–2009: Déjà Vu All Over Again*
Benjamin Isitt, *From Victoria to Vladivostok: Canada's Siberian Expedition, 1917–19*
James Wood, *Militia Myths: Ideas of the Canadian Citizen Soldier, 1896–1921*
Timothy Balzer, *The Information Front: The Canadian Army and News Management during the Second World War*
Andrew B. Godefroy, *Defence and Discovery: Canada's Military Space Program, 1945–74*
Douglas E. Delaney, *Corps Commanders: Five British and Canadian Generals at War, 1939–45*
Timothy Wilford, *Canada's Road to the Pacific War: Intelligence, Strategy, and the Far East Crisis*
Randall Wakelam, *Cold War Fighters: Canadian Aircraft Procurement, 1945–54*
Andrew Burtch, *Give Me Shelter: The Failure of Canada's Cold War Civil Defence*
Wendy Cuthbertson, *Labour Goes to War: The CIO and the Construction of a New Social Order, 1939–45*
P. Whitney Lackenbauer, *The Canadian Rangers: A Living History*
Teresa Iacobelli, *Death or Deliverance: Canadian Courts Martial in the Great War*

Graham Broad, *A Small Price to Pay: Consumer Culture on the Canadian Home Front, 1939–45*

Peter Kasurak, *A National Force: The Evolution of Canada's Army, 1950–2000*

Isabel Campbell, *Unlikely Diplomats: The Canadian Brigade in Germany, 1951–64*

Richard M. Reid, *African Canadians in Union Blue: Volunteering for the Cause in the Civil War*

Andrew B. Godefroy, *In Peace Prepared: Innovation and Adaptation in Canada's Cold War Army*

Nic Clarke, *Unwanted Warriors: The Rejected Volunteers of the Canadian Expeditionary Force*

David Zimmerman, *Maritime Command Pacific: The Royal Canadian Navy's West Coast Fleet in the Early Cold War*

Cynthia Toman, *Sister Soldiers of the Great War: The Nurses of the Canadian Army Medical Corps*

Daniel Byers, *Zombie Army: The Canadian Army and Conscription in the Second World War*

J.L. Granatstein, *The Weight of Command: Voices of Canada's Second World War Generals and Those Who Knew Them*

Colin McCullough, *Creating Canada's Peacekeeping Past*

Douglas E. Delaney and Serge Marc Durflinger, eds., *Capturing Hill 70: Canada's Forgotten Battle of the First World War*

Brandon R. Dimmel, *Engaging the Line: How the Great War Shaped the Canada–US Border*

Geoffrey Hayes, *Crerar's Lieutenants: Inventing the Canadian Junior Army Officer, 1939–45*

Meghan Fitzpatrick, *Invisible Scars: Mental Trauma and the Korean War*

Frank Maas, *The Price of Alliance: The Politics and Procurement of Leopard Tanks for Canada's NATO Brigade*

Patrick M. Dennis, *Reluctant Warriors: Canadian Conscripts and the Great War*

STUDIES IN CANADIAN MILITARY HISTORY
Published by UBC Press in association with the Canadian War Museum.